This book develops deterministic chaos and frac[...] point of iterated maps, but the method of ana[...] emphasis make it very different from all other books in the field. It is written to provide the reader with an introduction to more recent developments, such as weak universality, multifractals, and shadowing, as well as to older subjects like universal critical exponents, devil's staircases and the Farey tree.

The book is written especially for those who want clear answers to the following sorts of questions: How can a deterministic trajectory be unpredictable? How can one compute nonperiodic chaotic trajectories with controlled precision? Can a deterministic trajectory be random? What are multifractals and where do they come from? What is turbulence and what has it to do with chaos and multifractals? And, finally, why is it not merely convenient, but also necessary, to study classes of iterated maps instead of differential equations when one wants predictions that are applicable to computation and experiment?

Throughout the book the author uses a fully discrete method, a 'theoretical computer arithmetic', because finite (but not fixed) precision is a fact of life that cannot be avoided in computation or in experiment. This approach leads to a more general formulation in terms of symbolic dynamics and to the idea of weak universality. The author explains why continuum analysis, computer simulations, and experiments form three entirely distinct approaches to chaos theory. In the end, the connection is made with Turing's ideas of computable numbers and it is explained why the continuum approach leads to predictions that are not necessarily realized in computations or in nature, whereas the discrete approach yields all possible histograms that can be observed or computed.

This algorithm approach to chaos, dynamics, and fractals will be of great interest to graduate students, research workers, and advanced undergraduates in physics, engineering, and other sciences with an interest in nonlinear science.

*Chaos, dynamics,
and fractals*

Cambridge Nonlinear Science Series 2

Series editors
Professor Boris Chirikov, *Budker Institute of Nuclear Physics,
Novosibirsk, Russia*
Professor Predrag Cvitanović, *Niels Bohr Institute, Copenhagen*
Professor Frank Moss, *University of Missouri–St Louis*
Professor Harry Swinney, *Center for Nonlinear Dynamics,
The University of Texas at Austin*

Titles in this series

Chaos, dynamics, and fractals

An algorithmic approach to deterministic chaos

Joseph L. McCauley

Department of Physics, University of Houston

CAMBRIDGE UNIVERSITY PRESS

Published by the Press Syndicate of the University of Cambridge
The Pitt Building, Trumpington Street, Cambridge CB2 1RP
40 West 20th Street, New York, NY 10011-4211, USA
10 Stamford Road, Oakleigh, Melbourne 3166, Australia

First published 1993
First paperback edition (with corrections) 1994
Reprinted 1994

Printed in Great Britain by Athenæum Press Ltd, Gateshead, Tyne & Wear

A catalogue record for this book is available from the British Library

Library of Congress cataloguing in publication data available

ISBN 0 521 41658 2   hardback
ISBN 0 521 46747 0   paperback

KW

For: Chrissy, Katie, Marianne, and Hans

Contents

Preface

The study of deterministic chaos by iterated maps goes back to the mathematician H. Poincaré, but did not become a part of theoretical physics until after M. Feigenbaum's discovery, and analysis by a renormalization group method, of universal critical exponents at the transition to chaos in a class of one-dimensional maps (one-dimensional maps had also been studied as paradigms of chaos in higher-dimensional systems by Lorenz and Grossmann). Since the discovery of universality at transitions to chaos, and the observation of period-doubling sequences by A. Libchaber and his co-workers in fluid mechanics experiments, much has been written about deterministic chaos and fractals. However, one thing must be stated in the beginning: although this book is primarily about iterated maps, the method of analysis and choice of emphasis make it very different from all of the others. It is written for those who not only want an introduction to modern developments in nonlinear dynamics and fractals, but also want to understand the following questions: How can a deterministic trajectory be unpredictable? How can nonperiodic chaotic trajectories be computed? Is information loss avoidable or necessary in a deterministic chaotic system? Are deterministic chaotic orbits random? What are multifractals, and where do they come from? Why do we study iterated maps instead of differential equations?

Physicists like to begin with the laws of energy and momentum for a mechanical system (typically formulated in terms of differential equations), and describe how their consequences (solutions of differential equations) govern the motion of the different parts of the system. It is relatively easy to write down the coupled nonlinear equations that one should, in principle, begin with. The Navier–Stokes equations with boundary conditions for the flow past a sphere, e.g., are easy to derive, and look deceptively simple (the coefficients are all analytic), but their solutions are apparently *not* all simple: some of them, at high enough Reynolds numbers, presumably contain information that describes fluid turbulence. At present, we do not understand how to analyze the Navier–Stokes equations *analytically* in order to see whether, or in what form, the information that describes the energy-eddy cascade in fully developed turbulence (the phenomenology of the energy-eddy cascade is described in Chapter 10) is included in solutions at high Reynolds numbers.

The Navier–Stokes equations, although simple in appearance, apparently can generate much more *information* than we understand how to organize, either analytically or with the help of the best available high-speed digital computers. Not even with the fastest available digital computer can we make truncations of the original, nonlinear partial differential equations that we can justify (or even believe in) as telling us clearly or convincingly how turbulence follows from the laws of physics.

The inability to analyze mathematically the systems of differential equations that are of direct interest in experimental physics and engineering led at first to the neglect of nonlinear dynamics problems in physics, but later to the study of iterated maps. Iterated maps *have* been used successfully to analyze and to predict nonlinear phenomena that occur in nature: period doubling is only one example. The reason for the lack of a systematic quantitative connection between chaotic phenomena, as described theoretically by maps or as observed in nature, and the traditional language and ideas of physics is *symbolized* by our inability to discover the range and complexity of information that can be generated by the Navier–Stokes equations and other systems of coupled nonlinear partial differential equations.

The study of maps has the redeeming feature that we finally end up with mathematical models where we *can* see the chaos and can begin to learn to analyze and understand it both qualitatively and quantitatively. Furthermore, from the study of iterated maps, one sees clearly how nonlinear equations that are very much simpler in form than the Navier–Stokes equations can generate complicated behavior completely deterministically, eliminating the need for assumptions of randomness as a basis for complex phenomena in mathematics, and perhaps also for macroscopic phenomena in nature.

The situation whereby we should try to understand chaotic behavior from the differential equations approach is, in reality, much worse than we have admitted so far. Suppose that you want to try to understand some aspect of the earth's weather system from a theorist's standpoint. You should start with the Navier–Stokes equations including the heat equation (physics) and could try to go forward by making a 'mode expansion' (a mathematical method for replacing a system of partial differential equations by infinitely many ordinary differential equations), which you would find necessary to truncate arbitrarily, because (if the mode expansion converges) there is as much information in infinitely many nonlinearly coupled modes as there is in the original partial differential equations. This is not a justification for truncation, but serves only to emphasize the enormous difficulty of solving the problem. Suppose that you keep only three particular modes, which cannot contain enough information to describe the weather, but whose range of predicted behavior you hope to be able to understand *mathematically*, because the resulting 'information production system' is now reduced to three coupled nonlinear ordinary differential equations (the Lorenz model provides an example). Even if we *can* see chaos *qualitatively* in the geometric study of the flow in the model's three-dimensional phase space (as one can in the Lorenz model), we still do not understand how to integrate our chaotic model correctly *numerically*. No one has yet figured out how to do it the right way, although anyone with a desktop computer can 'integrate' every three-dimensional chaotic set of coupled differential equations incorrectly by ignoring the errors that accumulate rapidly. Error accumulation is a problem because chaotic equations generate information too fast for arbitrary truncation methods (neglect of all derivatives beyond a certain fixed order) and floating-point arithmetic (fixed-precision) to be able to handle. Unable to use 'standard' computation methods to generate results that are adequate for more than a superficial understanding of chaos in coupled systems of nonlinear ordinary differential equations, one *then* wants to understand how the study of the phase space flow can be replaced by the study of iterated maps. Digital machines operate very well with algorithms that require only *iteration*, so that this shift of emphasis is in the right direction for machine-based studies of chaos, but it still does not solve the problem of error accumulation. Digital computers can iterate integer- or string-valued maps perfectly, but standard programming methods that use floating-point arithmetic introduce uncontrollable errors into the iteration of a chaotic map of a *continuous* variable, which is why standard attempts to integrate chaotic differential equations numerically on a computer are subject to error pileup in the first place. The explanation how and why this problem arises amounts to a description of the essence of deterministic chaos and can be found in Chapter 2.

On the one hand, computers are used as tools in an attempt to discover properties of chaotic systems empirically. On the other hand, there are theorems in mathematics, at least for restricted classes of iterated discrete maps of continuous variables and flows, that predict certain special statistical properties to occur 'with measure one'. The question of whether appropriately coarsegrained versions of the measure-theoretic predictions should be expected to be reflected in machine calculations, much less in experiment, has not been adequately analyzed in any current book on nonlinear dynamics. Machine (or any) computation, even theoretically, takes place under circumstances of finite precision and finite time on a measure zero subset of the mathematicians' continuum. The probabilities that are generated by machine studies, and in real experiments and observations of nature, are simply the *frequencies of occurrence* of finitely many different discrete events. Analytical properties of an iterated map $z_{n+1} = f(z_n)$, in contrast, are derived on the basis of the presupposition of infinite precision (z_n is allowed to vary continuously) and infinite time limits (leading to ergodicity or mixing, and to probabilities that are defined as *set theoretic measures*, as opposed to frequencies of occurrence of finitely many events), often combined with a *hidden* assumption that it makes sense to assume that we can draw numbers randomly from the mathematical continuum. In theoretical physics, this discrepancy between the continuous and the finite usually does not pose a serious problem, but deterministic chaos is precisely the part of physics and mathematics where the geometric complexity of a dynamical trajectory is such as to defy its long-time predictability *unless* one is able to devise a calculation method that *completely* avoids the introduction of errors into the computation in the first place, because single trajectories *are* the sources of correct statistics. The result of trying to iterate chaotic maps of continuous variables while using floating-point arithmetic on a machine is that errors are introduced into the iteration scheme that can make it difficult to understand what the resulting numbers have to do with the statistics that *can* be generated mathematically by the different trajectories of the particular map. Even if that confusion should be eliminated (usually, it is not), one obtains at *best*, by that method, only a very special class of statistics that is compatible with (or is dictated by) the machine's roundoff/truncation algorithm. In contrast (see Chapter 9), even the *simplest* chaotic maps can in principle generate *every* imaginable statistical distribution that can be empirically constructed as a histogram. The extent within which this diversity of behavior can be realized in practice hinges upon the availability of algorithms that generate either initial conditions or symbol sequences. Furthermore, chaotic maps produce the different classes of statistics for sets of initial conditions that cannot be mimicked at all by iterating a map forward on a computer in the floating-point mode. There is an enormous and dangerous gap between the language and results of pure mathematics and the language and results of digital computers, and there is in the other available texts on chaos no bridge between the two. We use a language of 'theoretical computer arithmetic' to provide a formulation of chaos theory that computers can be programmed to handle exactly, in finite *but not fixed* precision, without the introduction of even a single error. The formulation is not merely theoretical, but can be carried out in practice. Our formulation also contains within it the possibility of all of the limits of a purely analytic mathematical approach based upon the continuum, if one mathematically takes the computationally and experimentally unattainable limits of infinite precision and infinitely many iterations, where events that occur with 'measure zero' could then be distinguished from events that occur with 'measure one'.

Because finite (but not fixed) resolution is an unavoidable fact of life in experiment and computation, all predictions that can be formulated in finite precision are equally treated in this text, including the 'measure zero' motions. Prejudice in favor of special metric properties of maps (which are, at any rate, nonuniversal) is avoided by an approach that is based upon symbolic dynamics. We also emphasize that experiment, computer simulation, and continuum analysis are three completely distinct fields of investigation of deterministic chaos whose regions of overlap were not previously made clear. We hope that this book contributes toward reducing the confusion over the overlap of computer-based and purely analytic studies of deterministic chaos. Deciding what is important for the description of nature is a much harder *unsolved* problem.

We begin in Chapter 1 with geometry, with flows in phase space, but we progress rapidly to iterated maps where we study maps as automata that process digit strings. Chaos theory is the deterministic theory of disorder. For a physicist, disorder is visualized discretely as lattice disorder, as in the classical statistical mechanics of discrete spin systems. The nearest numerical analog of a lattice is a digit string or a symbol sequence. This book describes deterministic disorder (which is just as good a phrase as deterministic chaos) mainly in terms of digit strings and their processors, automata, and also in terms of very useful *universal* disordered 'lattice configurations' called symbol sequences. The reason for our chosen formulation is that it is the simplest and, perhaps, the only one that yields a clear picture of what *is* deterministic chaos *within the context of computation*. Our digital method of analysis generates hierarchies of coarsegrained versions of the measure-zero properties (including the so-called multifractal measures) that are ignored in the usual treatment based upon iterated maps of continuous variables, as well as the measure-one behavior that is normally emphasized in other texts.

This book was written while the author was teaching modern deterministic dynamics in courses and lectures that were given over an eight-year period at the University of Houston, Universitetet av Oslo, Instituttet for energiteknikk (Kjeller, Norway), and Universität-Gesamthochschule-Siegen. The reading level is that of an advanced graduate-level physics course at the University of Houston. There the course was attended by engineering and physics students with standard backgrounds in mechanics and differential equations. Some parts of the text require more advanced mathematics knowledge for a complete understanding. For example, we have already assumed that the reader understands (or can learn) what the mathematician means by a measure, although the digital method of analysis that is developed in the text can be followed by a reader who has absolutely no knowledge of measure theory; in fact, measure theory is the *last* prerequisite that we would choose for a course on deterministic chaos. In Chapter 1 we introduce the time evolution operator and develop the classical connection between one-parameter continuous groups and the flows in phase space. Poisson brackets are introduced quite generally in Chapter 1, following the treatment of Arnol'd, and are used to discuss conservative systems in Chapter 3. The time evolution operator is used formally to discuss stroboscopic maps, which are studied further in Chapter 5. Chapter 3 requires advanced knowledge of Hamiltonian mechanics, and the necessary elementary introduction to Lie groups can be found in Chapter 1 and in Appendix 3.A at the end of Chapter 3. Deterministic chaos is introduced in Chapter 2 in the context of the Lorenz model, and the essential ideas are analyzed and systematically developed in terms of one-dimensional maps of the interval. In Chapter 4, maps are used to generate fractals, and the idea of generalized dimensions is introduced as a way to characterize singular probability distributions. In Chapters 6, 7, and 8 the reader is expected to understand Boltzmann's statistical definition of entropy. The terminology of critical phenomena, borrowed from statistical mechanics, is explained qualitatively and is used in Chapter 6 in order to set Feigenbaum's discovery of universality for the period-doubling transition to chaos into some historical perspective, although the mathematics in that chapter can be followed without the benefit of the historical perspective. Multifractals, based upon the formalism of the Chicago and NORDITA Schools, are introduced and discussed in Chapters 7, 8, and 9. Chapter 8 requires a physicist's background in classical statistical mechanics. In that chapter, we discuss the equivalence of the statistical equilibrium of one-dimensional, fully chaotic dynamics with one-dimensional lattice systems in statistical mechanics at both positive and negative temperatures, an advanced topic in multifractals. The phenomenology of the eddy cascade in fully developed turbulence is discussed in Chapter 10, along with nonfractal, fractal, and multifractal models of the cascade. The recent experimental results on intermittence in energy dissipation at small scales in turbulence are also discussed along with older results on spatial intermittence in the inertial range, but other experimental results are not discussed elsewhere in the text. For the experimental papers, we recommend P. Cvitanović's reprint collection (in the second edition) entitled *Universality in Chaos*. Finally, in Chapter 11, the theoretical basis for our fully discrete approach to chaos theory based upon the notion of algorithms for computable irrational numbers is given. For those without advanced background in physics, the book can be read by omitting Chapter 8. It is also possible to omit Chapter 3 if the reader is interested primarily in dissipative systems. Chapter 11 was written with the intention of making the reader reflect more carefully upon certain ideas about chaos in the context of computation, and what is possible in computation, and to provide an interesting point of view that cannot be found in other texts.

The book is intended for self-study, but can also be used in one of several different chapter-sequences for a one-semester course: 1, 2, 4, 6, 7, and 9 (chaos, fractals, transition to chaos by period-doubling, multifractals, and universal statistics in chaos); 1, 2, 4, 7, 9, and 10 (chaos, fractals, multifractals, universal statistics in chaos, and turbulence); 1, 2, 4, 5, and 6 (chaos, fractals, and transitions to chaos); 1, 2, 3, 4, and 5 (chaos in dissipative and conservative systems and the transition to chaos by destruction of quasiperiodic orbits); 1, 2, 3, 6, 7, and 8 (chaos, transition to chaos, multifractals, and statistical mechanics of chaos); or 1, 2, 4, 7, 9, and 11 (flows, iterated maps, symbolic dynamics, and automata).

I am extremely grateful for frequent guestfriendship to my good friends Tormod Riste and Arne Skjeltorp at the Norwegian Institute for Energy Technology (IFE), where forerunners of several of the book's chapters were written in 1985 in a stimulating but calm and friendly atmosphere that could hardly have been more conducive to thinking, preparing lectures, and writing. I spent the second half of my 1985–6 free year at the University of Siegen where I lectured and continued my research on the computability of chaos. For that experience, I am indebted to my good friend Dyt Schiller, who became infected by my enthusiasm for chaos theory in Houston in 1984, where I iterated maps on a computer, having deferred asking the hard question: what do the numbers mean? Later, in 1984, Herbert Wagner visited Houston and made us aware of the idea of cellular automata. In contrast with the clarity of the input and output of integer maps (or rules) like cellular automata, none of the articles in the literature on chaos theory explained clearly or correctly how one should interpret the pseudo-orbits that follow from every attempt to iterate a discrete chaotic map of a *continuous* variable on a machine while using floating-point arithmetic. The most widely quoted papers at that time, by Benettin and coworkers, and by Rannou, fell (honorably) very short of the mark, and Joe Ford's stimulating article in *Physics Today* ('How random is a coin toss?') raised many more questions for me than it answered. My ideas suddenly gelled while reading the brief, clear, stimulating and astounding description of the idea of computable numbers in Andrew Hodge's biography of Alan Turing. The reading of that same passage by my colleague Julian Palmore was the beginning of our collaboration on the formulation and study of deterministic chaotic maps as digit string processors. I am especially grateful to Julian for many telephone conversations, and for a fruitful collaboration from 1985–9 that produced research papers that form the basis for the digital method of this text. The parallel-processing method that is introduced in Chapter 2, for example, was explained to me by Julian on a cold, late January night near Oslo during my academic free year.

I am grateful to the University of Houston Physics Department for help with the preparation of early drafts of the manuscript which were typed, in large part, by Sandi White. At IFE, Gerd Jarrett was of enormous help with an even earlier version of parts of the manuscript. Final drafts were written at the University of Houston on a MacIntosh SE/30 that was purchased on an ONR grant, thanks to Mike Schlessinger, whom I met in Norway. I am grateful to Y. C. Chae and J. Kim for a careful reading of the first few chapters of early drafts of the manuscript and to many other graduate students in my classes since 1983, especially those from mechanical engineering at the University of Houston, for good questions and stimulating discussions. Finally, I have benefited from hearing a set of informal lectures presented in 1988 by Predrag Cvitanović at the University of Oslo (where the topological universality of unstable periodic orbits was emphasized). I am grateful to Fazle Hussain, who encouraged me in 1984 to transform my lectures into a book, and to Simon Moss, who also encouraged me to continue my work. My friend and constant companion, Cornelia Küffner, a writer and former editor for the Schwarzwälder Bote, helped me to focus on the content of the Preface by playing the role of a hypothetical antagonistic reader who is uninterested in hearing new ideas about old subjects. Finally, I am grateful to many other colleagues and students for discussions, including Preben Alstrøm, Eric Aurell, Tomas Bohr, Gemunu Gunaratne, Bambi Hu, Hua Tang, Jiri Müller, Harry Thomas, and Armand Yethiraz. I am especially grateful to Gemenu Gunaratne, Jiri Müller, Harry Thomas, and Dyt Schiller for reading parts of the manuscript and making suggestions for corrections and improvements. Several of the drawings for Chapters 1 to 6 were generated by computer by Hua Tang at the University of Houston and by Tuulike Storeboe at Kjeller.

Forerunners of Chapters 1, 2, 3, 5, and 11 can be found in *Physica Scripta* **T20** (1989), which represents the lectures given during my 1985–6 free year at the Universities of Oslo and Siegen. Former versions of Chapters 7, 8, and 10 constitute *Physics Reports* **189**, No. 5 (1990), which started as a set of informal lectures given at Kjeller in the summer of 1987. Chapter 9 and the last part of Chapter 10 are based upon the article in *Zeitschrift für Physik* **81**, 215 (1990), which ties together ideas of multifractals and computability.

At different stages of writing, teaching, and research, this work was supported by The American–Scandinavian Foundation (ASF), Norges Allmenvitenskapelige Forskningsråd (NAVF), the Nordisk Institutt for Atomfysikk (NORDITA) and the Office of Naval Research (ONR).

Joseph L. McCauley
Houston, Texas

Og eg gjorde både etter deira bøn og råd, og eg granska mykje i alle desse rødene med ettertenksamt minne, og så sette eg alle desse rødene i ei bok; ikkje berre til gaman for øyro og kvikk tidtrøyte for dei som høyrde, men heller til mangfaldig nytte for alle dei som med rett åthug eignar til seg denne boka og følger vel alt det som blir påbodi i boka. Og boka er såleis laga at ein vil tykkje ein har både laerdom og hugnad, og dessuten stort gagn, dersom det blir laert og påakta vel, det som er skrivi i henne. Men den som har fullt og rett skjøn, han blir var at det trengst ei storre bok til uttydning enn den som er skriven.

Kongsspegelen

Introduction

Since the beginning of physics, the study of theoretical mechanics has been synonymous with the study of the gross *regularities* of nature. Regularity of occurrence either in space or in time depends implicitly upon an underlying *stability* in the development of chains of events whereby small changes in starting conditions do not produce the need for a statistical description: nearby starting conditions must yield patterns of trajectories that are similar in appearance at later times. At the other extreme lie phenomena where attempts at reproducibility fail. They fail in the sense that 'repeated identical experiments', experiments where the starting conditions are as close to one another as experimental error will permit, yield very different, apparently unrelated patterns of trajectories as time goes on. This kind of behavior we call statistical. It has been widely assumed that the cause of statistical behavior is randomness, which, if one takes the word literally, is the same as saying that there is no identifiable cause at all, because the explanation of phenomena in terms of cause and effect requires a *deterministic* description of chains of events.[1]

Irregularities in the form of nonequilibrium disordered phenomena were once presumed to be describable only with the aid of assumptions of randomness: a deterministic description was thought either to be impossible or else was too complicated, in principle, to be practical. One

---

[1] In quantum theory, the equation of motion for the probability amplitude is deterministic, but there is no deterministic description of the space–time trajectory of a particle. In classical mechanics, there is no corresponding breakdown in the deterministic description of the motion in phase space. We expect that quantum effects are negligible at large enough length scales (fluid turbulence provides an example), except where there is macroscopic quantum coherence (superfluidity is an example). To the extent that quantum effects are negligible macroscopically, the analysis of this text will be correct on coarsegrained levels that are large compared with interatomic spacings (see Chapters 4 and 10 for the systematic coarsegraining of phase space).

reason why a deterministic description of disorder was thought to be impractical was the incorrect but widely held belief that apparent randomness in deterministic equations required for its appearance a large number of degrees of freedom. The realization that one-dimensional noninvertible and two- (and higher-) dimensional invertible maps can yield pseudorandom behavior, and that these same maps can describe, albeit incompletely, the behavior generated by higher-dimensional systems of deterministic nonlinear differential equations has changed the way that we think about the origin of phenomena in nature that require a statistical description.

The stability whereby the regularities of nature occur is explained *mathematically* by the well-known 'classical' attractors in the underlying dynamical system: stable equilibria, stable limit cycles, and stable tori in phase space.[2] Historically seen, concentration upon this knowledge led to the attempt to describe nature mathematically in terms of stable periodicity and stable quasiperiodicity. However, the solutions of problems where statistical behavior occurred far from thermodynamic equilibrium, with or without branching and fragmentation, were in no significant way advanced by this approach, even when arbitrary postulates of randomness were included. The most outstanding unsolved problems of this sort in classical physics were the phenomena of fluid turbulence, where the branching and fragmentation that define the eddy cascade require a statistical description, and the apparently irreversible approach of a conservative dynamical system to a state of statistical equilibrium that is described by a Gibbs distribution.

Most nonlinear problems in dynamics remained unsolved because there existed no *general* mathematical method of attack: each problem, in the absence of a universal method, seemed to be a law unto itself. In conservative mechanics, success was found through the use of Sophus Lie's method of continuous groups of transformations, but the solution of problems by symmetry methods, both in classical and quantum mechanics, led physicists only further along the path whereby problems were attacked from the standpoint of stable periodicity or stable quasi-periodicity (cf. Landau's theory of turbulence as an example), or else were declared to be random. Problems that are completely solvable by symmetry are the opposite of deterministic chaos, and the random

---

[2] We use the word attractor loosely here. Attractors and repellers occur only in dissipative systems. Strictly speaking, we should use the more correct phrase 'invariant set' in this introduction, but we defer that refinement until Chapter 2, where invariant sets are defined.

perturbation of stable orbits does not explain the statistics that arise from chaos far from thermal equilibrium.

Given that we understand dynamics near the 'classical' attractors and repellers (equilibria, limit cycles, and tori in phase space), the following intriguing question leads us to the archway beyond which deterministic chaos resides: what is the pattern made by a trajectory in a deterministic system where the motion is bounded, but all of the attractors of the classical variety are unstable inside the bounded region in phase space?[3] This intriguing question was not addressed by classical theoretical physics, but a dynamical system with exactly this property in a three-dimensional phase space was studied in 1963 by E. Lorenz, who attempted a numerical integration on a six-bit computer. *Because* of errors that were made by the computer's roundoff/truncation algorithm, he discovered what is now called sensitivity with respect to small changes in initial conditions: big changes in trajectory patterns occurred at later times owing to shifts in the last digits of the starting conditions. Because Lorenz had proven analytically that all orbits that started outside a certain ellipsoid in phase space had to enter that ellipsoid in finite time and are bounded by a concentric sphere that contains the ellipsoid, there was the *intuitive* idea of a stable *attractor* in spite of the instability of all *classical* attractors inside the bounding sphere. These two discoveries opened the door to something new, and led later to the introduction of the phrases 'sensitivity with respect to small changes in initial conditions' and 'strange attractor' in the work by the mathematicians Ruelle and Takens. Ruelle also contributed to the development of the formal mathematical connection between one-dimensional iterated maps and one-dimensional equilibrium statistical mechanics, where the symbol sequences of chaos theory become the disordered lattices of, e.g., one-dimensional classical spin systems.

The idea of a strange invariant set (attractors and repellers are invariant sets) is intimately connected with the appearance of statistical behavior in a deterministic system; in a dissipative system, fragmentation in phase space can also occur, leading to invariant sets that are strange attractors or strange repellers. How can one understand these 'nonclassical' phenomena in terms of equations describing the time evolution of the dynamical system? The path to further progress had already been suggested by H. Poincaré, who found it convenient to replace the study of a system of differential equations by the study of an iterated map. By using simple

---

[3] Here, the cause of the irregular motion cannot be accounted for by the 'random' perturbation of a stable periodic or stable quasiperiodic orbit. Instead, the disorderly conduct must be generated internally by the dynamics of the system itself.

iterated maps, it is easy to see how both fragmentation and statistical behavior develop purely deterministically from the rules of arithmetic. It is usually not practical to try to derive the correct map for a given set of differential equations. Instead, one studies relatively simple maps and then generalizes the results to include an entire class of topologically related maps by a universality principle. The main emphasis in chaos theory, by physicists, has been upon the discovery and implications of universality principles.

In this book, most of chaos theory is developed in terms of the simplest one- and two-dimensional maps, the best choice for a clear introductory presentation of the main ideas. We illustrate by an example in Chapter 2 that one- and two-dimensional maps cannot contain complete information about higher-dimensional dynamical systems, but they can provide a partial (incomplete) description of more complicated dynamical systems. Universality principles at transitions to chaos are discussed but we also discuss a different kind of universality whereby the different classes of statistics that can be generated by an entire class of fully chaotic maps, including asymmetric logistic maps that generate fractals, are explained and predicted in complete detail from the properties of the binary tent map in binary arithmetic. This is the reason for the emphasis placed in the text upon understanding the details of the binary tent map as a digit string processor.

Early in the book, we begin to use what one can call a language of theoretical computer arithmetic in order to systematically replace discrete maps of continuous variables by integer maps called automata. The automata are discrete processors of strings of digits. We develop this formulation of iterated maps of continuous variables by using expansions for all numbers and algebraic symbols in a definite base of arithmetic. As was stated by von Neumann, decimal expansions are an application of information theory, and our main job here is to keep track of the information that is generated extremely rapidly by a chaotic dynamical system. An advantage of this approach is that we can make as clear as is possible, in the context of computation, what deterministic chaos is, as well as what it is not. In the traditional continuum-analytical approach, the reader can easily become lost in the confusion of trying to understand how a system can be simultaneously deterministic *and* unpredictable, and is likely to end up with an inadequate idea of what happens when he programs a computer in an attempt to compute a chaotic trajectory. There are purely analytic theorems from mathematics that predict properties of chaotic orbits for certain well-known systems, for a class of initial conditions that occur with 'measure one' in the phase space. In the traditional approach, where the gap between continuum analysis (measure-one results) and computation via algorithms (measure-zero results) is not taken seriously, the 'measure-zero' nonperiodic orbits are ignored. We discuss the evidence that this viewpoint is inadequate. Our conclusion is that theorems from measure theory cannot be relied upon to tell us the trajectories of a chaotic system that are most likely to occur in experiment, nor can continuum analysis be used to weigh the relative importance of the different sorts of statistics that are generated by chaotic orbits in computations, where initial conditions are fixed precisely and algorithms can be used to avoid error pileup. We therefore develop a method of analysis that treats the coarsegrained versions of both the measure-zero and measure-one properties on an equal footing, and wherein the idea of a pattern of a chaotic orbit is seen to be equivalent to a pattern in a string of digits (symbolic dynamics), or the pattern of letters in a 'word' constructed from a finite alphabet.

One of our final observations is that chaos theory is the part of theoretical physics where Turing's ideas of computable numbers and functions enter naturally and nontrivially. The formulation of chaotic problems in terms of algorithms for digit strings of numbers (the automaton, or algorithmic formulation) is extremely useful whenever one wants to compute chaotic trajectories with controlled precision (i.e., if one wants to know what the numbers mean) or if one wants to know generally what are the possible applicable results of mathematical chaos theory. In the book's last chapter, we present arguments, based upon Turing's ideas, in favor of a fully discrete approach to chaos theory. A consequence of the use of algorithms is that one does not fall into what was called by von Neumann the 'state of sin' of believing that randomness can be produced from arithmetic. A consequence of reading this book is that one will learn how to use algorithms to generate the different statistical distributions of chaos that are not analyzed at all in other treatments of the subject.

1
Flows in phase space

1.1 Determinism, phase flows, and Liouville's theorem

We begin with the general idea that the state of a dynamical system is defined by n variables at time t, $X(t) = (x_1(t), \ldots, x_n(t))$. Geometrically, this is a point in an n-dimensional space called phase space. By determinism, we mean that there is a definite rule whereby, if the state $X(t)$ occurs, then it arose by a sequence of steps from a definite state $X(t_0)$ at an earlier time $t_0 < t$. In other words, if $X(t)$ occurs then $X(t_0)$ had to have occurred earlier. As can happen in a nonlinear system, nonuniqueness can arise (one initial condition $X(t_0)$ may give rise to two or more distinct solutions),[1] but this is only a complication. The main point is that there is no randomness or possibility of free choice in the time evolution from one state to another: the future is fixed once and for all by the past. Whether the time evolution rule is given in the form of a set of coupled differential equations, iterated maps, machines, or mathematical models of machines (automata) is unimportant for the definition at this stage, but we begin this chapter with deterministic systems of differential equations and end with discrete maps. The link between differential equations, iterated maps, and automata is introduced by examples in Chapter 2.

It is no accident that the historical path in physics has led to the attempt to understand nature through the study of differential equations. There is an underlying belief in continuity in classical dynamics that is based upon Euclidean geometry, and takes the form that an undisturbed body follows a straight line, at constant speed, in an inertial frame of reference

---

[1] An example is given by $dy/dt = y^{1/2}$ with $y(0) = 0$. Three possibilities follow: $y = 0$ for all t, $y = t^2/4$ for $t \geq 0$ (with $y = 0$ for $t < 0$), or $y = 0$ for $0 \leq t \leq t_0$, but $y = (t - t_0)^2/4$ for $t \geq t_0$.

(Newton's first law). That this is true within the classical regime, whereby quantum and relativistic effects can be ignored, is not questioned, but the generalization of this way of thinking leads us to attempt to explain all the macroscopic phenomena observed in nature as solutions of differential equations: smooth, continuous curves – smooth variations on the straight line theme. The continuum approach accounts very well for the motion of the earth about the sun, because the inverse square law of gravity yields a regular geometric object for the orbit, an ellipse. In contrast, there is a large collection of problems that can be formulated as solutions of differential equations, problems of growth of objects of irregular shape, for example, where the observed natural evolution of the system is very difficult to predict and to understand clearly from the point of view of smooth solutions of deterministic systems of differential equations. For example, the flow of a fluid is easily *formulated* as a coupled set of nonlinear partial differential equations, the Navier–Stokes equations, but fully developed fluid turbulence shows fragmentation combined with apparently statistical behavior. The fragmentation consists of the formation of a large eddy (or several eddies) that is unstable and fragments into several smaller, also unstable eddies. The branching into more and smaller eddies continues until viscosity dominates and damps the smallest eddies (Fig. 1.1). Statistics come into play because the observed space–time patterns of eddies in the cascade can deviate widely from one another when small changes are made in the initial conditions. It is hard to obtain solutions where such fragmentation can be understood within a way of

Fig. 1.1 Instability and fragmentation of larger vortices into smaller ones in a low-Reynolds number shear layer due to letting an ink droplet fall into a container filled with water (photo by A. Skjeltorp).

thinking whose basic building blocks are straight lines, ellipses, and hyperbolae, in spite of the fact that a 'marked particle' in the flow appears, to the unaided eye, to follow a continuous trajectory (but the pattern made by the trajectory is not a simple one). This is not to say that such problems cannot be understood in terms of differential equations, but it is fair to raise the question whether physics can be studied in terms of discrete maps and fully digital maps called automata, rather than in the language that seventeenth-century thinkers found advantageous for the description of an entirely different class of mechanics problems. At any rate, while our study begins with differential equations and flows in phase space, we shall rapidly progress to the study of maps, where the phenomena of deterministic chaos are much more transparent. The computation of chaotic orbits of maps will lead us naturally to the study of maps as automata. To summarize, by beginning with deterministic differential equations as flows in phase space, we shall be led to the study of iterated maps that generate both statistical behavior and fragmentation in phase space – all within a completely deterministic framework. The desire to compute chaotic orbits in a clear, completely unambiguous way will lead us to replace maps of continuous variables by integer maps, automata.

If we define our dynamical system by a system of differential equations,

$$\dot{x}_i(t) = V_i(x_1(t), \ldots, x_n(t), t, \mu), \tag{1.1}$$

with μ a control parameter (μ can also denote a set of control parameters), determinism is implicit in the fact that the solutions depend upon the initial conditions as well as upon t and μ:

$$x_i(t) = \Psi_i(x_1(t_0), \ldots, x_n(t_0), t_0, \mu). \tag{1.2}$$

Given that we start with differential equations, it is natural to try to think of the collection of trajectories in phase space as a flow analogous to the flow of a fluid. For the time being, we assume that the right-hand side of (1.1) is independent of t, which means that we restrict our considerations to self-regulating ('autonomous') systems, whereby the solution (1.2) depends only upon the time *difference* $t - t_0$. The extension to systems with a contribution to the right-hand side that varies periodically with t is given in section 1.4. A streamline of the flow is a trajectory of the dynamical system, and the velocity field $V = (V_1, V_2, \ldots, V_n)$ is everywhere tangent to the streamlines. The analog of the flow of a finite mass of fluid is the tube of streamlines traced out by the time development, starting from a finite phase space volume element $\delta\Omega(t_0) = \delta x_1(t_0) \cdots \delta x_n(t_0)$.

Every point in this volume element is an initial condition of (1.1), and so from each phase point issues forth a trajectory as t increases. The collection of these trajectories forms a cylinder, or tube, in phase space (Fig. 1.2). In nature, we can never specify the location of a point with infinite precision, so that a finite value of $\delta\Omega(t_0)$ geometrically represents the finite ignorance that is present at the start of every experiment or sequence of observations. In computation, we must write decimal expansions in order to compute, or else we must use rational numbers with larger and larger denominators, so that $\delta x_i(t_0)$ can be regarded as representing the remainder in the decimal expansion for an initial condition. It is sometimes possible to start a computation with $\delta\Omega(t_0) = 0$ by choosing rational numbers as initial conditions, but we can never avoid a finite $\delta\Omega(t_0)$ in nature. We shall see in Chapters 3 and 4 that a qualitative difference between the evolution of trajectories of chaotic and nonchaotic systems lies in the way that the *shape* of $\delta\Omega(t_0)$ changes as time goes on.

Let us, for the time being, set aside the question of how $\delta\Omega(0)$ might differ significantly in shape from $\delta\Omega(t)$, and concentrate instead upon the easier question: How does the size of the volume element change

Fig. 1.2 Tube of streamlines in phase space emanating from a ball of initial conditions.

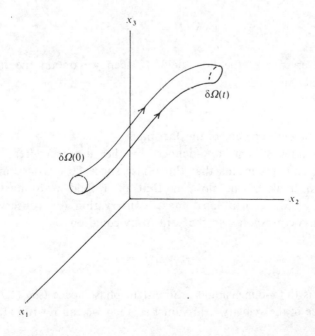

with time? Because

$$\delta\Omega(t) = J(t)\,\delta\Omega(0), \qquad (1.3)$$

where we take $t_0 = 0$, and where

$$J(t) = \frac{\partial(x_1(t), \ldots, x_n(t))}{\partial(x_1(0), \ldots, x_n(0))} \qquad (1.4)$$

is the Jacobi determinant of the transformation (1.2) (we want to think of (1.2) as a one-parameter coordinate transformation with parameter t), we need only study the derivative

$$\frac{dJ}{dt} = \frac{\partial(\dot{x}_1(t), x_2(t), \ldots, x_n(t))}{\partial(x_1(0), x_2(0), \ldots, x_n(0))} + \cdots + \frac{\partial(x_1(t), \ldots, \dot{x}_n(t))}{\partial(x_1(0), \ldots, x_n(0))}. \qquad (1.5)$$

By the chain rule for Jacobi determinants, we can write

$$\frac{\partial(x_1(t), \ldots, \dot{x}_i(t), \ldots, x_n(t))}{\partial(x_1(0), \ldots, x_n(0))} = \frac{\partial(x_1(t), \ldots, \dot{x}_i(t), \ldots, x_n(t))}{\partial(x_1(t), \ldots, x_n(t))} \cdot J(t), \qquad (1.6)$$

and this reduces to

$$\frac{\partial(x_1(t), \ldots, \dot{x}_i(t), \ldots, x_n(t))}{\partial(x_1(0), \ldots, x_n(0))} = \frac{\partial \dot{x}_i(t)}{\partial x_i(t)} J(t), \qquad (1.6b)$$

because the variables $(x_1(t), \ldots, x_n(t))$ are independent. If we observe that

$$\nabla \cdot V = \sum_{i=1}^{n} \frac{\partial \dot{x}_i}{\partial x_i} \qquad (1.7)$$

is the divergence of the flow field V, then we obtain the first-order differential equation,

$$\dot{J} = \nabla \cdot VJ, \qquad (1.8)$$

for the time development of the Jacobian.

Conservative systems are defined by the divergence-free condition $\nabla \cdot V = 0$, and this means that the phase space flow is incompressible: if we can think of the flow as that of a fluid with mass density $\rho(x_1, x_2, \ldots, x_n, t)$, then local mass conservation (or conservation of probability) corresponds to the continuity equation

$$\frac{\partial \rho}{\partial t} + \nabla \cdot \rho V = 0, \qquad (1.9)$$

where ∇ is the n-dimensional gradient in phase space (see (1.7)). If the divergence of the velocity field vanishes, then we can rewrite (1.9) in the form

$$\dot{\rho} = \frac{\partial \rho}{\partial t} + V \cdot \nabla \rho = 0, \tag{1.10}$$

which is exactly the condition for $d\rho/dt = 0$, or $\rho = $ constant in time along a streamline, because $\dot{\rho} = d\rho/dt$ is the time derivative along a streamline in the flow. That is, the flow is incompressible. A Hamiltonian system, the 'canonical' form of a conservative set of differential equations, is, when it exists, one with $n = 2f$ 'canonically conjugate' variables $X = (q_1, \ldots, q_f; p_1, \ldots, p_f)$ and a function $H(q_1, \ldots, p_f)$, the Hamiltonian, where

$$\dot{q}_i = \frac{\partial H}{\partial p_i}, \qquad \dot{p}_i = -\frac{\partial H}{\partial q_i}, \tag{1.11}$$

which can be regarded as a generalization to $2f$ dimensions of the idea of a stream function in the two-dimensional hydrodynamics of ideal fluids. It follows that

$$\nabla \cdot V = \sum_{i=1}^{f} \left(\frac{\partial^2 H}{\partial p_i \, \partial q_i} - \frac{\partial^2 H}{\partial q_i \, \partial p_i} \right) = 0 \tag{1.12}$$

for Hamiltonian systems. Therefore, we label a Hamiltonian system conservative whether or not H depends explicitly upon t. Liouville's theorem is usually quoted as the statement that the phase space density ρ has no explicit time dependence for Hamiltonian systems, but this is the weaker version of Liouville's theorem because there are probability distributions for which no density exists. The condition that $J = 1$ is a more general definition of a conservative system; in particular, it is directly applicable to iterated maps.

In order to illustrate that one needs only the differential equations (or iterated map) but not the solution in order to decide whether a given system is dissipative or conservative, consider any differential equation of the form $d^2x/dt^2 + f(x) = 0$. With $x = x_1$ and $dx/dt = x_2$, we obtain $\nabla \cdot V = 0$, so that all such second-order differential equations are conservative. However, from the equation $d^2x/dt^2 + g(dx/dt) + f(x) = 0$, we obtain $\nabla \cdot V = -\partial g/\partial x_2$, representing dissipative systems.

Hamiltonian systems are a subset of all possible conservative systems and were used by Gibbs to form the basis for the microcanonical ensemble in statistical mechanics, where a completely isolated dynamical system is considered. In that case, the entropy is given by $k \ln \Omega(t)$ and is constant in time (k is Boltzmann's constant). In contrast, the second law of thermodynamics demands that the entropy should increase monotonically until a state of equilibrium is reached. The question of how to resolve this apparent contradiction has been a fundamental problem in statistical mechanics since the time of Boltzmann. Gibbs suggested how the shape of $\Omega(t)$ should change with time in order to simulate an irreversible approach to statistical equilibrium ('mixing'), but it was not until much later that examples of conservative systems that could be shown to be mixing were identified. We discuss Gibbs' idea of mixing in Chapter 3, in conjunction with the interesting question of how the shape of the phase volume element changes with time in a chaotic dynamical system (there is a connection with the spreading of the ink droplet shown in Fig. 10.2).

It is typical in a nonconservative system that $\nabla \cdot V$ can change sign in different regions of phase space, but an example where this is not true is provided by the Lorenz model, which is introduced in section 1.3 below. For a purely dissipative system, $\nabla \cdot V < 0$ everywhere, so that phase space volumes $\delta\Omega(t)$ always contract during the flow.

In what follows, we shall introduce the classical attractors: stable equilibria, stable limit cycles, and stable periodic and quasiperiodic orbits on tori. Stable equilibria are zero-dimensional, while limit cycles are closed, one-dimensional curves in phase space. Each classical attractor has a definite dimension and not all attractors have the same dimension.

If one starts with a two-dimensional phase space, e.g., then the approach to zero of $\delta\Omega(t)$ for a dissipative system is easy to visualize in the classical case, because all the points in $\delta\Omega(t)$ eventually flow into one point or else contract onto a curve, so that $\delta\Omega(t) \to 0$ as $t \to \infty$ owing to a reduction in dimension. More generally, as the time goes to infinity in a dissipative system, the asymptotic motion takes place on a set of lower dimension than that of the full phase space owing to contraction of the volume element to zero size. In Chapter 2, the interesting question we will ask is: What happens when the motion in phase space is bounded, but the orbit is condemned to wander forever because all the available equilibria, limit cycles and quasiperiodic orbits are unstable? The answer to this question leads to the study of point sets that are called 'strange' because, among other properties, they usually have the cardinality of the continuum and can have a nonintegral fractal dimension. Fractals are introduced and discussed in Chapter 4. It is not easy to give simple examples that make clear in an elementary way how an attractor with a nonintegral dimension develops in a nonlinear system of differential equations, but simple mathematics can be used to show how fractal attractors and repellers are generated by discrete maps. At the end of this chapter, we shall describe one method whereby a certain kind of map, the stroboscopic map, can be obtained from differential equations. In Chapters, 2, 3, and 4, we show by example how behavior that can be called 'strange', in comparison with classical attractors and repellers, is generated by simple iterated maps.

We have introduced the idea that it is convenient to think of the solutions of a system of differential equations as a flow in phase space, but we have not discussed the conditions under which this is possible. In the study of canonical transformations in classical mechanics, one meets the notion that the Hamiltonian is the infinitesimal generator of the flow in phase space, and that the solutions of the differential equations can be regarded as a continuous group of transformations (see Chapter 3 for details). The Hamiltonian generates the group, and the condition for the existence of the transformation group is the same as the condition that the solutions can be described as a phase flow. When the system is nonconservative, no Hamiltonian can be constructed, but the condition for the phase flow viewpoint to hold is, again, the condition that the solutions (1.2) of (1.1) form a one-parameter group of transformations where the time is the group parameter. For certain self-regulating systems, one can then write

$$x_i(t) = U(t - t_0)x_i(t_0), \tag{1.13}$$

where $U(t - t_0)$ is the time evolution operator, but this is just a different (from (1.2)) formal way of writing the solution. In order to emphasize the group property, consider the condition that one can use a phase point $(x_1(t_0), \ldots, x_n(t_0))$ at time t_0 as the initial condition to find the solution $(x_1(t_2), \ldots, x_n(t_2))$ at the later time t_2, and that one can also use any other phase position $(x_1(t_1), \ldots, x_n(t_1))$ at an intermediate time t_1, $t_0 < t_1 < t_2$. The use of (1.2) then yields the condition

$$\psi(x(t_0), t_2 - t_0) = \psi(\psi(x(t_0), t_1 - t_0), t_2 - t_1), \tag{1.13b}$$

where we have rewritten the *n*-component system (1.2) in the shorthand notation $x(t) = \psi(x(t_0), t - t_0)$. The condition (1.13b) is just one of the conditions, the closure condition, that the transformations (1.2) form a one-parameter continuous group with the time t as group parameter (see Ince, 1986). The same condition can also be written more formally as

$$U(t_2 - t_0) = U(t_2 - t_1)U(t_1 - t_0). \tag{1.13c}$$

Qualitatively, this is a condition for path independence of solutions in phase space. Whenever the time evolution operator exists, then the motion can be described geometrically as a flow in phase space. Here, we assume that the system is self-regulating; the special case where the velocity field depends periodically upon the time t is discussed in section 1.4 below.

The condition for the existence of the time evolution operator, and therefore for the description of the motion as a phase flow, is that the solutions (1.2) of (1.1) exist for all *finite* times; that is, that the solutions are free of finite-time singularities. It is easy to see by example that this flow property is not guaranteed without further restrictions on (1.1): an example of a Hamiltonian system where there is no time evolution operator, hence no phase flow, is that of one-dimensional motion in a potential $U(x) = -x^4/2$, where the Hamiltonian is given by $H = \dot{x}^2/2 + U(x)$. Here, H is constant because $U(x)$ is time-independent (energy conservation), and, for the particular case where $H = 0$ ($H \leq 0$ is required), it follows that

$$\dot{x} = \pm x^2, \tag{1.14}$$

whose solutions have the form

$$x(t) = \frac{x_0}{1 \pm x_0(t - t_0)} \tag{1.14b}$$

and therefore have finite-time singularities that depend upon the choice of the initial condition x_0! In contrast, when $U = x^4/2$, or $U = -x^2$, then no singularities occur, and the phase flow picture is correct. Therefore, given a set of differential equations, it is first necessary to determine whether solutions are bounded at all finite times in order to be able to study the phase space motion as a flow. We shall perform such an analysis for the Lorenz model at the beginning of Chapter 2. In addition, in the case of nonlinear equations, one is not guaranteed uniqueness of solutions without first satisfying certain restrictions (see Arnol'd, 1973).

In the case of linear equations,

$$\frac{dX(t)}{dt} = AX(t), \tag{1.15}$$

the analysis becomes much easier. In particular, if A is a constant $n \times n$ matrix, then the time evolution operator is given explicitly by

$$U_t = e^{tA} = I + tA + t^2A^2/2! + \cdots + t^nA^n/n! + \cdots, \tag{1.16}$$

where I is the identity matrix, as is often shown implicitly in elementary texts on ordinary differential equations.

We can begin now with the classification of invariant sets in phase space. Attractors and repellers are examples of invariant sets that occur in dissipative systems. If all orbits in a definite region of phase space approach a particular limiting point set as $t \rightarrow \infty$, then that point set is called an invariant set (mathematically, one must take the closure of the infinite-time trajectory). The set of initial conditions yielding such orbits is called the basin of attraction for the attractor under consideration. We turn now to two of the three classical attractors, stable equilibria and stable limit cycles. Examples of the third case, that of stable quasiperiodic orbits on tori in phase space, are discussed later in Chapter 3 in the special context of Hamiltonian systems (periodic and quasiperiodic orbits on tori can also occur in dissipative systems).

1.2 Equilibria, linear stability, and limit cycles

Consider the self-regulating ('autonomous') system

$$\dot{x}_i = V_i(x_1, x_2, \ldots, x_n; \mu). \tag{1.1b}$$

At an equilibrium point, the velocity vector $V = (V_1, \ldots, V_n)$ vanishes. Denote such a point by $\bar{X} = (\bar{x}_1, \ldots, \bar{x}_n)$. This point is stable if it attracts nearby trajectories, unstable if it repels them.

In order to analyze the stability locally, we expand (1.1) about $\bar{X}$,

$$V_i(x_1, \ldots, x_n) = V_i(\bar{x}_1, \ldots, \bar{x}_n) + \delta x_j \frac{\partial V_i}{\partial x_j}\bigg|_{X=\bar{X}} + \cdots, \tag{1.17}$$

where $\delta x_j = x_j - \bar{x}_j$ and summation convention is used (sum over repeated indices). If we keep only the linear term in the displacement δX_i from equilibrium, we obtain the variational equation

$$\delta \dot{X} = A \, \delta X, \tag{1.18}$$

where A is the (constant) Jacobi matrix with entries

$$A_{ij} = \frac{\partial V_i}{\partial x_j}\bigg|_{X=\bar{X}} \tag{1.19}$$

Since A is a constant matrix,

$$\delta X(t) = \begin{pmatrix} \delta x_1(t) \\ \vdots \\ \delta x_n(t) \end{pmatrix} \tag{1.20}$$

is the displacement from equilibrium, $U_t = e^{tA}$ and so

$$\delta X(t) = e^{tA} \, \delta X(0), \tag{1.21}$$

where $\delta X(0)$ is the initial displacement from equilibrium.

If we can diagonalize A, then

$$A\hat{e}_i = \lambda_i \hat{e}_i, \tag{1.22}$$

where λ_i is the eigenvalue corresponding to the eigenvector $\hat{e}_i$. When the n eigenvectors $\hat{e}_i$ are linearly independent, then from

$$\delta X(0) = \sum_{i=1}^{n} c_i \hat{e}_i \tag{1.21b}$$

and from (1.21) and (1.21b) we obtain

$$\delta X(t) = \sum_{i=1}^{n} c_i e^{\lambda_i t} \hat{e}_i \tag{1.21c}$$

by using (1.21) and (1.22) combined with the fact that $A^n \hat{e}_i = \lambda_i^n \hat{e}_i$ in the power series expansion for e^{tA}.

The equilibrium point $\bar{X}$ is an attractor if $\mathrm{Re}\, \lambda_i < 0$ for all i, but is, except for special trajectories, a repeller if $\mathrm{Re}\, \lambda_i > 0$ for at least one value of the index i. In the former case the equilibrium point is called stable, whereas it is unstable in the latter case. If $\mathrm{Re}\, \lambda_i \leq 0$ for all i, the equilibrium point is also stable in linear approximation.

In the two-dimensional phase plane, the possible equilibria are determined by the different possible eigenvalue problems (1.21), and there are only six that are distinct. Let us take $\bar{X} = (0, 0, \ldots, 0)$ so that $\delta x_i = x_i$ in the variational equation (1.18). We shall not write the matrix A in its most general possible form, but shall instead illustrate the six different possibilities.

Consider first the case where $\dot{x}_1 = x_2$, $\dot{x}_2 = -x_1$. Here,

$$A = \begin{pmatrix} 0 & 1 \\ -1 & 0 \end{pmatrix} \tag{1.23}$$

and the eigenvalue equation $\det |A - \lambda I| = 0$ yields $\lambda = \pm i$. It follows from $x_1 \dot{x}_1 + x_2 \dot{x}_2 = 0$ that $x_1^2 + x_2^2$ is constant, so the streamlines in the (x_1, x_2) plane are concentric circles (rotating clockwise) about the origin (Fig. 1.3). The equilibrium point at the origin is called an elliptic point because the local phase portrait, the plot of the velocity field, describes concentric ellipses. In the more general case, a pair of purely imaginary eigenvalues corresponds to an elliptic point or center.

If next we consider the case $\dot{x}_1 = x_2$, $\dot{x}_2 = x_1$, then $(0, 0)$ is again an equilibrium point,

$$A = \begin{pmatrix} 0 & 1 \\ 1 & 0 \end{pmatrix}, \tag{1.24}$$

and $\lambda = \pm 1$. Hence, $\delta X(t) = c_1 e^t \hat{e}_i + c_2 e^{-t} \hat{e}_2$, so that $\hat{e}_1$ is an unstable direction whereas $\hat{e}_2$ is stable. Since $x_1^2 - x_2^2 = $ constant, these directions are the asymptotes of a hyperbola (Fig. 1.4) and the origin is called a hyperbolic point. In general, a pair of real eigenvalues with opposite sign correspond to a hyperbolic point (saddle point).

We now generalize to include the damped harmonic oscillator, $\dot{x}_1 = x_2$, $\dot{x}_2 = -\beta x_2 - x_1$, with friction coefficient $\beta > 0$ (when $\beta < 0$, then energy is pumped into the system). In this case,

$$A = \begin{pmatrix} 0 & 1 \\ -1 & -\beta \end{pmatrix}, \tag{1.25}$$

and it follows that the eigenvalues of A are

$$\lambda = \frac{-\beta \pm \sqrt{\beta^2 - 4}}{2}. \tag{1.26}$$

If $|\beta| > 2$ then both eigenvalues are real. If both eigenvalues are negative, then the origin is an attractor and is called a stable node, or streamline

Fig. 1.3 Local phase portrait in the neighborhood of an elliptic point.

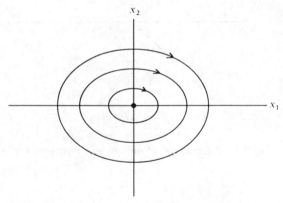

Fig. 1.4 Streamlines near a hyberbolic point.

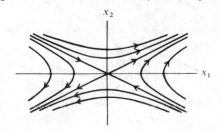

sink (Fig. 1.5). If both eigenvalues are positive, then the origin is a repeller and is called an unstable node (streamline source–simply reverse the flow direction in Fig. 1.5). In either case, the flowlines radiate into or out of the origin, but not necessarily isotropically: for example, if $\lambda_i < 0$ for $i = 1, 2$, and $|\lambda_2| > |\lambda_1|$ then $\delta X(t) \sim c_1 e^{-|\lambda_1|t} \hat{e}_1$ as $t \to \infty$ and the streamlines enter the origin tangent to the direction of the eigenvector $\hat{e}_1$ (Fig. 1.6).

When $|\beta| < 2$, it follows that

$$\lambda = (-\beta \pm i\Omega)/2 \qquad (1.27)$$

where $\Omega = \sqrt{4 - \beta^2}$ is an oscillation frequency. When $0 < \beta < 2$,

Fig. 1.5   Isotropic flow into a sink.

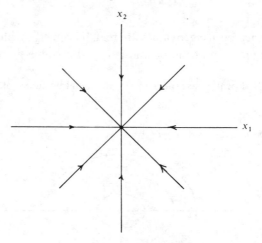

Fig. 1.6   An example of anisotropic flow into a sink.

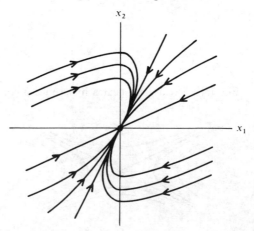

Re $\lambda_{1,2} < 0$ and the equilibrium point $(0,0)$ is an attractor that is called a stable focus (Fig. 1.7): the local phase portrait is a spiral that winds into the origin.

If $-2 < \beta < 0$, the streamlines diverge from the origin as they wind about it, and we have the case of an unstable focus (reverse the flow direction in Fig. 1.7).

In the general nonlinear case there may be many equilibria (solutions of $V = 0$) distributed over phase space. Each local flow diagram (local phase portraits) will correspond to one of the six possibilities discussed above, but the full nonlinear vector field $V = (V_1, \ldots, V_n)$ is needed in order to plot the streamlines (flow diagram, or global phase portrait) between and far from the equilibria. When $n = 2$, it is easy to plot $dx_2/dx_1 = V_2/V_1$. Clearly, if a physical system has several different stable equilibria, then the observed equilibrium state will depend upon the basin of attraction from which the choice of initial conditions is made. Given an attractor the *basin of attraction* of a particular attractor is the set of initial conditions from which the motion asymptotically approaches the attractor. Stable equilibria are zero-dimensional attractors, but there are also higher-dimensional attractors and repellers.

All of the attractors and repellers studied so far are zero-dimensional. There are also one-dimensional attractors and repellers in driven non-conservative systems, closed curves that are periodic solutions called limit cycles. Consider as an example the system

$$\ddot{x} + (x^2 + \dot{x}^2 - 1)\dot{x} + x = 0. \tag{1.28}$$

Fig. 1.7   Stable spiral.

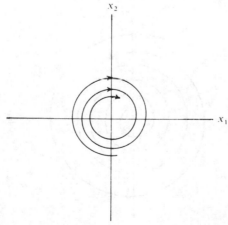

The energy $E = (\dot{x}^2 + x^2)/2$ changes according to

$$dE/dt = -\dot{x}^2(x^2 + \dot{x}^2 - 1), \tag{1.29}$$

and from $\dot{E} = -f\dot{x}$ it follows that the friction force is $f = \dot{x}(x^2 + \dot{x}^2 - 1)$. In other words, there is an energy flow through the system yielding normal damping when $r^2 = x^2 + \dot{x}^2 > 1$, but the system gains energy if $r^2 < 1$, so that the phase plane is divided into two distinct regions, of energy gain and energy loss. At $r = 1$, $\dot{E} = 0$ and we have the steady state: energy gain is balanced by dissipation. By inspection, the circle $x^2 + \dot{x}^2 = 1$ is a solution of (1.28) and this is an example of a limit cycle (Fig. 1.8). If we transform to polar coordinates $r^2 = x^2 + \dot{x}^2$, $\tan \theta = \dot{x}/x$, it follows that

$$\dot{r} = r(1 - r^2) \sin^2 \theta,$$
$$\dot{\theta} = (1 - r^2) \sin \theta \cos \theta - 1, \tag{1.30}$$

so that the cycle is stable ($\dot{r} > 0$ if $r < 1$, $\dot{r} < 0$ if $r > 1$) and $\dot{\theta} \sim -1$ when $r \sim 1$, so that the direction of rotation is clockwise.

The reader should check that there is an unstable spiral at the origin. Notice also that the amplitude of oscillation ($r = 1$) is completely independent of the initial conditions, in contrast to linear oscillations.

In the above example, the nature of the equilibrium point in the linear approximation persists under inclusion of the full nonlinear terms. It is easy to give an example that shows that this is not always true. Consider

$$\dot{x}_1 = x_2 + x_1 r(1 - r),$$
$$\dot{x}_2 = -x_1 + x_2 r(1 - r), \tag{1.31}$$

Fig. 1.8   Flow near a stable limit cycle whose shape is the unit circle. An unstable spiral lies inside the circle, with an unstable equilibrium point at the origin.

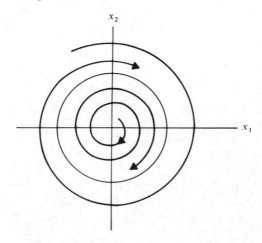

where $r^2 = x_1^2 + x_2^2$. In the linear theory, $\dot{x}_1 = x_2$, $\dot{x}_2 = -x_1$, and the origin is an elliptic point. In polar coordinates (r, θ), with $\tan \theta = x_2/x_1$, the original system becomes

$$\dot{r} = r^2(1 - r),$$
$$\dot{\theta} = -1,$$
(1.31b)

and has a stable limit cycle at $r = 1$. The origin, however, is not an elliptic point: it is an unstable focus! The conclusion is that when periodic solutions occur in a linearized theory, there is no guarantee that they persist under even the weakest nonlinear perturbation.

We can understand why, in a conservative system, all equilibria are either elliptic or saddle points: nodes and foci are impossible. The proof is simple: consider a finite phase volume $\delta\Omega(0)$ within the basin of attraction of a stable node or a stable spiral. Then $\delta\Omega(t) \to 0$ at $t \to \infty$, because all points in $\delta\Omega(0)$ are attracted by the equilibrium point. This result is inconsistent with Liouville's theorem, which demands that $\delta\Omega(t)$ is constant in a conservative system.

In a two-dimensional self-regulating system, attractors are limited to zero and one dimension, stable equilibria and stable limit cycles (Poincaré–Bendixson theorem). This eliminates the possibility of deterministic chaos. To find chaos, you need at least three coupled nonlinear differential equations in a nonconservative system (see the Lorenz model in Chapter 2), or at least a four-dimensional phase space in a Hamiltonian system (see the Hénon–Heiles model in Chapter 3). However, as we show in Chapter 2, one needs only one dimension in order to obtain chaotic behavior from a nonlinear discrete map.

We turn now to the path from regular to completely unstable motion via bifurcations.

1.3 Change of stability (bifurcations)

We begin as usual with the nonlinear system

$$\dot{x}_i = V_i(x_1, \ldots, x_n; \mu),$$
(1.1b)

but now the emphasis is on how the stability of an equilibrium point $\bar{x}_i = x_i(\mu)$ (a solution of $V(\bar{x}_1, \bar{x}_2, \ldots, \bar{x}_n; \mu) = 0$) is affected when the control parameter μ is varied. Near the equilibrium point $\bar{X} = (\bar{x}_1, \ldots, \bar{x}_n)$,

$$\delta X(t) = e^{tA} \delta X(0),$$
(1.21)

where $\delta X(t) = X(t) - \bar{X}$ is a small deviation from equilibrium. Consider

in particular the case where there is a critical value μ_c, such that
Re $\lambda_i(\mu) < 0$ for all i when $\mu < \mu_c$, Re $\lambda_j(\mu_c) = 0$ for at least one i, $i = j$,
and Re $\lambda_j(\mu) > 0$ when $\mu > \mu_c$. In other words, μ_c is found by solving
Re $\lambda_j(\mu_c) = 0$.

As the simplest example of change of stability, consider

$$\dot{x} = \mu x - x^2. \tag{1.32}$$

The equilibria are given by $\bar{x} = 0$ and $\bar{x} = \mu$. Linearizing, with $\delta x(t) = x(t) - \bar{x}$, we obtain $\delta x(t) \sim e^{\mu t}$ when $\bar{x} = 0$, which is stable if $\mu < 0$,
unstable if $\mu > 0$. $\mu_c = 0$ is the bifurcation point. If $\bar{x} = \mu$ then $\delta X(t) \sim e^{-\mu t}$,
so that $\bar{x} = \mu$ is unstable if $\mu < 0$, stable when $\mu > 0$. The bifurcation
diagram is shown in Fig. (1.9), a plot of the equilibrium solution $\bar{x}(\mu)$ vs μ.

A branching of solutions that takes the shape of a pitchfork is shown
qualitatively in Fig. (1.10). There, the solution that is stable for $\mu < \mu_c$
becomes unstable for $\mu > \mu_c$ where two new stable solutions spontaneously
appear and grow from the critical point at μ_c. We consider next examples
of pitchfork and Hopf bifurcations. A Hopf bifurcation occurs when the
real parts of two complex conjugate eigenvalues λ and λ^* pass through
zero as a control parameter is varied, so that Re $\lambda < 0$ if $\mu < \mu_c$ (stability)
and Re $\lambda > 0$ if $\mu > \mu_c$ (instability), and where d Re $\lambda/d\mu > 0$ at μ_c.

As an example of a Hopf bifurcation, consider the model where (in
polar coordinates)

$$\dot{r} = r(\mu - r), \qquad \dot{\theta} = -1. \tag{1.32b}$$

When $\mu < 0$ then the origin is a stable focus. When $\mu > 0$, the focus is

Fig. 1.9   Bifurcation diagram showing exchange of stability: plot of
equilibrium solutions as the control parameter is varied. The solid
lines represent stable equilibria, the dashed ones unstable equilibria.

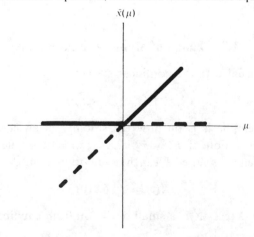

unstable but there is also a stable limit cycle at $r = \mu$. In Cartesian coordinates, $x_1 = r \cos \theta$ and $x_2 = r \sin \theta$. The equations of linear stability for the equilibrium point at the origin yield the two complex conjugate eigenvalues

$$\lambda = \mu + i, \qquad \lambda = \mu - i \tag{1.32c}$$

that describe the change of stability of the focus. The Hopf bifurcation occurs at $\mu = 0$.

In an inverse Hopf bifurcation, an initially unstable limit cycle encloses a stable focus. As the inverse Hopf bifurcation point is approached, the limit cycle shrinks down onto the stable spiral and finally gives birth to an unstable focus. A simple model with this behavior is given by replacing $\mu - r$ by $\mu + r$ in (1.32b).

The Lorenz model follows from a certain truncation of the Navier–Stokes equations, and is defined by the system

$$\dot{x} = \sigma(y - x),$$

$$\dot{y} = \rho x - y - xz, \tag{1.33}$$

$$\dot{z} = -\beta z + xy,$$

where σ, β, and ρ are control parameters.

We take σ, β, and ρ to be positive, so that with

$$\nabla \cdot V = \frac{\partial \dot{x}}{\partial x} + \frac{\partial \dot{y}}{\partial y} + \frac{\partial \dot{z}}{\partial z} = -\sigma - \beta - 1 < 0 \tag{1.34}$$

Fig. 1.10 Example of a pitchfork instability: a stable equilibrium point (solid line) becomes unstable (dashed line) as the control parameter is increased, and two new stable equilibria (solid lines) are born.

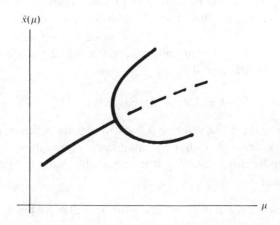

the Lorenz model is dissipative, with an exponential contraction rate $\delta\Omega(t) \sim e^{-(\sigma+\beta+1)t}$ for volume elements $\delta\Omega(t) = \delta x\,\delta y\,\delta z$ in phase space.

There are three distinct equilibria: $\bar{x} = \bar{y} = \bar{z} = 0$ and $(\bar{x}, \bar{y}, \bar{z}) = (\pm\sqrt{\beta(\rho-1)}, \pm\sqrt{\beta(\rho-1)}, \rho-1))$. Note that when $\rho < 1$ there is only one equilibrium point, the origin. We shall see how the other two equilibria grow into existence in the phase space as a pitchfork bifurcation.

Linear stability is determined by the Jacobi matrix

$$A = \begin{pmatrix} -\sigma & \sigma & 0 \\ \rho - \bar{z} & -1 & -\bar{x} \\ \bar{y} & \bar{x} & -\beta \end{pmatrix}, \tag{1.35}$$

evaluated at an equilibrium point.

With $\bar{x}, \bar{y}, \bar{z} = (0, 0, 0)$, we obtain

$$A = \begin{pmatrix} -\sigma & \sigma & 0 \\ \rho & -1 & 0 \\ 0 & 0 & -\beta \end{pmatrix} \tag{1.35b}$$

and the eigenvalue equation

$$(\beta + \lambda)[(\sigma + \lambda)(1 + \lambda) - \rho\sigma] = 0, \tag{1.36}$$

which has the three roots $\lambda = -\beta$ and

$$\lambda = [-(\sigma + 1) \pm ((\sigma + 1)^2 + 4\sigma(\rho - 1))^{1/2}]/2.$$

The root $\lambda = [-(\sigma + 1) + ((\sigma + 1)^2 + 4\sigma(\rho - 1))^{1/2}]/2$ changes from a negative to a positive value as ρ increases through 1, so that there is a bifurcation at $\rho = 1$, where $(0, 0, 0)$ loses its stability. Because the two new equilibria also come into play at $\rho > 1$, we have an example of a pitchfork bifurcation.

Next, we observe that a Hopf bifurcation takes place at a critical value of ρ_h: we consider the equilibria

$$(\bar{x}, \bar{y}, \bar{z}) = (\pm\sqrt{\beta(\rho-1)}, \pm\sqrt{\beta(\rho-1)}, \rho - 1))$$

and ask whether there is a value of ρ such that there is one real, negative eigenvalue λ, combined with two imaginary eigenvalues $\lambda_2 = i\alpha$ and $\lambda_3 = -i\alpha$. Denote this value of ρ, if it occurs, by ρ_h. It follows from the eigenvalue equation

$$\lambda^3 + \lambda^2(\sigma + \beta + 1) + \lambda\beta(\sigma + \rho) + 2\sigma\beta(\rho - 1) = 0, \tag{1.37}$$

by equating coefficients, that

$$\lambda_1 = -\sigma - \beta - 1,$$

$$\alpha^2 = \beta(\sigma + \rho_h), \qquad (1.38)$$

$$-\lambda_1\alpha^2 = 2\sigma\beta(\rho_h - 1).$$

These three equations require $\rho_h = (\sigma(\sigma + \beta) + 3\sigma)/(\sigma - \beta - 1)$, and we assume that $\sigma > \beta + 1$. In addition, to show that $\text{Re}\,\lambda$ crosses the imaginary axis at $\rho = \rho_h$ and does not get stuck there, you must show that $d\lambda(\rho_h)/d\rho > 0$.

If you now make an expansion in the small parameter $\delta\rho = \rho - \rho_h$ and compute the approximate eigenvalues to lowest order in $\delta\rho$, you can see that λ_1 remains negative regardless of whether $\delta\rho < 0$ and that $i\alpha$ is approximately the imaginary part of a complex eigenvalue λ where $\text{Re}\,\lambda < 0$ if $\delta\rho < 0$ but $\text{Re}\,\lambda > 0$ if $\delta\rho > 0$. Therefore, we have a Hopf bifurcation at $\rho = \rho_h$ with two unstable spirals unwinding from the points $(\pm\sqrt{\beta(\rho - 1)}, \pm\sqrt{\beta(\rho - 1)}, \rho - 1))$ when $\delta\rho > 0$. We defer the discussion of the flow in this region until Chapter 2.

The question of what the local flow is when $1 < \rho < \rho_h$ is less interesting but deserves to be answered, because one can learn that there is yet another bifurcation in this region. The main point is that the foci that become unstable at $\rho = \rho_h$ could not have come into existence at $\rho = 1$: at $\rho = 1$ where the cubic equation in λ has two negative roots and one vanishing root, all of these roots must survive slightly displaced in the complex λ-plane when $\delta\rho = \rho - 1$ is slightly positive. Therefore, the only way that the stable foci can be born is for two of the three real roots to coalesce at some larger value $\rho_1 > 1$ of ρ, so that for $1 < \rho < \rho_1$ there are three real eigenvalues (it is left to the reader to prove that these are all negative). When $\rho = \rho_1$, two of these roots collapse into one, and there is one real eigenvalue λ_1 along with two complex conjugate ones λ and λ^* when $\rho_1 < \rho_h$.

In order to make a picture of all this, look at the cubic polynomial

$$F(\lambda) = \lambda^3 + (\sigma + \beta + 1)\lambda^2 + \lambda\beta(\sigma + \rho) + 2\beta\sigma(\rho - 1), \qquad (1.39)$$

whose zeros are our sought-after eigenvalues. When there are three real, negative roots $\lambda_1, \lambda_2, \lambda_3$, then the function $F(\lambda)$ has the qualitative shape shown in Fig. 1.11a–c. When $\rho > 1$ all the coefficients are positive, so that the number of negative roots is either one or three. And because there are three *distinct* real roots when $\rho = 1$ (one lies at the origin), there must be three distinct *negative* roots when ρ is slightly greater than one (all roots have finite velocity $d\lambda/d\rho$ in the complex λ-plane as ρ varies). As ρ increases, λ_2 and λ_3 move toward one another until at $\rho = \rho_1 > 1, \lambda_2 = \lambda_3$, and $F(\lambda_2) = 0$. It follows that $3\lambda_2^2 + 2\lambda_2(\sigma + \beta + 1) + \beta(\sigma + \rho_1) = 0$, and the larger (less negative) of the solutions is our root λ_2:

$$\lambda_2 = (-(\sigma + \beta + 1) + ((\sigma + \beta + 2) - 3\beta(\sigma + \rho_1))^{1/2})/3. \quad (1.40)$$

Because $\lambda_1 = -2\lambda_2 - \sigma - \beta - 1$, we can also compute λ_1. Then, you have to plug (1.40) back into $F(\lambda_2) = 0$ in order to compute ρ_1. We leave this to the reader. The main point is that as ρ increases through ρ_1, λ_2 gives birth to two complex conjugate eigenvalues and therefore to the birth of the two stable foci that we found in existence at $\rho \leq \rho_h$.

We end this section with an example of a symmetry-breaking pitchfork bifurcation in a Hamiltonian system. Consider the case of a bead sliding without friction on a hoop of radius a. The hoop rotates with angular speed ω about a direction parallel to gravity (Fig. 1.12). The Lagrangian is

$$L = ma^2\dot{\theta}^2/2 + ma^2\omega^2(\sin^2\theta)/2) + mga\cos\theta, \quad (1.41)$$

and with $P = \partial L/\partial\dot{\theta} = ma^2\dot{\theta}$, the Hamiltonian $H = P\dot{\theta} - L$ is

$$H = P^2/2ma^2 - mga\cos\theta - ma^2\omega^2(\sin^2\theta)/2. \quad (1.41b)$$

With the substitutions $t^2 \rightarrow t^2 g/a$ and $H \rightarrow H \cdot mga$ we obtain

$$H = \dot{\theta}^2/2 + U(\theta), \quad (1.42)$$

where $\rho = \dot{\theta}$ and $U(\theta)$ is an effective potential

$$U(\theta) = -\cos\theta - \mu(\sin^2\theta)/2, \quad (1.42b)$$

with dimensionless control parameter $\mu = \omega^2 a/g$.

Equilibria are extrema of (1.42), $\sin\bar{\theta}(1 - \mu\cos\bar{\theta}) = 0$ and there are

Fig. 1.11   (a) Three different stable equilibria are guaranteed by the three distinct negative roots ($\rho < \rho_1$). (b) Two stable equilibria have merged ($\rho = \rho_1$). (c) The polynomial (1.39) now has two complex conjugate roots with negative real part corresponding to the birth of two stable foci, and one real, negative root ($\rho > \rho_1$).

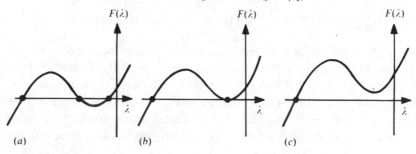

(a)   (b)   (c)

two possibilities, $\sin\bar{\theta} = 0$ or $\cos\bar{\theta} = 1/\mu$. The first, $\sin\bar{\theta} = 0$, yields $\bar{\theta} = 0$ or $\bar{\theta} = \pm\pi$. The latter occur at maxima of $V(\theta)$ and are therefore hyperbolic. The solution $\bar{\theta} = 0$ is an elliptic point if $\mu < 1$ because it occurs at a minimum of $U(\theta)$. However, the qualitative picture changes as μ increases through one: at $\mu = 1$, $U''(\bar{\theta}) = \cos\bar{\theta} - \mu\cos 2\bar{\theta}$ vanishes when $\bar{\theta} = 0$ and becomes a local maximum of $U(\theta)$ when $\mu > 1$ (Fig. 1.13a, b),

Fig. 1.12   Bead on a frictionless, uniformly rotating hoop.

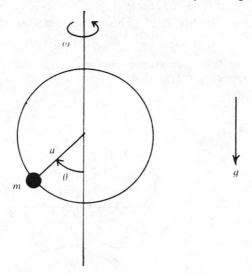

Fig. 1.13   (*a*) A single potential minimum corresponds to a single elliptic point. (*b*) Two new minima, corresponding to two new elliptic points in phase space, emerge owing to a symmetry-breaking pitchfork bifurcation.

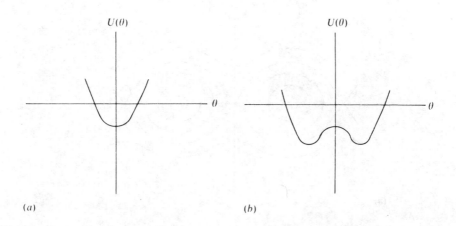

(*a*)   (*b*)

so that $\bar{\theta} = 0$ is then hyperbolic. At μ slightly greater than one, the new elliptic points

$$\bar{\theta} = \pm \cos^{-1}(1/\mu) \qquad (1.43)$$

begin to branch out from the origin, so that we have another example of pitchfork bifurcation. It is easy to check, for example, that $U''(\bar{\theta}) = (\mu^2 - 1)/\mu$. Because $H(\theta) = H(-\theta)$, the solutions (1.43) break reflection symmetry. So, we have a simple example of a symmetry-breaking pitchfork bifurcation. The phase portraits are shown as Figs 1.14a and b.

It is easy to show in (1.43) and $\bar{\theta} = \pm\pi/2$ as $\mu \to \infty$. The bifurcation diagram is shown in Fig. 1.15. This concludes our introduction to

Fig. 1.14   (*a*) Phase portrait corresponding to Fig. 1.13*a*. (*b*) Stream-lines corresponding to Fig. 1.13*b*.

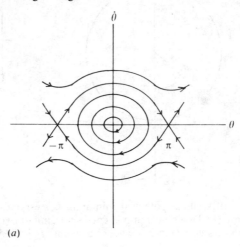

(*a*)

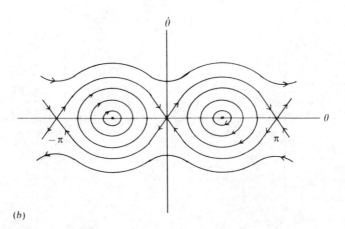

(*b*)

bifurcations, although we have not covered all possible cases of interest here (the saddle-node bifurcation is a case in point). Next, as preparation for Chapter 5 and more generally for orientation, we discuss the replacement of a periodically driven system by an iterated map.

1.4 Periodically driven systems and stroboscopic maps

The main purpose of this section is to show how the study of a system of differential equations can be replaced by the study of a discrete map. Unfortunately, there is no *unique* replacement of a system of differential equations by a map, except for the case where the dynamical system is periodically driven, in which case the stroboscopic map follows uniquely. We begin with that case, because the reader needs it as background for Chapter 5. In general, the fact that the replacement of a differential system by a map is typically not unique can be seen as motivation for the study of universality principles.

To begin, we rewrite (1.1) symbolically as

$$\dot{X} = V(X, t); \tag{1.1d}$$

that is, the velocity field $V(X, t) = (V_1, \ldots, V_n)$ depends explicitly upon the time. In this case, you cannot plot phase trajectories independently of the time, so that you cannot think of a flow in phase space either. Another way to say it is that the time evolution operators $U(t_2, t_1)$ and $U(t_1, t_0)$ for different time intervals do not commute. This means that they do not form a one-parameter group, because all elements in an Abelian group commute with each other (see Ince, 1986).

Fig. 1.15   Bifurcation diagram for a bead on a frictionless rotating hoop.

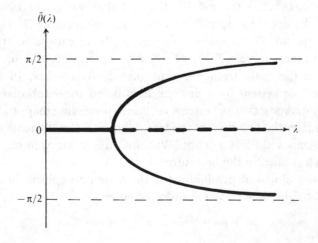

In the case of periodic forcing with period τ, $V(X, t + \tau) = V(X, t)$ so that one also obtains the same velocities at times, $t + 2\tau, t + 3\tau, \ldots$. If we set $t = 0$ and let n be an integer, then because $U(n\tau + s, n\tau) = U(s, 0)$ it follows that

$$U(n\tau + s, 0) = U(n\tau + s, n\tau)U(n\tau, 0) = U(s, 0)U(n\tau, 0). \quad (1.44)$$

With $s = \tau$, we get

$$U(n\tau + \tau, 0) = U(\tau, 0)U(n\tau, 0) = AU(n\tau, 0) \quad (1.45)$$

where we have written $A = U(\tau, 0)$. By induction, we then obtain

$$U(n\tau, 0) = A^n, \quad (1.46)$$

which yields the desired group property because A commutes with itself.

There is an immediate important consequence of this, which we shall state abstractly and then illustrate by an example. If we write

$$x_1(t) = x_{1,N}, x_2(t) = x_{2,N}, \ldots, x_n(t) = x_{n,N},$$

and

$$x_1(t + \tau) = x_{1,N+1}, x_2(t + \tau) = x_{2,N+1}, \ldots,$$

then it follows that we have an n-dimensional iterated discrete map,

$$\begin{aligned}
x_{1,N+1} &= G_1(x_{1,N}, \ldots, x_{n,N}; \tau) \\
x_{2,N+1} &= G_2(x_{1,N}, \ldots, x_{n,N}; \tau) \\
&\vdots \\
x_{n,N+1} &= G_n(x_{1,N}, \ldots, x_{n,N}; \tau),
\end{aligned} \quad (1.47)$$

called the stroboscopic map because (i) $Ax_i(t)$ is a function of all n variables $x_1(t), \ldots, x_n(t)$ and (ii) the fact that we go to times $t + \tau$, $t + 2\tau, \ldots$ by applying $A, A^2, \ldots$ means that the functions $G_1, G_2, \ldots, G_n$ *are the same for all iterations* $N = 1, 2, \ldots$. If we choose to strobe the system at time intervals $T, 2T, 3T, \ldots$ with $T \neq \tau$, then the functions G_i will not be the same from one time interval to another. In this way we replace the system by a discrete map called the stroboscopic map. Furthermore, conservative systems yield area-preserving maps (if $\nabla \cdot V = 0$ for (1.1d) then the Jacobian of the transformation (1.47) equals one) and linear systems yield linear maps. We turn now to an example that has been much studied in the literature.

The damped-driven pendulum (or the driven Josephson junction) is described by the driven second-order equation

$$\alpha\ddot{\theta} + \beta\dot{\theta} + \gamma \sin \theta = A + K \cos \omega t, \quad (1.48)$$

and there are two competing frequencies, ω and $\sqrt{\gamma/\alpha}$. Let us now try to construct the general form of the stroboscopic map. If we set $\theta = \theta$, $r = \dot{\theta}$, $\tau = 2\mu/\omega$, $\theta(t) = \theta_n$, $\theta(t+\tau) = \theta_{n+1}$, $r(t) = r_n$ and $r(t+\tau) = r_{n+1}$, then the time evolution operator A is defined formally by the power series (with coefficients obtained directly by differentiation of the original differential equation),

$$\theta_{n+1} = \theta_n + (A\tau^2/2 - \beta A\tau^3/3! + \omega K\tau^3/3!)$$
$$- \gamma(\tau^2/2 - \beta\gamma\tau^3/3!) \sin \theta_n - \gamma\tau^3 r_n \cos \theta_n/3!$$
$$+ (\tau - \beta\tau^2/2 + \beta^2\tau^3/3!)r_n + \cdots, \tag{1.49}$$

$$r_{n+1} = r_n(1 - \beta\tau + \beta^2\tau^2/2) + (A - A\beta\tau^2 + k\omega\tau^2/2)$$
$$- (\gamma\tau + \gamma\tau^2\beta/2) \sin \theta_n + \gamma\tau^2 r_n \cos \theta_n/2 + \cdots +, \tag{1.49b}$$

which may converge for small enough τ (large ω). A partial resummation of the series yields the stroboscopic map

$$\theta_{n+1} = G_1(\theta_n, r_n),$$
$$r_{n+1} = G_2(\theta_n, r_n) \tag{1.49c}$$

in the form

$$G_1(\theta, r) = \theta + \Omega + h(\theta, r) + (1 - J)r/b,$$
$$G_2(\theta, r) = Jr + \Omega' + g(\theta, r), \tag{1.50}$$

where h and g are periodic in θ with period 2π, Ω and Ω' are constants, and $J = e^{-\beta\tau/\alpha}$ is the Jacobian of the map (1.49) and is the same as the Jacobian

$$J = \frac{\partial(\theta(t), \dot{\theta}(t))}{\partial(\theta(0), \dot{\theta}(0))} = e^{-\beta t/\alpha} \tag{1.51}$$

of the differential equation (1.48) evaluated at $t = \tau$. This is an example of how the flow of a periodically driven system of nonlinear differential equations can be replaced by a stroboscopic map.

In principle, it is convenient in the study of nonlinear dynamical systems to replace the differential equations by a discrete map called a Poincaré map. As we have mentioned above for the case of stroboscopic maps, except for the case of a periodically driven system, the replacement by a map is not unique: there is no unique prescription for forming a Poincaré map, for example. Nor is it, in most cases, easy to derive a map analytically that accurately represents a system of differential equations over long time intervals. It is much easier to speak in general theoretical terms about Poincaré maps than it is to construct examples of maps that accurately represent a nonlinear dynamical system.

In practice, a system of differential equations is replaced by a definite discrete map (one that is *not* a Poincaré map) every time that one discretizes the system for the purpose of computation. In a chaotic system, this kind of replacement (discretization for computation) can lead to uncontrolled errors after only very short time intervals. In order to understand how and why this happens, one should first understand what is meant by chaos for the simplest discrete maps, where the property called deterministic chaos can easily be seen to be intimately connected with the idea of rapid error accumulation in computation (cf. Chapter 2). In this connection one should note that derivatives of *all orders* are required in order to use differential equations to construct stroboscopic maps whenever τ is finite ($\omega < \infty$). Only in the special case of infinitesimal time steps (or very few iterations, when τ is finite but small enough) can one justify keeping *only* the first few derivatives in the construction of the map (1.47). In practice, no one yet knows the correct closed form of the stroboscopic map for the system (1.49). Instead, in the literature, arguments of universal behavior have been used to motivate the study of relatively simple discrete maps in an attempt to understand the system (1.49) near the transition to chaos (Chapter 5). In the next chapter, we follow a procedure that was used originally by Lorenz to discover a method whereby one can, by systematically ignoring certain information that is contained in the complete solutions of the dynamical system, replace the study of a complicated dynamical system (not periodically driven) by the study of a lower-dimensional map. In practice, it is usual to study certain simple maps and then to argue that all maps in the same universality class will exhibit similar behavior. Examples of universality principles are presented in Chapters 5, 6, and 9.

The main point here and illustrated in the next chapter is that there is no known generally applicable analytic prescription that takes one from a given set of differential equations to a discrete map that is given in closed algebraic form or even as a convergent series, that accurately represents those differential equations for arbitrarily long time intervals. Instead, universality principles are sought after and are then used to justify the study of certain algebraically simple discrete maps in order to understand the properties of entire classes of more complicated dynamical systems.

1.5 Continuous groups of transformations as phase space flows

In the beginning of this chapter, we have introduced the idea of a one-parameter group of transformations as a flow in phase space. The transformations are provided by the solutions of a set of differential equations for different initial conditions and depend upon one parameter, the time t. The group is Abelian (commutative), because every continuous one-parameter group is commutative. However, the point of view can be generalized to include multiparameter groups and therefore non-commuting flows. The generalization includes dissipative as well as conservative dynamical systems and is presented below, although, in this text, we use the multiparameter transformation group perspective explicitly only in the analysis of conservative systems in Chapter 3.

In what follows, a flow is characterized by a certain vector field. The vector field is everywhere tangent to the streamlines of the flow, and provides the right-hand side of the differential equations that describe the flow *locally*. Through integration, the flow is described *globally* by a group operator that is analogous to the time evolution operator. Whenever the group parameter is the time, then the group operator is the time evolution operator. In general, distinct flows do not commute because the local velocity fields have a nonvanishing Poisson bracket, i.e. the flows are not locally compatible. However, under certain well-defined conditions, a collection of r noncommuting flows can generate an r-parameter group, where $r > 1$, where each separate flow is a one-dimensional subgroup of the larger group of transformations. In order to illustrate all of this, we begin with a discussion of the group parameter.

Consider a flow that is defined by the continuous transformations

$$x(a) = \psi(x(0), a), \tag{1.52}$$

where a is a parameter, x and ψ stand for $(x_1, \ldots, x_n)$ and $(\psi_1, \ldots, \psi_n)$, and the phase space is n-dimensional. The time t is an example of a group parameter, but we make no restriction to that specific case. The identity transformation is given by setting $a = 0$ (this choice is convenient, but not necessary), inverses exist, and every combination of transformations is again a transformation (group closure):

$$x(c) = \psi(x(0), c), \tag{1.53}$$

$$x(c) = \psi(x(a), b), \tag{1.54}$$

so that

$$\psi(\psi(x(0), a), b) = \psi(x(0), c). \tag{1.55}$$

Therefore there must be a definite combination law for the group parameter
$$c = \phi(a, b), \tag{1.56}$$

where $b = 0$ yields the identity $a = \phi(a, 0)$. An infinitesimal transformation $x + dx = f(x, \delta a)$ follows from setting $b = \delta a$, whereby $c = a + da$ and $x(c) = x(a) + dx$, so that

$$dx_i = V_i(x)\, \delta a, \tag{1.57}$$

where $V_i(x) = \partial \psi_i(x, b)/\partial b$ is to be evaluated at $b = 0$. From $a + da = \phi(a, \delta a)$ we get

$$da = \delta a v(a), \tag{1.58}$$

where $v(a) = \partial \phi(a, b)/\partial b$ is to be evaluated at $b = 0$. A *canonical* parameter is *additive*, like the time. In general, however, ϕ, is not linear in a. The canonical parameter θ for our Abelian group is given by integrating $d\theta = da/v(a)$. As an example, in special relativity the relative velocity of two inertial frames provides the noncanonical group parameter, whereas the canonical parameter is an 'angle' in the hyperbolic space–time. Another example is provided by three-dimensional Euclidean rotations, where the Euler angles are not canonical parameters, but the rotation angles in the three mutually perpendicular yz, xz, and xy planes provide canonical parameters for the three subgroups. In what follows, we shall always assume that the canonical parameter has been found because the differential equations of the flow take on their simplest form in that case.

An infinitesimal transformation $x + dx$ induces, in any scalar phase space function, the transformation $F \to F + dF$, where

$$dF/d\theta = LF \tag{1.59}$$

and the linear operator L (called the infinitesimal generator of the group) is determined by the flow field $V(x)$:

$$L = \sum_1^n V_i(x) \frac{\partial}{\partial x_i}. \tag{1.60}$$

In particular, the differential equations that describe the flow locally are given, as expected, by

$$dx_i/d\theta = V_i(x). \tag{1.61}$$

Here, we have put the cart before the horse: we have started with the idea of the global flow (the solution of the differential equations) and have worked backward to obtain the differential equations that describe the local aspects of the flow. In principle, if one could know all possible flows in advance, then there would not be any reason to begin an analysis of a dynamics problem with differential equations. We normally cannot do that, because one needs relatively little information in order to describe a flow locally (vector field), whereas an enormous amount of information can be required in order to characterize a flow globally, *especially* when a deterministic system is chaotic. The point is made in order to encourage the reader to think about the reason why we have been tied, historically, to *differential equations* as a starting point for formulating physics.

Globally, the flow is characterized by the solutions, which follow from the application of the group operator to different initial conditions. The group operator U, in turn, is generated by the differential operator L (or vector field V). If $F(x)$ can be expanded in a Taylor series

$$F(x(\theta)) = F(x) + \theta \, dF(x)/dx + \theta^2 \, d^2F(x)/dx^2/2 + \cdots, \quad (1.62)$$

where $x = x(0)$, then by successive differentiations of $dF/d\theta = LF$ with respect to θ, we can substitute L for $d/d\theta$, resume the power series and obtain $F(x(\theta)) = F(U(\theta)x) = U(\theta)F(x)$, where $U(\theta) = \exp(\theta L)$. Here we have also used the fact that the infinitesimal generator L is invariant under the group that it generates,

$$L = \sum_1^n V_i(x(a)) \frac{\partial}{\partial x_i(a)} = \sum_1^n V_i(x(0)) \frac{\partial}{\partial x_i(0)}, \quad (1.63)$$

and have also assumed that the power series converges for all values of θ and for all initial conditions. Normally, this assumption does not hold, but when the group is compact then the exponentiation of L to obtain U is possible for all values of θ (there is no convergence problem). For example, for rotations in the xy-plane, and using Cartesian coordinates, we have the infinitesimal transformation $dx = y \, d\theta$, $dy = -x \, d\theta$, so that the vector field is $V = (y, -x)$. The infinitesimal operator is $L = y \, \partial/\partial x - x \, \partial/\partial y$, and it follows that $x(\theta) = \exp(\theta L)x = x \cos \theta - y \sin \theta$, $y(\theta) = \exp(\theta L)y = x \sin \theta + y \cos \theta$, which hold for all values of θ.

In general, *if* there were no convergence problem, *then* every nonlinear system of self-regulating differential equations could be solved in the form

$$x_i(t) = \exp\left(t \sum_1^n V_i \, \partial/\partial x_i\right) x_i = x_i + tV_i(x) + t^2 \sum_1^n V_j(x) \frac{\partial V_i}{\partial x_j}\bigg/2 + \cdots,$$

$$(1.64)$$

but power series typically do not converge for all times for arbitrary solutions of differential equations (the radii of convergence are generally limited by singularities in the complex plane, even if there are no singularities on the real-time axis). In particular, the behavior at large times cannot usually be discovered from a power series about $t = 0$. When the power series solution about $t = 0$ has a finite radius of convergence, then one can, in principle, proceed by analytic continuation to later times. Even when the power series would converge for all times, more and more terms in the series would be required in order to discover the long-time behavior of the solution.

We now generalize to the case of two or more flows. Consider r different flows $U_k(\theta_k)$, each with its own canonical parameter θ_k and with vector field $V^k(x)$, $k = 1, \ldots, r$:

$$V_k^i(x) = \frac{\partial U_k}{\partial \theta_k} x_i \quad \text{at} \quad \theta_k = 0. \tag{1.65}$$

The corresponding infinitesimal operator for the kth flow is given by

$$L_k = \sum_{i=1}^{n} V_k^i(x) \frac{\partial}{\partial x_i}, \tag{1.66}$$

and the differential equations that describe the flow locally are $dx_i/d\theta_k = V_k^i(x)$. Generally, two distinct flows are not compatible in the sense that they do not commute: $U_1(\theta_1)U_2(\theta_2)$ usually differs from $U_2(\theta_2)U_1(\theta_1)$. The degree of incompatibility is measured by the noncommutativity of the infinitesimal generators L_1 and L_2:

$$\frac{\partial^2}{\partial \theta_1 \, \partial \theta_2} \{F(U_1 U_2 x) - F(U_2 U_1 x)\} = (L_2, L_1)F(x) \quad \text{when } \theta_1 = \theta_2 = 0, \tag{1.67}$$

where $(L_2, L_1) = L_2 L_1 - L_1 L_2$ is called the commutator of L_2 and L_1. What is immediately interesting is that the commutator is not, as it would superficially appear, a second-order differential operator; rather,

$$(L_2, L_1) = \sum_{i,k=1}^{n} \left(V_2^i \frac{\partial V_1^k}{\partial x_i} - V_1^i \frac{\partial V_2^k}{\partial x_i} \right) \frac{\partial}{\partial x_k}, \tag{1.68}$$

so that $(L_2, L_1) = L_3$ is the infinitesimal operator of a *third* flow whose vector field V_3 is given by the *Poisson bracket* $V_3 = [V_1, V_2]$ of the vector fields V_1 and V_2:

$$V_3^i = \sum_{k=1}^{n} \left(V_2^k \frac{\partial V_1^i}{\partial x_k} - V_1^k \frac{\partial V_2^i}{\partial x_k} \right). \tag{1.69}$$

That is, the infinitesimal generator of the new flow is given by

$$L_3 = \sum_{k=1}^{n} [V_1, V_2]_k \frac{\partial}{\partial x_k}. \tag{1.70}$$

Because the Jacobi identity

$$[V_1, [V_2, V_3]] + [V_2, [V_3, V_1]] + [V_3, [V_1, V_2]] = 0 \tag{1.71}$$

is satisfied by the Poisson brackets of the vector fields (the reader is invited to prove this), it follows that the infinitesimal generators also satisfy the Jacobi identity:

$$(L_1, (L_2, L_3)) + (L_2, (L_3, L_1)) + (L_3, (L_1, L_2)) = 0. \tag{1.72}$$

Our infinitesimal operators, under commutation, generate a linear algebra that is called a Lie algebra. One can also say that the vector fields of the flows, under the Poisson bracket operation, generate the same Lie algebra.

As an example of a Lie algebra, consider real rotations in three dimensions. The three vector fields that define the flow locally are $V_1 = (z - y)$, $V_2 = (-x, z)$, and $V_3 = (y, -x)$. The corresponding infinitesimal generators obey the 'angular-momentum' commutation rules

$$(L_i, L_j) = \varepsilon_{ijk} L_k, \tag{1.73}$$

where we have used summation convention (sum over repeated index k) and ε_{ijk} is the completely antisymmetric three-index symbol. An arbitrary flow has the form

$$U(\boldsymbol{\alpha}) = \exp\left(\sum_{i=1}^{3} \alpha_i L_i \right), \tag{1.74}$$

and this means that any combination of three generally noncommuting rotations $U_1(\theta_1) U_2(\theta_2) U_3(\theta_3)$ is equivalent to a *single* rotation through an angle α about a *single* axis $\hat{\alpha}$ (Euler's theorem on rigid body motion follows, for example).

In our example above, there is a property that was not assumed in the more formal discussion that preceded it: we started with the assumption that each flow U_k is a one-dimensional rotation group, but observed that an arbitrary product of the two or more noncommuting rotation operators is *again* a rotation operator (see (1.74)), so that the three separate one-dimensional flows combine to form a larger group, the group of rotations in three dimensions. This is reflected locally by the fact that the three infinitesimal operators L_i form a complete set: every commutator of any two of them is given by the third. This brings us to the next important question: what does the Lie algebra look like in the general case of an r-parameter group, where every product $U_k(\theta_k)U_l(\theta_l)$ of two (generally noncommuting) flows is again a flow in the group? Each group element is an element of a one-parameter subgroup, but these r subgroups can combine to form a larger group that is a multiparameter Lie group. The conditions for a set of transformations

$$x^i = \psi^i(x^1, \ldots, x^n; a^1, \ldots, a^r), \tag{1.75}$$

depending continuously and differentiably upon r parameters $(a^1, \ldots, a^r)$ to form a group, are: (i) that the group combination law $\psi(\psi(x, a), b) = \psi(x, c)$ is possible because there is a combination law for the parameters

$$c^\sigma = \phi^\sigma(a^1, \ldots, a^r; b^1, \ldots, b^r); \tag{1.76}$$

(ii) there is an identity, $x = \psi(x, a_0)$, where, without loss of generality, we can take $a_0 = 0$; and (iii) each transformation has a unique inverse $x(a) = \psi(x(c), \bar{c})$.

Lie's idea was to study the flows locally in the tangent space near the identity, which is determined by the n vector fields V_σ^i:

$$dx^i = V_\sigma^i \, \delta a^\sigma. \tag{1.77}$$

We must now pay attention to whether the vector is covariant (lower indices) or contravariant (upper indices), summation convention (sum over repeated indices) is used, and the required metric g_{kl} normally follows from writing down a kinetic energy $T = g_{kl} \, dx^k \, dx^l / 2 \, dt^2$. Correspondingly, the group parameter combination law yields

$$da^\sigma = v_\rho^\sigma \, \delta a^\rho. \tag{1.78}$$

Consider the matrix v with entries v_ρ^σ, and let λ denote its inverse; both v and λ depend upon $a = (a^1, \ldots, a^r)$. Then

$$\delta a^\rho = \lambda_\sigma^\rho \, da^\sigma, \tag{1.79}$$

and so it follows that

$$dx^i = V_\sigma^i \lambda_\tau^\sigma(a) \, da^\tau. \tag{1.80}$$

The transformation that is induced in any scalar function $F(x)$ is therefore given by writing

$$dF = \delta a^\sigma X_\sigma F, \tag{1.81}$$

where

$$X_\sigma = V_\sigma^i(x) \frac{\partial}{\partial x^i} \tag{1.82}$$

is the infinitesimal generator that is determined by the velocity field whose components are given by $V_\sigma^i(x)$. We are now prepared to turn to Lie's integrability condition.

In order that the transformations are path-independent, the integrability conditions

$$\frac{\partial^2 \psi^i}{\partial a^s \, \partial a^t} = \frac{\partial^2 \psi^i}{\partial a^t \, \partial a^s}, \tag{1.83}$$

must be satisfied for all i and for all s, t. This condition yields, after a little manipulation, that

$$[V_k^i(x), V_\gamma^i(x)] = \left(\frac{\partial \lambda_\rho^\tau}{\partial a^\sigma} - \frac{\partial \lambda_\sigma^\tau}{\partial a^\rho} \right) v_\kappa^\rho v_\gamma^\sigma V_\tau^i(x), \tag{1.84}$$

which is the closure relation for the Lie algebra expressed in terms of Poisson brackets of the vector fields. If we write

$$c_{\kappa\gamma}^\tau = \left(\frac{\partial \lambda_\rho^\tau}{\partial a^s} - \frac{\partial \lambda_\sigma^\tau}{\partial a^\rho} \right) v_\kappa^\rho v_\gamma^\sigma, \tag{1.85}$$

then the quantities $c_{\kappa\gamma}^\tau$ are independent of a, because there is no explicit a-dependence on the left-hand side of the Poisson bracket relation (1.84) above. The constants $c_{\kappa\gamma}^\tau$ are called the structure constants of the Lie algebra, and the closure relation

$$[V_\kappa^i(x), V_\gamma^i(x)] = c_{\kappa\gamma}^\tau V_\tau^i(x), \tag{1.86}$$

for the Lie algebra means simply that the Poisson bracket of any two vectors in the tangent space is itself a linear combination of the r linearly independent vector fields that define the Lie algebra (the corresponding infinitesimal generators satisfy the commutation rules $(X_\sigma, X_y) = c_{\sigma y}^\tau X_\tau$). Therefore, the vector fields of the Lie algebra provide r linearly independent (but not necessarily orthogonal) basis vectors in the tangent space near the group identity, in terms of which any other vector in the tangent space can be expanded. The Poisson bracket relation is the starting point for the development of the theory of Lie algebras and Lie groups. The modern theory relies heavily upon the contributions of Cartan's classification of Lie algebras by using root diagrams.

The Poisson bracket formalism will be applied to conservative systems in Chapter 3, but no restriction to conservative systems is made in the development above. In fact, one can discuss the Navier–Stokes equations in terms of Lie group generators (see Birkhoff, 1950).

2
Introduction to deterministic chaos

2.1 The Lorenz model, the Lorenz plot, and the binary tent map

We begin with the way that deterministic chaos was discovered numerically in a system of three coupled nonlinear differential equations by the meteorologist E. Lorenz, in an attempt to integrate the model on a computer. In addition, Lorenz showed analytically that orbits are attracted to enter a certain phase space volume from which escape is impossible, but which contains no stable classical attractors (equilibria, limit cycles, and tori). Consequently, he discovered numerically that nearby initial conditions yielded trajectories with entirely different spatial patterns.

The bifurcation sequence for the Lorenz model was discussed in the first chapter, where we assumed without proof that the model generates a flow in phase space. In order that the vector field

$$V = (\sigma(y - x), (\rho - z)x - y, xy - \beta z)$$

in the Lorenz model defines a flow for all positive times, it is necessary that the solutions $X(t) = (x, y, z)$ are bounded at all $t < \infty$, i.e. that no singularities develop at *finite* times. The consequence of uniformly bounded solutions at all finite times is that one is led to conclude that the model has an attractor, even when all of the classical attractors have become unstable by bifurcations. That is, we are led to conclude that the Lorenz model has a 'strange' attractor. The use of the word strange was appropriate until examples of nonclassical attractors became known, but the use of the word persists in more modern discussions. Let us see how the existence of a nonclassical attractor in the Lorenz model can be inferred.

In order to study the boundedness of solutions, and to prove that the model defines a flow in phase space, start with any sphere of radius R

centered at $(0, 0, \rho + \sigma)$ and consider the time rate of change of the radius

$$R^2 = x^2 + y^2 + z'^2, \tag{2.1}$$

where $z' = z - \rho - \sigma$. Along a trajectory,

$$R\dot{R} = x\dot{x} + y\dot{y} + z'\dot{z}', \tag{2.2}$$

where dR/dt is the radial component of the flow and is therefore normal to the surface of the sphere. In the new variables (x, y, z') the Lorenz model becomes

$$\dot{x} = \sigma(y - x),$$

$$\dot{y} = -x(\sigma + z') - y, \tag{2.3}$$

$$\dot{z} = xy - \beta z' - \beta(\sigma + \rho),$$

so that

$$R\dot{R} = -\sigma x^2 - y^2 - \beta z'^2 - \beta(\sigma + \rho)z'. \tag{2.2b}$$

The radial component of the flow vanishes whenever

$$R\dot{R} = -\sigma x^2 - y^2 - \beta z'^2 - \beta(\sigma + \sigma)z' = 0, \tag{2.4}$$

and the points where this occurs are given by the intersection of the sphere with the ellipsoid

$$\sigma x^2 + y^2 + \beta(z' + (\sigma + \rho)/2)^2 = \beta((\sigma + \rho)/2)^2, \tag{2.5}$$

so that

$$R\dot{R} = \beta((\sigma + \rho)/2)^2 - (\sigma x^2 + y^2 + \beta(z' + (\sigma + \rho)/2)^2). \tag{2.4b}$$

From this result, we conclude that the radial component of the velocity is negative (points toward the origin) everywhere outside the ellipsoid, is positive inside the ellipsoid, and vanishes along any intersection of the sphere with the ellipsoid. We therefore obtain the following qualitative information about the flow: consider a sphere with a radius so large that the ellipsoid lies entirely inside it. The radial component of the flow at the surface of the sphere points everywhere inward, and every streamline that begins outside the sphere must enter the sphere within a finite time. Because $dR/dt < 0$ everywhere outside the ellipsoid, a streamline must eventually intersect the ellipsoid, and when this happens then the radial component of the velocity vanishes. At such a point, the velocity vector is entirely azimuthal and so either points into the ellipsoid or does not. If it does, then the streamline enters the ellipsoid. If it does not, then in the next instant the streamline picks up a radial component of velocity

$dR/dt < 0$ and so must intersect the ellipsoid again at a later finite time. The same reasoning applies to that intersection, so that the streamline must eventually, perhaps after many intersections, enter the ellipsoid. Once inside, it can leave again (unless it goes into a sink, it *must* leave again, because $dR/dt > 0$), but as soon as it exits, $dR/dt < 0$ holds again, so that the streamline eventually must re-enter the ellipsoid. The flow pattern can be quite complicated, but we can say that, given any sphere that contains the ellipsoid completely, all streamlines must enter and then are confined within that sphere forever. Hence, all solutions $X(t)$ are bounded by any such sphere, even as $t \to \infty$, so that the time evolution operator exists and the phase flow picture is thereby confirmed. It is an interesting exercise to use three-dimensional computer graphics to plot the flow field on a lattice within any sphere that contains the ellipsoid.

More generally, for the forced dissipative equations

$$\dot{x}_i = a_{ijk} x_j x_k - b_{ij} x_j + c_j, \tag{2.6}$$

where $i = 1, 2, \ldots, n$ (and summation convention is used), if $a_{ijk} x_i x_j x_k = 0$ and b_{ij} is a positive definite matrix (all eigenvalues are positive), then one can prove that all solutions are bounded as $t \to \infty$ (see Ghil and Childress, 1987). Equations of this form are typical of the Navier–Stokes equations when mode expansions are arbitrarily truncated. The Lorenz model is an example of one such truncation (see Schuster, 1984 or Ghil and Childress, 1987). Here, our viewpoint is that the Lorenz model is a 'mathematics laboratory' for discovering deterministic chaos; therefore we do not reproduce its purely formal derivation from the hydrodynamics equations. Especially in the turbulent regime, one does not know whether the mode expansions converge, and the justification of the replacement of the Navier–Stokes equations by finitely many ordinary differential equations would require sufficiently rapid convergence of the mode expansion. Next, we formally generalize our elementary ideas of attractors and repellers.

The idea of an attractor, as introduced in the first chapter, is that of a point set in phase space, a curve or a torus, that draws the orbit toward it as $t \to \infty$, whenever an initial condition is chosen that belongs either to the attracting point set or to the attractor's 'basin of attraction'. Other possible attractors, including the so-called 'strange' attractors, also can be defined analytically, purely formally, as the point set that is the closure of a 'generic', meaning 'typical', orbit of the system asymptotically as the time goes to infinity. The idea of a flow is a purely *geometric* notion, but the use of the word 'typical' immediately introduces the idea of *probable* behavior, which means that the opportunity to make choices of something or other must somehow arise. So far, we have discussed only geometry, because statistics do not arise when one sticks to the classical attractors and repellers. In order to generalize our idea of an attractor in a useful way, we must leave the domain of pure geometry and introduce considerations based upon specific classes of different initial conditions in phase space, because that is where differences in statistical behavior of chaotic orbits arise, as we shall demonstrate. A 'typical' orbit is defined from the *analytic* or *continuum* standpoint as one that occurs 'with measure one' in a hypothetical random draw of initial conditions (the famous choice-problem) from the continuum that constitutes the basin of attraction (this includes all points that lie on the attractor).

Geometrically, strange attractors are made up of a continuum of points in a way that is qualitatively and nontrivially different from smooth curves and tori: they have a fragmented structure that can be like that of a Cantor set (Chapter 4), and the motion on the Cantor-like set is either completely unstable in the sense that nearby initial conditions yield orbits that develop very unsimilar spatial patterns as a function of time (Chapter 4), or else it is only marginally stable at a boundary of chaos (Chapters 5 and 6). In addition, different subsets of a strange attractor can be defined formally as the closure of qualitatively different orbits that start from different *classes* of initial conditions on the attractor (cf. Chapters 7 and 9 for discussions of classes of initial conditions), so that when one leaves behind equilibria, limit cycles, and tori, which are the classical attractors, then the description of the flow becomes both qualitatively and quantitatively much more difficult. Let us proceed with the reasoning that leads to the conclusion that the Lorenz model has a nonclassical attractor.

In Chapter 1, we discovered three different zero-dimensional attractors of the Lorenz model: when $0 < \rho < 1$, the origin is an attractor (it is a repeller when $\rho > 1$). When $1 < \rho < \rho_h$, the points

$$(\pm\sqrt{\beta(\rho - 1)}, \pm\sqrt{\beta(\rho - 1)}, \rho - 1)$$

are attractors – again, zero-dimensional ones. The Lorenz model also has one-dimensional attractors: stable limit cycles. Stable quasiperiodic motion on a torus in phase space can also be an attractor. This is all consistent with another fact: because $J(t) = e^{-(\sigma + \beta + 1)t} \to 0$ as $t \to \infty$, any initial phase volume $\delta x\, \delta y\, \delta z$ contracts to zero asymptotically, and this is consistent with the confinement of the $t \to \infty$ solutions to any subspace of dimension less than three, 0-, 1-, or 2-dimensional, for example.

Although it would be typical that the Hopf bifurcation at ρ_h would yield stable limit cycles for $\rho_h < \rho < \rho_c$, where ρ_c is some critical value where the limit cycles become unstable, this does not happen in the Lorenz model. There, when $\rho > \rho_h = \sigma(\sigma + \beta + 3)/(\sigma - \beta - 1)$, there are no stable equilibria (sinks of the flow) or stable limit cycles, nor are there stable quasiperiodic orbits. Yet, the fact that the orbits are attracted irreversibly to enter and evolve forever inside any sphere that entirely contains the ellipsoid suggests intuitively the idea of an attractor somewhere within the sphere's volume. But what is the nature of an attractor under the conditions that no rest, no stable periodic of quasiperiodic motion is possible, when a phase point is condemned to wander for ever inside a bounded volume? We expect that, when it is bounded by the sphere, a trajectory must wander into and out of the ellipsoid infinitely many times as the time goes to infinity. By studying only the classical mechanics textbooks of our quantum-mechanical era, physicists had no preparation to ask or to analyze this sort of question. It was E. Lorenz, a theoretical meteorologist, who was led to the discovery of sensitivity with respect to small changes in initial conditions by paying careful attention to the results of an attempted numerical integration of his model for the parameter values $\sigma = 10$, $\beta = \frac{8}{3}$, and $\rho = 28$ ($\rho_h = \frac{470}{19}$ on this case). We have chosen these same parameter values in order to generate Fig. 2.1 to 2.3.

Here is not exactly what Lorenz observed, but it is computationally equivalent to what he observed: discretize the Lorenz equations for computation – this means that you must replace them by a three-dimensional iterated map that represents your integration scheme. It is not a Poincaré map (that is a more sophisticated idea), but it is a discrete map. Now, with $\sigma > \beta + 1$ and ρ slightly greater than ρ_h, choose an initial condition (x_0, y_0, z_0) and iterate the map on a computer – do not worry at first about the numerical accuracy of your integration scheme; it is not and cannot be accurate if you iterate more than a few times, it turns out, but we will come to that later (a check on the accuracy of your integration scheme is to integrate backward and require that you recover your initial condition to within several digits accuracy). Now, choose a new initial condition $(x_0 + \delta x_0, y_0 + \delta y_0, z_0 + \delta z_0)$ close to, but different from, the old one. The shift $(\delta x_0, \delta y_0, \delta z_0)$ represents a change in what we would like to call the *least*-significant digits of all or of any one of the values of (x_0, y_0, z_0). What happens is not what we were taught to expect: the two solutions start near each other, but diverge exponentially fast from each other as you iterate the map. This is called 'sensitivity with respect to small changes in initial conditions' or, more poetically, 'the butterfly effect'. It is instructive to make two different plots, using color graphics on a computer in order to compare the projections of the computed trajectories onto one and two dimensions for two slightly differing initial conditions. One should plot: (i) the time series for the variable $z(t)$ vs t

(Fig. 2.1*a*, *b*), and (ii) and $z(t)$ vs $y(t)$ (Fig. 2.2). In both plots, it is interesting to mark the locations of the projections of the unstable equilibria and the ellipsoid.

The Lorenz model can be called chaotic in the sense that there is effectively no such thing as a least-significant digit in an initial condition: big changes grow from small differences as the bounded trajectory loops many times about the unstable fixed points. It should not escape the reader's notice that this defining property of chaos suggests that chaotic orbits are not easy to compute. In practice, sensitivity with respect to small changes in initial data means that statistics are required for the description of such systems in nature. Imagine a thought experiment where a closed, isolated system that obeys the Lorenz model is prepared in a definite initial state. Data are collected experimentally in the form of a time series for some physical variable, $z(t)$, for example. Next, one attempts an identical repetition of the same experiment: the time series for $z(t)$ is recorded for the same length of time, starting from the same

Fig. 2.1   Time series for the coordinate $z(t)$ resulting from a typical attempt at numerical integration of the Lorenz model on a VAX for initial conditions (x_0, y_0, z_0) equal to (*a*) (1, 1, 1) and (*b*) (1.1, 1.1, 1.1). Note that the two series are approximately the same for times earlier than about $t = 11$, but then diverge strongly from each other (butterfly effect).

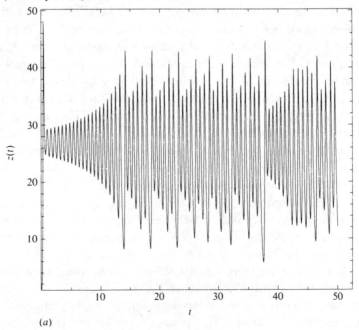

(*a*)

initial state. However, in the laboratory, the 'same' initial conditions can *at best* be prepared to within a *finite* accuracy $(x_0 + \delta x_0, y_0 + \delta y_0, z_0 + \delta z_0)$. While this finite accuracy of preparation may be reduced, by using more accurate measuring apparati and by taking more care in state-preparation, it can never, under *any* circumstances, be reduced to zero. Infinite precision is quite impossible in the laboratory, if not in the mind of a mathematician. Therefore, *no* initial condition can ever be exactly repeated in a real physical system in nature, and that a small error is made means that the two experiments will yield similar time series for a short while, but the pattern of the second series will eventually diverge quantitatively from the pattern of the first one (see Fig. 2.1*a, b*). If one performs many such 'repeated, identical experiments' then one can define an empirical statistical distribution that describes how often $z(t)$ takes on a value that lies within a given range of values. Therefore, deterministic chaotic systems have the capability to explain natural phenomena that appear to be 'random' in that they are necessarily described by the collection of statistical data *but may actually have an underlying deterministic cause and description*. One is, with good reason, excited by this possibility.

What Lorenz really did was to stop a computation and then to restart it on the same machine at a later time. The shift in initial conditions

Fig. 2.1—continued.

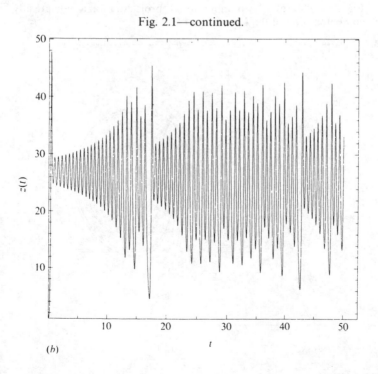

(*b*)

occurred because the machine did use as input exactly the same number that he typed into the machine after having read that number as output. In other words, because of the way that the roundoff/truncation algorithm (which is typically unknown to the programmer), treats digits the machine represented *in two slightly different ways* the same number displayed upon the computer screen, depending upon whether the display represented what he typed as *input* for an initial condition, or read from the screen as *output* from a calculation.

The main point is now that there is a familiar idea of stability combined with an entirely different idea of instability. First, there is the suggestion of stability in the sense of an attractor: all orbits that start outside a certain ellipsoid enter it in finite time and are trapped there forever. However, at least on a subset of the phase space inside the ellipse, there is an instability in the sense of sensitivity with respect to small changes in initial data. At first sight this confluence of stability and instability seems hard to explain. It will turn out to be not hard to understand when we finally abandon the attempt to understand all of the details of the flow, and concentrate instead upon iterated maps that represent certain aspects of the flow. We

Fig. 2.2 Plot of a two-dimensional projection of a numerically computed orbit of the Lorenz model.

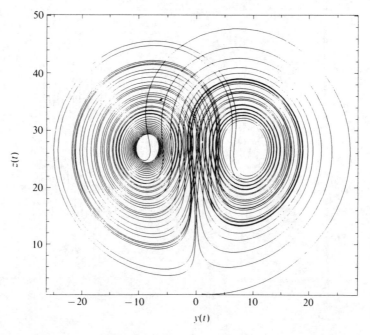

have already seen an example in Chapter 1 where a flow can be replaced by one kind of map, a stroboscopic map.

Lorenz's studies were motivated by the question whether the weather is predictable in the long run, and led to the discovery of the butterfly effect (sensitivity with respect to small changes in initial conditions) in a completely deterministic system of three coupled nonlinear equations. By the butterfly effect is meant that the weather in Eidfjord could be changed by the flapping of a butterfly's wings in Mittenwald, if the weather system is deterministic-chaotic. We shall use the phrase butterfly effect simply to mean sensitivity with respect to small changes in initial conditions. In order to most clearly understand the mathematical reason for the butterfly effect, it is helpful to replace the original system of differential equations by a simpler iterated map than the one provided by Lorenz's numerical integration scheme. Poincaré knew, around the turn of the century, that what we now call 'strange' behavior could occur in perfectly deterministic systems of nonlinear differential equations, and he was the first to propose the theoretical study of nonlinear systems by replacing the original

Fig. 2.3   From the data of Fig. 2.1a: a plot of successive maxima of $z(t)$ against each other.

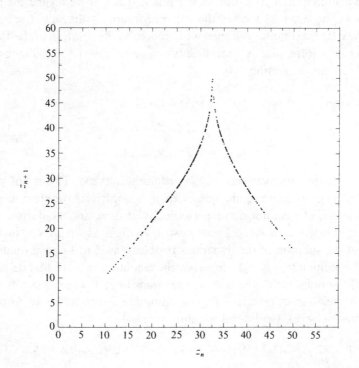

equations by a so-called 'return map'. Here is Poincaré's idea: imagine a (hyper-)surface in phase space. Every time the orbit passes through the surface in a given region of phase space (Fig. 2.1), you record the location $x_i = (x_{1i}, x_{2i}, \ldots, x_{ni})$ of the intersection for $i = 1, 2, \ldots, N$ intersections. Then, plot x_{i+1} vs x_i. Geometrically, this means that you begin to study an iterated map of n variables. It is sometimes possible, by ignoring a certain amount of information that the differential equations contain, to extract a discrete map that has *lower* dimension than the original system of equations. Lorenz concentrated his attention upon certain partial information produced by his numerical integration scheme by constructing the following plot, now called a Lorenz plot: He iterated his integration scheme numerically, and then plotted successive maxima of the coordinate $z(t)$ against one another. If we denote the successive maxima by $z_1, z_2, \ldots, z_n, \ldots$, then one plots z_{n+1} vs z_n. Lorenz's result is similar to that shown as Fig. 2.3. The points appear to fall on a single curve, suggesting approximately an underlying one-dimensional map $z_{n+1} = f(z_n)$.

Under what set of circumstances could the Lorenz plot yield a lower (than three-)dimensional map *analytically*? Let x_n and y_n denote the values of $x(t)$ and $y(t)$ for which the nth maximum z_n occurs, beginning with some definite initial condition in the phase space. If we regard the phase point (x_n, y_n, z_n) that locates the nth maximum as the initial condition for the next maximum z_{n+1}, then we can discuss the solution of the Lorenz model in a form that is superficially suggestive of a three-dimensional iterated map (cf. section 1.4),

$$x_{n+1} = G_1(x_n, y_n, z_n; \tau_n),$$

$$y_{n+1} = G_2(x_n, y_n, z_n; \tau_n), \qquad (1.47b)$$

$$z_{n+1} = G_3(x_n, y_n, z_n; \tau_n),$$

where τ_n is the time between the two adjacent maxima. This is not yet an iterated map, because τ_n changes with n, a nontrivial problem, because the maxima of z cannot occur periodically in time, and we do not know how τ_n changes with n. To know how τ_n changes as n varies is to know part of the solution of the dynamics problem, as is to know a qualifying initial condition (x_0, y_0, z_0). However, the condition that the first derivative of $z(t)$ vanishes each time that a maximum in $z(t)$ occurs permits us to use the relation $z_n = x_n y_n - \beta z_n$ to eliminate either x_n or y_n from the equations (1.47b), yielding (if we eliminate x_n)

$$\beta G_3(\beta z_n/y_n, y_n, z_n, \tau_n) = G_2(\beta z_n/y_n, y_n, z_n, \tau_n)G_1(\beta z_n/y_n, y_n, z_n, \tau_n), \quad (1.47c)$$

which, in principle, should determine τ_n in terms of (x_n, y_n, z_n). If this solution for τ_n as a function of the phase point is possible, then τ_n can be eliminated from the lower two of equations (1.47b) to yield a *two-dimensional iterated map*

$$y_{n+1} = \bar{G}_2(y_n, z_n),$$
$$z_{n+1} = \bar{G}_3(y_n, z_n).$$
(1.47d)

This two-dimensional map *cannot* contain complete information about the flow in the three-dimensional phase space: we arrived at it by ignoring all questions about that flow *except* for the question where the maxima of $z(t)$ occur. However, in contrast with two coupled differential equations, two-dimensional iterated maps can yield *chaotic* motion, so that our iterated map can contain partial information about the Lorenz attractor. In fact, as we shall see below, deterministic chaos even occurs in *one*-dimensional iterated maps. Unfortunately, a further reduction in dimension is possible *analytically* only on the basis of an assumption that we cannot justify: if the variable y_n would change rapidly relative to z_n, yielding approximately the 'slaving' condition $y_n \approx \bar{G}_2(y_n, z_n)$, then the elimination of y_n from the equation for z_n in (1.47d) would produce a one-dimensional iterated map,

$$z_{n+1} = f(z_n),$$
(1.47e)

to describe approximately the Lorenz plot. It is absolutely clear that the Lorenz plot can be at best *approximately* one-dimensional, because the required one-dimensional map is noninvertible (it has a two-valued inverse, two values of z_n for each single value of z_{n+1}), while the solution of the Lorenz differential equations is invertible (because the Jacobian is finite), as is any map that represents the Lorenz flow completely accurately. The main point is that a one-dimensional map *cannot* provide more than partial information about a higher-dimensional attractor and the three-dimensional flow near or on that attractor, in spite of claims made to the contrary in several contemporary textbooks. One-dimensional maps that describe bounded chaotic motion are by necessity noninvertible (must have at least a two-valued inverse), whereas the flow and the return maps that can describe it in detail are necessarily uniquely invertible (the Jacobi determinants that are required for inversion and backward motion in time are always finite).

There are features of trajectories that have their analogs in iterated maps. For example, a stable limit cycle in phase space corresponds to a stable fixed point of the return map, whereas an unstable limit cycle corresponds to an unstable fixed point of the map. For a one-dimensional map (1.47e), the fixed point satisfies $z = f(z)$. Stability follows from linearizing about the fixed point, yielding, with $\delta z_n = z_n - z$, the linearized equation and its solution $\delta z_n = f'(z)\,\delta z_{n-1} = (f'(z))^{n-1}\,\delta z_0$. If the magnitude of $f'(z)$ is smaller than unity then the fixed point is stable, whereas if the magnitude of the slope is greater than unity then the fixed point is unstable. In general, a map may have several or no stable equilibria (bifurcation sequences for two particular classes of one-dimensional maps are discussed in Chapters 5 and 6). Furthermore, if the time series $z(t)$ is chaotic, then so is the map. Clearly, the numerical Lorenz return map cannot be not understood from the standpoint of stable fixed points, and to understand the chaos in the flow it is best to begin by understanding how a one-dimensional map can exhibit self-generated chaotic behavior.

There are other ways to try to obtain an iterated map that has a lower dimension than the flow. The Rössler model is defined by a set of three coupled nonlinear equations and has been studied by the following procedure: a trajectory is projected onto the xy-plane, and x_{n+1} is plotted against x_n, where x_n is recorded every time that y vanishes with $\mathrm{d}y/\mathrm{d}t > 0$. The numerical calculation produced a one-dimensional map because of the information that was systematically ignored in order to define the numbers $\{x_n\}$.

In this text, we abandon the aim to derive specific maps analytically from systems of differential equations and concentrate instead upon the study of low-dimensional maps, especially one-dimensional maps, with the expectation that the systematic neglect of enough detailed information in higher-dimensional systems might, at least in some cases, yield low-dimensional maps *if one had enough information about the solution to carry out the approximation procedure systematically and correctly*. The approximation of the two-dimensional Lorenz plot by a one-dimensional map gives us *partial* information about the Lorenz model: we can use the one-dimensional map to model both the geometry of the set on which the maxima of $z(t)$ are distributed, as well as the distribution itself on that set. The fact that the Lorenz model has an attractor leads to the fact that the points z_n fall into a bounded interval. We want to understand how the iteration of a noninvertible map can produce 'strange' motion, motion that is fully deterministic, yet is 'apparently' random (i.e., generates statistical behavior completely deterministically), on a bounded interval.

The study of differential equations leads to the study of flows in phase space, where specific trajectories and specific initial conditions are not of direct interest, but the study of systems with nonclassical attractors leads us to the study of iterated maps. The study of chaotic maps, in turn, will lead us back to the consideration of both initial conditions and arithmetic, because the statistics that are generated by a chaotic map on a strange invariant set (the set defined by the closure of a chaotic trajectory in phase space) turn out to depend strongly upon the choice of initial conditions. How can one use the Lorenz plot to illustrate that the motion is chaotic, to understand that the invariant set is nonclassical, and that the motion on that set generates statistics that in some sense mimic randomness? It all goes back to the butterfly effect, which arises whenever there is bounded motion without stable classical attractors, irrespective of whether one studies differential equations or iterated maps.

In order to understand the butterfly effect in detail, and to understand how it leads to statistical behavior, we can follow Lorenz and replace the Lorenz plot by a one-dimensional map that is partly similar to the numerical Lorenz return map, but has a form simple enough for us to be able to analyze it completely by using binary arithmetic. In this way, we can come to appreciate in complete detail what is meant by sensitivity with respect to small changes in initial conditions. The Lorenz model is not an accurate approximation to the original hydrodynamics equations that describe the weather system, but it is a good mathematics laboratory for the study of analytic properties of chaos. Likewise, there is no need here to study *exactly* the Lorenz map: it is mathematically convenient to replace it at this introductory stage by a qualitatively similar but simpler map that can be thoroughly understood both analytically and numerically. The goal here is to understand deterministic chaos with as little mathematical complication as possible. Furthermore, as we shall illustrate in Chapter 9, there are *universality principles* that make the exact choice of map unimportant: whenever a map belongs to a certain class of maps, one need study only the simplest map in the universality class in order to understand how the others behave. We turn now to the study of the binary tent map, which looks deceptively simple but defines the universality class that describes 'fully developed chaos' (Chapter 9).

The binary tent map is defined by the equations (Fig. 2.4)

$$z_{n+1} = \begin{cases} 2z_n, & z_n \in [0, \frac{1}{2}), \\ 2(1 - z_n), & z_n \in (\frac{1}{2}, 1], \end{cases} \tag{2.7}$$

and its dynamics can be easily understood in base-2 arithmetic where we write all numbers and parameters as binary strings,

$$z_n = \sum_{j=1}^{\infty} \varepsilon_i(n)/2^j = 0.\varepsilon_1(n)\varepsilon_2(n)\cdots\varepsilon_N(n)\cdots, \tag{2.8}$$

where $\varepsilon_i(n) = 0$ or 1. The map is exactly equivalent to a binary shift map defined by the simple automaton

$$\varepsilon_i(n + 1) = \begin{cases} \varepsilon_{i+1}(n), & \varepsilon_1(n) = 0, \\ 1 - \varepsilon_{i+1}(n), & \varepsilon_1(n) = 1, \end{cases} \tag{2.9}$$

so that initial conditions that are written as finite binary strings

$$z_0 = \sum_{i=1}^{N} \varepsilon_i(0)/2^i = 0.\varepsilon_1(0)\varepsilon_2(0)\cdots\varepsilon_N(0) \tag{2.10}$$

truncate to zero in finite time $n = N$. More generally, if we write $z_0 = P/Q$ where P and Q are both integers, then the subsequent iterates z_n can take on only one of Q different possible values $0, 1/Q, 2/Q, \ldots, (Q-1)/Q$. Therefore, after at most Q iterations of the map, the initial condition is repeated and you have periodicity. Rational numbers, numbers of the form $z_0 = P/Q$ with P and Q both integers, have binary expansions that either truncate after a finite number of bits or else consist of an infinitely repeating block of bits. For example $\frac{1}{4} = 0.0100000\cdots$, but $\frac{1}{3} = 0.010101\cdots$. Hence, all rational initial conditions of the tent map yield periodic orbits. However, all of these orbits are unstable, and one way that deterministic chaos manifests itself is that (with P and Q relatively prime and with Q prime relative to 2), a small increase in Q typically yields an entirely different orbit of longer period. For example, if you start with $z_0 = \frac{1}{3}$ then the map hops onto the unstable fixed point $z_1 = \frac{2}{3}$, whereas $z_0 = \frac{2}{9}$, which differs from the original initial condition by only $\frac{1}{9}$, yields the unstable

Fig. 2.4   The binary tent map.

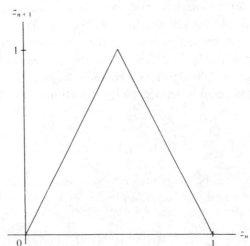

3-cycle made up of the points $\frac{2}{9}$, $\frac{4}{9}$, and $\frac{8}{9}$. If we think of z_n vs n as a discrete time series, then while the two time series start very near to each other, the long-time behavior of one (1-cycle) is very different from that of the other (3-cycle). In stark contrast, a slight change in the initial conditions for a *stable* periodic orbit, e.g. of the simple harmonic oscillator, produces an orbit that is geometrically similar to the original orbit and has the same period as the original orbit. For a stable limit cycle (the periodic orbits of the tent map are discrete versions of *unstable* limit cycles), a small change in the initial condition has no effect at all as the time goes to infinity so long as the initial condition is inside the basin of attraction of the stable limit cycle. Therefore, the system of unstable periodic orbits of the tent map is completely unfamiliar to us from traditional studies of mechanics and differential equations: because rational numbers are dense in the real number system, the tent map's unstable periodic orbits are dense in the phase space.

Nontruncating periodic binary strings for z_0 yield unstable periodic orbits by virtue of the fact that rationals are made up of blocks of bits that repeat infinitely often (see section 2.4). Irrational initial conditions, on the other hand, yield unstable nonperiodic orbits because there is no periodicity in the binary string of an irrational number. Where is the instability in this case? If one changes the infinite string of 'least-significant bits' $\varepsilon_N\varepsilon_{N+1}\cdots$ in an irrational initial condition, then the result is an entirely different nonperiodic orbit after N iterations of the map. That is, the spatial *patterns* of the two orbits are completely different: each spatial pattern reflects the discrete bit-pattern in the infinite-length tail $\varepsilon_N\varepsilon_{N+1}\cdots$ in its own initial condition. For the tent map, the reason for this 'sensitivity with respect to small changes in initial conditions', for both periodic and nonperiodic orbits, can be understood as follows: in z_0, the information stored as *any* bit, say $\varepsilon_j(0)$, is propagated to the left as the map is iterated, and, after j iterations, determines the information in the *leading* bit $\varepsilon_1(n)$. You can understand this most easily by computing the output of the automaton (2.9) for a binary initial condition, e.g. the orbit for $z_0 = \sqrt{2} - 1 = 0.0110101000001001\cdots$. The point is that there is no such thing as a 'least-significant bit' that can be neglected if you iterate the map long enough – the 'least-significant bits' *are* the reason for the complicated spatial patterns of deterministic chaotic orbits. This is in direct contradiction to our experience with the stable periodic and quasiperiodic orbits of nonchaotic dynamical systems. Both the periodic and quasiperiodic orbits of the tent map can be called chaotic because they have information flow, right to left in digit strings, as the map is iterated forward, at the rate of one bit per iteration. That is, these orbits exhibit the butterfly effect at the rate of one bit per discrete time step. In what follows, we shall see that the information flow property of one bit per iteration (right to left) on binary arithmetic can be summarized by saying that the tent map has a positive Liapunov exponent $\lambda = \ln 2$. Whenever a map exhibits information flow at the average rate of one digit per iteration (right to left) in base-μ arithmetic, then the map's trajectory has an average Liapunov exponent equal to $\ln \mu$. A word of caution: maps with variable slope, especially asymmetric maps, typically have a continuum of Liapunov exponents, where the expected Liapunov exponent depends strongly upon the class of initial conditions that is chosen (see Chapter 7).

What is 'strange' about the pattern of a trajectory of the tent map? Its 'strangeness' lies in the fact that a single orbit, originating from a definite irrational initial condition, has a closure that is either the entire unit interval or else is a finite subset of the unit interval. In fact, a single, well-defined orbit generates a *statistical distribution* on the unit interval. In other words, determinism leads to *statistical regularity*, to the establishment of laws of statistics. To understand this, break up the unit interval into 2^n distinct subintervals, each of length 2^{-n}. Label the first interval by the n-bit binary word $000\cdots 0$, the second by $000\cdots 01, \ldots$ and the last by $111\cdots 1$. The frequency with which a definite n-bit word $\varepsilon_1\varepsilon_2\cdots\varepsilon_n$ occurs as the first n bits of the sequence of iterates z_n tells us how often the trajectory visits the particular interval that is labeled by that particular string, and the frequency with which this visitation occurs is completely determined by the bit distribution in the initial condition z_0. You can make the subdivision of the phase space as fine as you like, so, in the infinite precision limit, the iterates of the map have a closure that covers the unit interval densely for a large class of initial conditions – those that cover the unit interval *uniformly* are discussed later in this chapter. Here, the 'strange' invariant set is the entire unit interval (e.g., the binary tent map, with slope magnitude 2) or at least a finite-length subset of the unit interval (e.g., tent maps with slope magnitude between 1 and 2), so long as the initial condition is an irrational number.

We have discussed two different kinds of statistical behavior above. The first is that initially nearby orbits generate different spatial *patterns* (irreproducibility of nearly identically prepared experiments). The simplest discrete pattern is that of a one-bit symbol sequence: merely read and catalog the first bits $\varepsilon_1(1)\varepsilon_1(2)\cdots\varepsilon_1(n)$ of each iterate z_n of the map. The resulting discrete pattern of 0's and 1's tells us how often the trajectory visits the intervals $[0, \frac{1}{2}]$ and $(\frac{1}{2}, 1]$. For two different irrational initial conditions, one gets two different *nonperiodic* patterns of bits. The second is that *single orbits* generate statistics. The reason for the second phenomenon is the same as the reason for the first: information flow that turns least-significant digits into leading digits in a finite number of iterations, which means that trajectories can be computed only for a number of iterations that is consistent with the number of bits that have been calculated in the initial condition.

We have seen, by example, that even the very simplest chaotic systems, one-dimensional maps, can have invariant sets (attractors and repellers are examples of invariant sets) that have the cardinality of the *continuum*, that is, of *uncountable* sets of points, combined with the property of entirely different spatial patterns of orbits issuing forth from balls of nearby initial conditions. The *formal* definition of an attractor as the closure of an orbit (infinite time limit) is not useful for the description of either experiment or computation, because definitions that are based upon the need for '2^∞' data points, the cardinality of the continuum (as expressed from the standpoint of binary arithmetic), in order to define an orbit that has a geometrically complex pattern, cannot be implemented *in practice*. Our position is that theoretical definitions that cannot be verified or used in either experiment or computation should not be adopted as a basis for the development of chaos theory, especially when they can lead to a viewpoint that is misleading. In the literature on chaos, the use of the formal definition of an attractor leads to the development of the prejudice that trajectories that are *measure-theoretically* generic are the preferred ones both in nature and in computation. The so-called generic trajectories are the ones that follow from a certain *uncountable* set of initial conditions that would *theoretically* occur with absolute certainty if one could make a truly *random draw* of numbers from the continuum. In order to begin to make these ideas clear, we shall illustrate in section 2.4 below an example of a generic orbit for the binary tent map from the measure-theoretic viewpoint. However, in what follows, we shall not prefer any particular set of initial conditions in our analysis.

By successive stages of simplification (by making approximations that are abstractly equivalent to discarding information),[1] Lorenz finally constructed a model that he could analyze, and found deterministic chaos (sensitivity with respect to small changes in initial conditions) at the very simplest level of nonlinearity, that of the binary tent map. One must then entertain the possibility that the earth's weather system is also

---

[1] Here we must admit that the word 'discard' is not quite appropriate, because we do not have such detailed information to begin with. If one had all the appropriate information (solutions for different initial conditions), one could start by systematically ignoring large amounts of detailed information about complicated trajectories.

deterministic-chaotic *because* it is governed by deterministic nonlinear equations of motion. In that case, sensitivity with respect to small changes in initial data would mean that changes in the 'least-significant' digits of an initial condition that describes the 'present state of the weather' would very quickly become transformed by the information flow property into changes in the leading digits of a weather prediction at later times. This fact would mean the defeat of any attempt to predict the weather over arbitrarily long times *even with a perfectly accurate mathematical model of the weather*, simply because one cannot know the required initial conditions from nature (the present state of the weather system) to within better than *finite* accuracy in practice. The other side of the coin is that, given an accurate mathematical model of the weather, the information flow property will permit us to predict the future state of the weather for a length of time that is determined by the accuracy within which we know the present state of the weather: present conditions are initial conditions for the future in any deterministic system. That is, if the weather system is deterministic rather than random, then *some* degree of short-range predictability may be possible.

We turn now to details of two *analytic* (as opposed to algorithmic, or digital) ways to describe deterministic chaos: (i) Liapunov exponents and (ii) invariant densities. The idea of a positive Liapunov exponent is the more basic one because it is just a way to describe the information flow property, which is the reason for sensitivity with respect to small changes in initial data, whereas continuous invariant densities can also be generated by stable, quasiperiodic motion on a torus (see Chapter 3). In what follows, we interest ourselves only in invariant densities that are generated by the action of a positive Liapunov exponent.

2.2 Local exponential instability of nearby orbits: the positive Liapunov exponent

In the automaton (2.9), the property equivalent to sensitivity with respect to small changes in initial conditions of (2.7) is that information is transferred from right to left in binary strings at the rate of one bit per iteration. We now introduce the analytic description of this same idea: the positive Liapunov exponent.

In order to find some conceptual ground to stand on, let us return for the moment to the idea of an unstable limit cycle. Such a curve has the property that it repels all nearby trajectories, so that they diverge from the limit cycle at an exponential rate. Consider now the possibility of a trajectory that is not a limit cycle but that repels all nearby trajectories in phase space at an exponential rate. Consider also the possibility that such repelling trajectories are dense in the system's phase space; such dense sets are provided in systems like the tent map (so-called 'hyperbolic systems') by the unstable periodic orbits. Then, for any two repelling trajectories that are separated initially by the distance $\delta x(0)$, the magnitude of the separation at a later time t will be

$$|\delta x(t)| \sim |\delta x(0)|\, e^{\lambda t}, \tag{2.11}$$

where λ is called the largest Liapunov exponent and depends in general upon the location of the initial condition $x(0)$ in phase space. In one dimension, the relation

$$\lambda = \frac{1}{t} \ln \frac{|\delta x(t)|}{|\delta x(0)|} \tag{2.11b}$$

for large t and small $|\delta x(0)|$ defines the Liapunov exponent, while a similar relation in higher dimensions yields only the largest Liapunov exponent. A positive Liapunov exponent is the essence of deterministic chaos: the basic idea is that small errors $\sim |\delta x(0)|$ in initial data are magnified exponentially, so that an orbit can be accurately computed only for times t such that

$$|\delta x(0)|\, e^{\lambda t} \ll L, \tag{2.11c}$$

or

$$t \ll \frac{1}{\lambda} \ln(L/|\delta x(0)|), \tag{2.11d}$$

where L denotes the extent of the phase space where the (bounded) chaotic motion occurs, and $\delta x(0)$ denotes the uncertainty in our knowledge of an initial condition $x(0)$.

The largest Liapunov exponent λ can be computed by linearizing the equations of motion about a particular unstable orbit, as we illustrate below for discrete maps in one and two dimensions.

In order to formulate useful expressions for the numerical computation of Liapunov exponents, let us begin with the case of a one-dimensional map. If one first understands the formulation for discrete maps then one can apply it to differential equations which are converted into discrete maps for the purpose of computation. We begin in one dimension with an iterated map

$$x_{n+1} = f(x_n), \tag{2.12}$$

where the iterations x_n are confined to a point set within some bounded interval $[a, b]$ as $n \to \infty$. The main idea is as follows: we compute an exact orbit (via a gedanken computation, if necessary) $x_0 \rightarrow x_1 \rightarrow x_2 \rightarrow \cdots$, beginning from some initial condition x_0, and then we linearize the map about this exact orbit for some nearby initial condition $x_0 + \delta x_0$:

$$x_n + \delta x_n = f(x_{n-1} + \delta x_{n-1}) \approx f(x_{n-1}) + f'(x_{n-1}) \, \delta x_{n-1} + \cdots. \quad (2.13)$$

We then obtain the linearized error propagation equation

$$\delta x_n \simeq f'(x_{n-1}) \, \delta x_{n-1}, \quad (2.14)$$

which we can solve exactly to obtain

$$\delta x_n \cong \prod_{i=0}^{n-1} f'(x_i) \, \delta x_0, \quad (2.15)$$

where δx_0 is the initial separation of the two orbits. With

$$|\delta x_n| \approx \sum_{i=0}^{n-1} |f'(x_i)| |\delta x_0| \quad (2.15b)$$

we can define the Liapunov exponent

$$\lambda = \lim_{n \to \infty} \frac{1}{n} \sum_{i=0}^{n-1} \ln|f'(x_i)| = \frac{1}{n} \ln|f^{(n)'}(x_0)| \quad (2.15c)$$

where $f^{(n)}(x_0) = x_n$ is the nth iterate of f and, except for special maps, λ depends upon x_0. We can then use (2.15c) to rewrite (2.15) in the form

$$|\delta x_n| \sim |\delta x_0| \, e^{\lambda n}, \qquad n \gg 1. \quad (2.16)$$

Naturally, one must take care in computation that, for large n, $|\delta x_0|$ is small enough for the neglect of quadratic and higher-order terms in $f(x_n)$ to be justified. It is typical that one will not in computation obtain *pointwise* convergence to a definite number λ, because there will be variations in λ whenever the map's slope f' is not constant along a trajectory.

For an arbitrary two-dimensional discrete map of the plane,

$$\begin{aligned} x_{n+1} &= f(x_n, y_n), \\ y_{n+1} &= g(x_n, y_n), \end{aligned} \quad (2.17)$$

we can also linearize about some exact orbit,

$$(x_0, y_0) \rightarrow (x_1, y_1) \rightarrow \cdots \rightarrow (x_n, y_n) \rightarrow \cdots,$$

to obtain the error propagation equation

$$\delta X_{n+1} = A_n \, \delta X_n, \quad (2.18)$$

where the vector

$$\delta X_n = \begin{pmatrix} \delta x_n \\ \delta y_n \end{pmatrix} \tag{2.19}$$

describes the uncertainty in the nth iterates that follows from the uncertainty in the initial conditions, and the Jacobi matrix

$$A_n = \begin{pmatrix} \dfrac{\partial f}{\partial x_n} & \dfrac{\partial f}{dy_n} \\ \dfrac{\partial g}{\partial x_n} & \dfrac{\partial g}{\partial y_n} \end{pmatrix} \tag{2.20}$$

is evaluated for each iteration at the point (x_n, y_n) on the exact trajectory, starting from the initial condition (x_0, y_0).

Consider next the linearized equation for error propagation

$$\begin{aligned} \delta X_n &= A_{n-1} A_{n-2} \cdots A_0 \, \delta X_0 \\ &= J_n \, \delta X_0 \end{aligned} \tag{2.21}$$

in terms of the matrix $J_n = A_{n-1} A_{n-2} \cdots A_0$.

The eigenvalues $\mu_i(n)$ of J_n vary with n, as do the eignvectors $e_i(n)$, and both are obtained by solving the equations

$$\begin{aligned} \det|J_n - \mu(n)I| &= 0, \\ J_n e_i(n) &= \mu_i(n) e_i(n). \end{aligned} \tag{2.22}$$

Note that the directions of the eigenvectors are typically not constant along the trajectory $(x_0, y_0) \to \cdots \to (x_n, y_n) \to \cdots$. If $e_1(n)$ and $e_2(n)$ (which we assume to be normalized, so that $\|e_i(n)\| = 1$) are linearly independent, then we can write

$$\delta X_0 = \sum_{i=1}^{2} c_i e_i(n) \tag{2.23}$$

in order to obtain an equation for the error δX_n in the nth iterate,

$$\delta X_n = J_n \, \delta X_0 = \sum_i c_i \mu_i(n) e_i(n). \tag{2.24}$$

Liapunov exponents λ_1 and λ_2 are then defined by

$$\mu_i(n) \sim e^{n\lambda_i} \tag{2.25}$$

for large n, so that

$$\lambda_i \sim \frac{1}{n} \ln \mu_i(n), \qquad n \gg 1, \qquad (2.26)$$

and these numbers λ_i will in general vary when the initial conditions (x_0, y_0) are varied. A positive value $\lambda_i > 0$ corresponds to a local expansion of small areas along the direction of e_i, while $\lambda_i < 0$ yields a local contraction of small areas along the direction $e_i(n)$. Areas transform as

$$\Delta a_n = |J_n| \, \Delta a_0, \qquad (2.27)$$

where Δa_0 is some initial element of area in phase space, and $|J_n| = \det J_n$ is the Jacobian determinant of the transformation $(x_0, y_0) \rightarrow (x_n, y_n)$. We obtain $\det J_n \equiv \mu_1(n)\mu_2(n)$ so that, for large n we can rewrite (2.27) as

$$\Delta a_n \cong e^{n(\lambda_1 + \lambda_2)} \, \Delta a_0. \qquad (2.28)$$

Hence, $\lambda_1 + \lambda_2 = 0$ corresponds to an area-preserving map, while $\lambda_1 + \lambda_2 < 0$ describes a dissipative map. With $\lambda_1 + \lambda_2 < 0$, deterministic chaos requires one positive exponent, whereas for regular ('dissipative') motion, $\lambda_i < 0$ for $i = 1, 2$. The condition $|\mu_1| = |\mu_2|$ is both necessary and sufficient for regular motion in an area-preserving map. The existence of non-zero Liapunov exponents provides a mechanism whereby filamentation to statistical equilibrium can occur in an area-preserving map, as we show in the next chapter.

We have seen that if one knows the state of a chaotic dynamical system to within finite resolution, which is at best the case in observations of nature, then one can only predict the time evolution of the system over a time interval that is given by (2.11d), where the uncertainty δx_0 represents the resolution of the experiment. But what about computation, which is certainly not identical with the observation of nature? Does it mean that chaotic orbits cannot be computed? What if one starts with an exact initial condition given in the form of an algorithm that generates as many of the digits $\varepsilon_i(0)$ in the string $x_0 = 0.\varepsilon_1(0)\varepsilon_2(0)\cdots\varepsilon_N(0)\cdots$ as you like? In this case, the digits take on any one of μ possible values $\varepsilon_i(0) = 1, 2, 3, \ldots, \mu - 1$; we must expand the number x_0 in some integral base-μ of arithmetic. We can arbitrarily choose to work in any base $\mu = 2, 3, 4, \ldots$, just as one can arbitrarily choose to study a partial differential equation in rectangular, spherical, elliptic, $\ldots$ coordinates. In this framework, to what extent can we or cannot we compute the first N digits of the iterates

$$x_n = 0.\varepsilon_1(n)\varepsilon_2(n)\cdots\varepsilon_N(n)\cdots \qquad (2.8b)$$

of a chaotic map

$$x_{n+1} = f(x_n)? \tag{2.12}$$

One can handle this problem very slowly by hand, or one can write a program for a high-speed computer whereby the digit strings are handled carefully by taking into account the information flow property that is determined by the Liapunov exponent for a particular initial condition. Given that this is true, what determines the limit of predictability, our ability to 'see' into the future in the case of a chaotic system? We have seen that, for the tent map, the limitation is that of computer time, the time required to compute the nth digit in the initial condition. We show now that, for arbitrary maps, the limitation is still that of computation time.

In order to understand the most fundamental limitation, we discretize the error propagation equation

$$|\delta x_n| \sim |\delta x_0|\, e^{\lambda n}, \qquad n \gg 1, \tag{2.16}$$

where we understand λ as the Liapunov exponent for the particular trajectory under consideration. Then by δx_0 we mean μ^{-N}, and since the maximum error means that the first digit of x_n is wrong, $\delta x_n \approx 1$, we obtain from (2.16)

$$1 \approx e^{n\lambda}\mu^{-N} \tag{2.16b}$$

or

$$n \approx N \ln \mu / \lambda \tag{2.16c}$$

as the number of iterations that one can perform without committing an error in the first digit of x_n, given the first N digits of x_0. This is the information flow property, the discretized statement of the butterfly effect. It tells us that the most fundamental limitation on the ability to compute the future from the past in a deterministic-chaotic system is the limitation of computer time, which is nontrivial due to the absolute necessity to increase the precision of the computation as one iterates the map in order to avoid the introduction of uncontrollable errors.

In experiment, in contrast, one cannot know initial conditions exactly, so we are also interested in the statistical properties of chaotic orbits that are shared by entire *classes* of initial conditions. For example, for a dynamical system where a class of initial conditions can be found that generates a distribution of iterates that can be described by a stationary density $p(x)$, all time averages can be replaced by phase space averages. Such a system (map *plus* a 'measure one' class of initial conditions) is called ergodic, and because λ is defined for $n \gg 1$ by the discrete time average

$$\lambda = \sum_{i=0}^{n-1} \ln|f'(x_i)|/n, \tag{2.29}$$

it follows for ergodic systems that the expected value of the Liapunov exponent in the infinite time limit is given by

$$\lambda = \int_a^b dx p(x) \ln|f'(x)| = \langle \ln|f'(x)| \rangle \tag{2.30}$$

as a phase space average (ensemble average). The meaning of ergodic is simply that the time averages can be replaced by a certain spatial average that is itself time-independent. Given the stationary density $p(x)$, also called the invariant density, one can also compute the expected root-mean-square fluctuation in λ,

$$\Delta \lambda \simeq \langle (\ln|f'(x)| - \lambda)^2 \rangle^{1/2} \tag{2.29b}$$

in order to estimate the variation in λ that occurs in the time-average whenever the map's slope varies.

We turn next to the equation of motion that gives rise to the stationary probability density whenever a density $p(x)$ can be defined. The stationary, or invariant, density follows only for mixing systems and even then only for a certain class of initial conditions of the system. Mixing is defined more carefully in Chapter 3, but we can state the essence of the idea as follows: consider an initial spread of uncertainty in x_0 that is defined, e.g., by the probability density $p_0(x)$. In the simplest case,

$$p_0(x) = \begin{cases} 1/\delta x_0, & x \in [x_0, x_0 + \delta x_0], \\ 0, & \text{otherwise,} \end{cases} \tag{2.31}$$

although the following argument does not require exactly this initial density. If we think of the segment δx_0 of the x-axis as a droplet of red dye and of the unit interval as an otherwise clear, one-dimensional body of water, then as $n \to \infty$ the chaotic map colors the water uniformly pink. That is what is meant by mixing. The large n limit is the limit of statistical equilibrium. Mathematically, it is the positive Liapunov exponent λ that stretches the initially localized droplet of dye into a long, thin filament and distributes it uniformly over the phase space.

2.3 The Frobenius–Peron equation (invariant densities)

We turn now to a statistical-mechanical description of deterministic chaos: the use of a 'master equation'. Consider a discrete map

$$x_n = f(x_{n-1}) \tag{2.32}$$

of the unit interval. Let $p_0(x)$ denote a probability density describing uncertainty in the initial condition x_0; that is, $p_0(x)$ is supposed to describe our ignorance of initial conditions, and we want to write down an equation that describes the propagation of this ignorance as the discrete time n increases. We can define a 'master equation' for $p_n(x)$, the probability density that describes the error in the nth iterate x_n, and this master equation is called the Frobenius–Peron equation:

$$p_n(x) = \int_0^1 dy\, p_{n-1}(y)\, \delta(x - f(y)). \tag{2.33}$$

Hence, we have a Markovian equation with a deterministic transition rate $\delta(x - f(y))$ that expresses the totally deterministic nature of the iterative process: there is no *ad hoc* assumption of randomness here. Since a chaotic one-dimensional map cannot have a single-valued inverse (this restriction does not appear in two or more dimensions), we assume for simplicity a two-valued inverse: let $y_1 = f^{-1}(x)$ and $y_2 = f^{-1}(x)$ denote the two points y_1 and y_2 that iterate to the same point x (qualitatively our map has a single maximum). By using the transformation rule for delta functions,

$$\delta(x - f(y)) = \sum_{i=1}^{2} \frac{\delta(y - y_i)}{|f'(y_i)|}, \tag{2.34}$$

we obtain from (2.33) the result

$$p_n(x) = \sum_{i=1}^{2} \frac{p_{n-1}(y_i)}{|f'(y_i)|}. \tag{2.35}$$

Note that equation (2.35) can also be written in the form

$$p_n(x) = \sum_{i=1}^{2} p_{n-1}(y_i)\, e^{-\lambda_i(y_i)}, \tag{2.36}$$

where $\lambda(y_i) = \ln|f'(y_i)|$ is the 'local' Liapunov exponent for one iteration of the map. It is the action of the Liapunov exponent that generates probability distributions that remind us superficially of random behavior, although the process under consideration is completely deterministic.

The invariant density $p(x)$ is defined by the limiting case of a certain statistical equilibrium, yielding the stationary density where

$$p(x) = p_n(x) \tag{2.37}$$

is independent of the discrete time n.

Consider as an example the class of tent maps, both symmetric and unsymmetric, given by

$$x_{n+1} = ax_n \quad \text{if } x_n < b/(a+b), \qquad b(1-x_n) \quad \text{if } x_n > b/(a+b), \qquad (2.38)$$

and where the largest value that x_{n+1} can take on, for $0 < x_n < 1$, is $ab/(a+b) \geq 1$. That is, the tent map peaks at or above unity. For $a = b = 2$, we retrieve our binary tent map. It is easy to write down the equation for the invariant density,

$$p(x) = \frac{1}{a}\left[p\left(\frac{x}{a}\right) + \frac{a}{b}p\left(1 - \frac{x}{b}\right)\right]. \qquad (2.38b)$$

If we ask whether the uniform density $p(x) = 1$ is a solution, we find that the condition is that $ab = a + b$. In other words, both the binary tent map ($a = b = 2$) and every *asymmetric* tent map that peaks exactly *at* unity has a *uniform* invariant density. The reason why tent maps that peak above unity do not follow suit is that they generate *fractal* invariant sets, and one cannot define a density *pointwise* on a fractal (cf. Chapters 4 and 7).

We introduced the invariant density from the viewpoint of uncertainty in initial data, a viewpoint that seems useful for experiment, but not for computation. In computation, where the initial condition is necessarily specified precisely, the invariant density still has a meaning: it describes the distribution of iterates $\{x_n\}$ of the map, starting from any *one* initial condition in a certain (so far unspecified) class of initial conditions, the class that would occur with absolute certainty if one could draw numbers randomly, with infinite precision, from the continuum. In other words, the invariant density, in principle, describes the statistics that are generated by a single trajectory of the dynamical system, for a single initial condition that must be chosen from the special *class* of initial conditions that yield ergodic motion. The proof of this for Bernoulli shifts can be found in the monograph by Niven. The proof is given for the binary tent map at the end of the next section; the proof for the Bernoulli shift can also be inferred from the one given for the binary tent map.

Another word of warning. The idea of the master equation, from the statistical mechanics viewpoint, is that initial nonequilibrium densities $p_0(x)$ should relax to the invariant density. This relaxation is illustrated below for the Bernoulli shift. However, there are nontrivial stationary probability distributions that we shall systematically discover (in Chapter 9) that cannot be described by a density, and do *not* relax to the infinite-time solution described by the invariant density. The so-called 'multifractal measures' fall into this category, as do other interesting stationary distributions that cannot be described by densities. To put it in a nutshell, the range of statistical behavior that is generated by the binary tent map (for all possible classes of initial conditions) is not confined to the uniform invariant density; it includes multifractal distributions and more. Invariant densities, and invariant distributions that are not described by a density, are considered in more detail in section 9.1.

2.4 Simple examples of fully developed chaos for maps of the interval

The binary tent map has many properties in common with the binary Bernoulli shift, but we begin with the analysis of the details of deterministic chaos for Bernoulli shifts which are simpler.

Consider the Bernoulli shift

$$x_{n+1} = Dx_n, \qquad \mathrm{mod}\ 1, \tag{2.39}$$

with $x_0 \in [0, 1]$. The solution is given in closed form by

$$x_n = D^n x_0, \qquad \mathrm{mod}\ 1, \tag{2.39b}$$

where x_0 is the initial condition. Consider first the case where $|D| < 1$. Since $x_n \to 0$ as $n \to \infty$, the origin is a stable fixed point. If $D > 1$, there are no stable fixed points, since with $\delta x_n = D^n \delta x_0$, all perturbations δx_0 iterate away from the origin (the origin is hyperbolic when $D > 1$). For $D = 2, 3, 4, \ldots$, we have the usual Bernoulli shifts. $\lambda = \ln D$ is the Liapunov exponent.

For $D = 2$, the binary shift, $x_{n+1} = 2x_n \bmod 1$ can be written as

$$x_{n+1} = \begin{cases} 2x_n & 0 \le x_n < \tfrac{1}{2}, \\ 2x_n - 1, & \tfrac{1}{2} < x_n \le 1. \end{cases} \tag{2.40}$$

In order to understand the discrete dynamics and to make contact with 'computer arithmetic', it is useful to do the arithmetic in some definite base a where the nth iterate is written as the a-ary expansion

$$x_n = \sum_{j=0}^{\infty} \varepsilon_j / a^j, \qquad \varepsilon_j = 0, 1, 2, \ldots, a - 1 \tag{2.41}$$

with a integral.

One can show quite generally that rational numbers have periodic expansions in *every* integral base a, and that this periodicity is the reason for the dense set of unstable periodic orbits generated by Bernoulli shifts.

It is important to understand this fact, because sets of unstable periodic orbits that are dense on an attractor or repeller are characteristic of fully developed chaos, which is the simplest case to analyze. Much harder to analyze are systems where some parts of phase space are chaotic while other regions are not. By fully developed chaos, we mean in this text that there are no regions of finite size in phase space where the Liapunov exponent is less than or equal to zero. The binary tent map provides an example of fully developed chaos, and in Chapter 4 we shall study the case where fully developed chaos occurs on a fractal repeller – the arithmetic description of the orbits on the fractal will require no new analysis, because the analysis that we develop for the binary tent map will carry over to the fractal case.

A rational number is a number $x_0 = p/q$, with p and q both equal to integers. Irrational numbers, e.g. numbers such as $\sqrt{2}$, e, or π, cannot be expressed as the ratio of two integers. Rational numbers are countable and dense but occupy no space in the unit interval. Irrational numbers occupy all the space between 0 and 1 and are uncountable. Given that the rationals are countable, it is easy to show that they occupy no space. To see this, we label the rationals by integers $1, 2, 3, \ldots$, and centered upon the nth rational we construct a one-dimensional interval size $\varepsilon/2^n$. The total length occupied by the rationals in $[0, 1]$ is then less than

$$L = \sum_{n=1}^{\infty} \varepsilon/2^n = -\varepsilon + \varepsilon/(1 - \tfrac{1}{2}) = \varepsilon, \qquad (2.42)$$

which can be made as small as we like because ε is arbitrary. Hence, $L = 0$ and all the space is occupied by the irrationals.

It is easy to show that a rational number is built of periodic blocks of digits in every base $a = 2, 3, 4, \ldots$ of arithmetic, and vice versa. To see this, let

$$x = \sum_{j=1}^{\infty} \varepsilon_j/a^j, \qquad \varepsilon_j = 0, 1, 2, \ldots, a - 1. \qquad (2.43)$$

Given x in the form p/q, we construct the base-a expansion for x as follows: first, we write

$$ap/q = [ap/q] + p_1/q, \qquad (2.44)$$

when $[ap/q]$ is the integer part of ap/q and p_1/q is the remainder. Then $\varepsilon_1 = [ap/q]$. We construct next

$$ap_1/q = [ap_1/q] + p_2/q, \qquad (2.44b)$$

where $\varepsilon_2 = [ap_2/q]$, and the rest of the pattern is easy to see. Because there are *at most* q choices for each p_i, one of the p_i must eventually repeat for $i \leq q$. Therefore, the resulting sequence of integers is periodic, so that we can write

$$p/q = 0.\varepsilon_\alpha \varepsilon_\beta \cdots \varepsilon_K \varepsilon_1 \varepsilon_2 \cdots \varepsilon_N \varepsilon_1 \cdots \varepsilon_N \cdots, \quad (2.45)$$

where $\varepsilon_1 \varepsilon_2 \cdots \varepsilon_N$ is the basic periodic block and $\varepsilon_\alpha \cdots \varepsilon_K$ represents an initial block of digits that is not repeated. In base-a, a factor $1/a^n$ in p/q will not be repeated, e.g. N is the 'period' of p/q in base a and the algorithm that was used to obtain the 'decimal' expansion is just the Euclidean algorithm, which forms the basis for ordinary division. It follows that irrational numbers are nonperiodic in *every* base of arithmetic; this is the origin of the unstable nonperiodic orbits of the Bernoulli shift and the binary tent map.

We now prove that if the expansion of x in base a is periodic, then the expansion necessarily represents a rational number. Assume that (with $x \in [0, 1]$)

$$x = \sum_{j=1}^{\infty} \varepsilon_j/a^j = 0.\varepsilon_1 \varepsilon_2 \cdots \varepsilon_N \cdots \quad (2.46)$$

is periodic with period N, and where $\varepsilon_j = 0, 1, 2, \ldots, a - 1$. Then

$$a^N x = \varepsilon_1 \varepsilon_2 \cdots \varepsilon_N . \varepsilon_1 \varepsilon_2 \cdots \varepsilon_N \cdots$$

$$= \varepsilon_1 \varepsilon_2 \cdots \varepsilon_N + 0.\varepsilon_1 \varepsilon_2 \cdots \varepsilon_N \cdots$$

$$= \alpha + x, \quad (2.47)$$

where

$$\alpha = \varepsilon_1 \varepsilon_2 \cdots \varepsilon_N = \sum_{j=0}^{n-1} \varepsilon_{N-j} a^j \quad (2.48)$$

is an integer, so that $x = \alpha/(a^N - 1)$ is the ratio of two integers, and that completes our proof.

The reader should notice that all finite-length binary strings iterate in finite time to the origin under the action of the binary Bernoulli shift, but this does not happen on a typical hand calculator. The reason for this is that hand calculators typically use base-10, whereas digital computers typically use binary arithmetic. It means that periodic orbits of the binary Bernoulli shift can be computed on the former but not on the latter machines.

For an arbitrary rational number $x_0 = p/q$ in base $a = 2$, we have

$$x_0 = 0.\varepsilon_\alpha \varepsilon_\beta \cdots \varepsilon_K \varepsilon_1 \varepsilon_2 \cdots \varepsilon_N \varepsilon_1 \cdots \varepsilon_N \cdots, \quad (2.49)$$

where N is the period. With this particular representation, the Bernoulli shift $x_{n+1} = 2x_n \bmod 1$ amounts to a bit-shift one place to the right, dropping the integral part: for example,

$$x_1 = 0.\varepsilon_\beta \cdots \varepsilon_K \varepsilon_1 \cdots \varepsilon_N \cdots,$$
$$x_2 = 0.\varepsilon_\gamma \cdots \varepsilon_K \varepsilon_1 \cdots \varepsilon_N \cdots, \tag{2.50}$$

and it is clear that the block $\varepsilon_1 \varepsilon_2 \cdots \varepsilon_N$ repeats systematically after an initial transient. Periodic orbits of binary shifts are dense in the unit interval because the rational numbers are dense, but are unstable because of exponential magnification of errors. If we know x_0 only to within an error δx_0, then the error after n iterations becomes

$$\delta x_n = 2^n \, \delta x_0, \qquad \bmod 1,$$
$$= 2^{n \ln 2} \, \delta x_0, \qquad \bmod 1, \tag{2.51}$$

where $\lambda = \ln 2$ is the Liapunov exponent. Clearly, the corresponding automaton is just $\varepsilon_j(n) = \varepsilon_{j+1}(n-1)$, and $\lambda = \ln 2$ represents information flow at the rate of one bit per iteration in the cellular automaton. If δx_0 is irrational, then when $\delta x_n \sim 1$, we shall have lost all information as to the location of x_n in $[0, 1]$, and the relaxation time for that loss of information is given by $1 \sim 2^n |\delta x_0|$ or

$$n \sim -\ln|\delta x_0|/\ln 2 = -\ln|\delta x_0|/\lambda. \tag{2.52}$$

When x_0 is irrational, the closure of the orbit as $n \to \infty$ (the 'attractor') is the entire unit interval. The question then arises: What is the probability density for the iterates?

The master equation for the binary Bernoulli shift is given by

$$p_n(x) = [p_{n-1}(x/2) + p_{n-1}((x+1)/2)]/2 \tag{2.53}$$

because the shift map has a two-valued inverse, so that there are two points $x_{n-1} = x_n/2 < \frac{1}{2}$ and $x_{n-1} = (x_n + 1)/2 > \frac{1}{2}$ that iterate to x_n.

The invariant density $p(x)$ is a fixed-point solution of the Frobenius–Peron equation. A stable fixed-point solution corresponds to an irreversible approach of the system to equilibrium. We can prove that an arbitrary smooth initial density $p_0(x)$ evolves irreversibly to $p(x) = 1$ by direct iteration of the master equation:

$$p_1(x) = (p_0(x/2) + p_0((x+1)/2))/2$$
$$p_2(x) = (p_0(x/2^2) + p_0((x+1)/2^2) + p_0((x+2)/2^2) + p_0((x+3)/2^2))/2^2$$
$$\vdots \tag{2.54}$$

so that

$$p_n(x) = \sum_{j=0}^{2^n-1} p_0((x+j)/2^n)/2^n \qquad (2.55)$$

is an average of the values of p_0 at 2^n different points distributed from $x/2^n \sim 0$ to $(x + 2^n - 1)/2^n \sim 1$, as $n \to \infty$. To go further, notice that the sum has the form of a Riemann integral: if we set $x_j = (x + j)/2^n$, then $\Delta x_j = \Delta j/2^n = \frac{1}{2}^n$ and the range of x_j (as $n \to \infty$) is from 0 to 1. Therefore, for $n \to \infty$,

$$p_n(x) = \sum_{j=0}^{2^n-1} \Delta x_j\, p_0(x_j) \to \int_0^1 p_0(x)\, \mathrm{d}x = 1. \qquad (2.56)$$

So, the averaging of $p_0(x)$ due to the chaotic map causes the system irreversibly to approach the statistical equilibrium represented by the uniform invariant density $p(x) = 1$. This uniform density describes the distribution of iterates of the map for a measure-one class of initial conditions in the unit interval, the so-called normal numbers that were studied by Borel (see Niven, 1956). Other classes of initial conditions yield other probability distributions, including the so-called multifractal measures that are the topic of Chapters 7 and 9.

The Bernoulli shift is the simplest example of deterministic chaos. There, one studies directly chaos at its most fundamental source: the deterministic patterns provided by the nonperiodic decimal expansions of irrational numbers. There are two ways whereby these studies form the foundation for the study of more complicated maps. First, the decimal expansions occur as initial conditions. Second, they occur as symbol sequences. The development of chaos theory in terms of nonperiodic symbol sequences is given in Chapters 8 and 9 for one-dimensional, fully chaotic maps.

Before graduating to the study of symmetric and asymmetric maps with either piecewise constant or continually variable slope, and where the motion can occur either on the unit interval or on a fractal subset of the interval, it is useful to understand more about the dynamics of the map that is only slightly more complicated than the Bernoulli shift, the binary tent map

$$z_{n+1} = \begin{cases} 2z_n, & z_n \in [0, \tfrac{1}{2}), \\ 2(1 - z_n), & z_n \in (\tfrac{1}{2}, 1]. \end{cases} \qquad (2.7)$$

Because $z_{n+1} < 1$ whenever $z_n < 1$, all iterates are confined to the unit interval. Every orbit has a Liapunov exponent $\lambda = \ln 2$, and this is reflected by the fact that the information transfer in binary representations of the map's iterates,

$$z_n = 0.\varepsilon_1(n)\varepsilon_2(n)\cdots\varepsilon_N(0)\cdots, \qquad (2.8)$$

where $\varepsilon_i(n) = 0$ or 1, takes place at the rate of one bit per iteration, right to left. We now prove by a different method than the one used in section 2.1 that rational initial conditions yield periodic orbits. The proof is useful as an exercise in digital thinking, which we shall use throughout this book.

Suppose that the initial condition $z_0 = 0.\varepsilon_1(0)\varepsilon_2(0)\cdots\varepsilon_N(0)\cdots$ is rational, which is to say that the binary string is periodic and has the form $z_0 = 0.\varepsilon_1\varepsilon_2\cdots\varepsilon_N\varepsilon_1\varepsilon_2\cdots\varepsilon_N\cdots$, where $\varepsilon_1\varepsilon_2\cdots\varepsilon_N$ is the basic repeated block. For example, $\frac{1}{3} = 0.0101\cdots$. It follows from the automaton

$$\varepsilon_i(n+1) = \begin{cases} \varepsilon_{i+1}(n), & \varepsilon_1(n) = 0, \\ 1 - \varepsilon_{i+1}(n), & \varepsilon_1(n) = 1, \end{cases} \qquad (2.9)$$

that $z_1 = 0.\varepsilon_2\varepsilon_3\cdots\varepsilon_N\varepsilon_1\cdots$ if $\varepsilon_1 = 0$ but

$$z_1 = 0.(1 - \varepsilon_2)(1 - \varepsilon_3)\cdots(1 - \varepsilon_N)(1 - \varepsilon_1)\cdots$$

if $\varepsilon_1 = 1$. Then, we obtain $z_2 = 0.\varepsilon_3\varepsilon_4\cdots\varepsilon_N\varepsilon_1\cdots$ if $\varepsilon_2 = 0$ and $\varepsilon_1 = 0$ or if $\varepsilon_2 = 0$ and $\varepsilon_1 = 1$, but $z_2 = 0.(1-\varepsilon_3)(1-\varepsilon_4)\cdots(1-\varepsilon_N)(1-\varepsilon_1)\cdots$ if $\varepsilon_2 = 1$ and $\varepsilon_1 = 1$ or if $\varepsilon_2 = 1$ and $\varepsilon_1 = 0$. From the continuation of this reasoning it follows that either $\varepsilon_N = 0$ and $z_N = 0.\varepsilon_1\varepsilon_2\cdots\varepsilon_N\cdots$ so that the initial condition is repeated and the period of the orbit is N, or else $\varepsilon_N = 1$ and $z_N = 0.(1 - \varepsilon_1)(1 - \varepsilon_2)\cdots(1 - \varepsilon_N)\cdots$. In the latter case it follows that $z_{N+1} = z_1$ and the period is still N. It also follows that nonperiodic binary strings yield nonperiodic orbits of the tent map.

The rule by which the sequence of bits in the initial condition determines the first bit of the orbit $z_n = 0.\varepsilon_1(n)\cdots$, hence the 'symbol sequence' $\varepsilon_1(0)\varepsilon_1(1)\cdots\varepsilon_1(n)\cdots$ is that $\varepsilon_1(n) = 0$ if $\varepsilon_{n-1}(0)\varepsilon_n(0) = 00$ or 11, whereas $\varepsilon_1(n) = 1$ if $\varepsilon_{n-1}(0)\varepsilon_n(0) = 01$ or 10. In other words, in contrast to the Bernoulli shift, *pairs* of bits in the initial condition determine the leading bit in the chaotic orbit. The general rule stated above follows from the application of the automaton (2.9) to a specific example; e.g., from looking at the iterates of $z_0 = \sqrt{2} - 1$ in binary arithmetic,

$$z_0 = 0.0110101000001001\cdots. \qquad (2.9b)$$

It is equally easy to show that the triplet of bits $\varepsilon_{n-2}(0)\varepsilon_{n-1}(0)\varepsilon_n(0)$ in z_0 determines the first two bits $\varepsilon_1(n)\varepsilon_2(n)$ in z_n, and so on. Therefore, the distribution of blocks of binary digits in the initial condition determines the statistical distribution of the map's iterates over the phase space: the frequency with which 0 and 1 occur as the first bit tells us how often the orbit visits the two intervals $[0, \frac{1}{2})$ and $(\frac{1}{2}, 1]$; the frequencies with which 00, 01, 10, and 11 occur as the first two bits of z_n tell us how often the intervals $[0, \frac{1}{4})$, $(\frac{1}{4}, \frac{1}{2})$, $(\frac{1}{2}, \frac{3}{4})$, and $(\frac{3}{4}, 1]$ are visited by the orbit, and so forth. We can see directly by this method how different statistical distributions are built up by iterating the map from initial conditions that have different distributions of blocks of bits.

We can now make an observation about the class of trajectories that generates the invariant density of the tent map. Consider the partition of the unit interval into 2^N intervals of equal size $l_i = 2^{-N}$. In binary arithmetic, the 2^N different combinations of m 0's and $N - m$ 1's that make up the first N-bits of the iterate $x_n = 0.\varepsilon_1(n)\varepsilon_2(n)\cdots\varepsilon_N(n)\cdots$ can be used as address labels for the 2^N intervals. For example, the N-bit word made of N 0's labels the interval that starts at the origin, while the N-bit word made entirely of 1's labels the interval that ends at $x = 1$. Whenever a particular N-bit word occurs as the first N-bits of x_n, then the map visits the interval that is labeled by that word. In this picture, uniform invariant density, integrated over the ith of the 2^N intervals, yields the same probability $P_i = 2^{-N}$ for the map to visit that interval or any one of the $2^N - 1$ other intervals of equal size (to be consistent, one can write the index i in binary as well). Therefore, the orbits that generate the invariant density are exactly the ones where every combination of m 0's and $N - m$ 1's occurs equally often in the first N-bits of the iterates, and where this is true for *every* value $N = 1, 2, 3, \ldots$, as n goes to infinity. That is, 0 and 1 occur equally often as the first bit of z_n; 00, 01, 10, and 11 occur equally often as the first two bits of z_n; and so on. If one uses the rule derived above that connects the first N-bits in z_n to a block of $N + 1$ bits in z_0, then it is easy to show that, in z_0, 0 must occur as often as 1; 00, 01, 10, and 11 must each occur a quarter of the time; and so on. A number z_0 with this property is called 'normal' (the word is due to Lebesgue), and the properties of these numbers were studied extensively by Borel. The resulting even distribution of both bits and iterates exhibits *statistical independence*, but we do not call such behavior 'random' because the resulting statistical independence is generated systematically by the deterministic system, that is, by rules of arithmetic. One can call this behavior 'apparently random' in the sense that it is pseudorandom. Finally, the initial conditions that generate the invariant density of the tent map are exactly the ones that the measure theorist would typically call generic. Because 'almost all' initial conditions (i.e., with 'measure-one') yield such orbits, one often hears the argument that a random draw of initial conditions from the continuum will yield one that, for the tent map, generates the uniform distribution of iterates *with certainty*. The idea that it makes sense to make a random draw of numbers from the continuum will be challenged in the last section of this chapter. In Chapter 9 we show how the tent map can be used to generate every possible statistical distribution that can be empirically constructed.

2.5 Maps that are conjugate under differentiable coordinate transformations

Consider a map

$$x_{n+1} = g(x_n) \tag{2.57}$$

and a related map

$$z_{n+1} = f(z_n) \tag{2.58}$$

which is obtained from the original map (2.57) by a coordinate transformation

$$z_n = h(x_n), \tag{2.59}$$

with $h' \neq 0$. That is, the transformation is presumed invertible for all x_n and z_n. We denote by λ and σ the Liapunov exponents of the maps (2.58) and (2.57) respectively, i.e.

$$\sigma \cong \frac{1}{n} \sum_{i=0}^{n-1} \ln|g'(x_i)|. \tag{2.60}$$

The question is: how is σ related to λ, where λ is the exponent for the corresponding trajectory of map f (trajectories map one-to-one)? From (2.57) we obtain

$$\frac{dx_{n+1}}{dx_n} = g'(x_n), \tag{2.61}$$

and, likewise,

$$\frac{dz_{n+1}}{dz_n} = f'(x_n) \tag{2.61b}$$

follows from (2.58), so that from $dz_n = h'(x_n)\, dx_n$ we obtain

$$\frac{dz_{n+1}}{dz_n} = \frac{h'(x_{n+1})}{h'(x_n)} \frac{dx_{n+1}}{dx_n}, \tag{2.62}$$

which becomes

$$f'(z_n) = \frac{h'(x_{n+1})}{h'(x_n)} g'(x_n) \tag{2.63}$$

when the above results are combined. Inserting the above results yields

$$\lambda \cong \frac{1}{n} \sum_{i=0}^{n-1} \left[\ln|g'(x_i)| + \ln\left|\frac{h'(x_{i+1})}{h'(x_i)}\right| \right], \tag{2.63b}$$

which can be rewritten in the form

$$\lambda = \sigma + \ln\left|\frac{h'(x_n)}{h'(x_0)}\right| \Big/ n. \tag{2.63c}$$

It follows that $\sigma \to \lambda$ as $n \to \infty$. In other words, for times $n \gg 1$, the Liapunov exponent, for trajectories that map onto each other, is invariant under *differentiable* coordinate transformations.

The ensemble average exponent for the asymmetric tent map with $ab = a + b$, computed by using the invariant density $\rho = 1$, is $\lambda = (b \ln a + a \ln b)/(a + b)$. This result differs from $\ln 2$, the exponent for the binary tent map. We therefore expect that the transformation from this map to the binary tent map is not differentiable.

For differentiable coordinate transformations, invariant densities also map onto each other. By the conjugacy operation, the map f is given by $f = hgh^{-1}$, i.e., $z_{n+1} = h(g(h^{-1}(z_n)))$. Likewise, if $p_g(x)$ denotes the invariant density of g (assuming that g has one), and $p_f(z)$ denotes f's invariant density, then because probability is conserved under coordinate transformations, $p_f(z)\, dz = p_g(x)\, dx$, it follows that

$$p_f(z) = p_g(h^{-1}(z))/h'(h^{-1}(z)). \tag{2.64}$$

This result leads us to suspect that the binary tent map can only be differentiably conjugate to maps that have nonuniform invariant densities.

As an example, consider the logistic map $f(x, D) = Dx(1 - x)$ for the case where $D = 4$. In this case, the invariant density was first calculated analytically by J. von Neumann. The trick is that when $D = 4$, the logistic map is conjugate to the binary tent map by a differentiable coordinate transformation. For the binary tent map

$$x_{n+1} = \begin{cases} 2x_n, & x_n < \tfrac{1}{2}, \\ 2(1 - x_n), & x_n > \tfrac{1}{2}, \end{cases} \tag{2.7}$$

we have already shown that the invariant density is uniform, $p(x) = 1$.

If we transform variables according to

$$x_n = \frac{2}{\pi} \sin^{-1} \sqrt{z_n}, \tag{2.59b}$$

then the equation $x_{n+1} = 2x_n$ becomes

$$\frac{2}{\pi} \sin^{-1} \sqrt{z_{n+1}} = \frac{4}{\pi} \sin^{-1} \sqrt{z_n}, \tag{2.65}$$

or

$$z_{n+1} = (\sin (2 \sin^{-1} \sqrt{z_n}))^2. \tag{2.66}$$

Using $\quad \theta = \sin^{-1} \sqrt{z}, \quad \sin\theta = \sqrt{z} \quad$ and $\quad \cos\theta = \sqrt{1-z}, \quad$ yields

$\sin 2\theta = 2 \sin \theta \cos \theta$, or

$$z_{n+1} = 4z_n(1 - z_n),\qquad(2.67)$$

which is a logistic map.

The invariant density of this logistic map therefore can be calculated from (2.64) with $p_f(x) = 1$, and the result is

$$p(x) = \frac{1}{\pi\sqrt{x(1-x)}}.\qquad(2.68)$$

Our logistic map provides another example of 'fully-developed chaos' (the binary tent map provides the first example): the entire phase space, in this case the unit interval, is densely filled with unstable orbits. The binary tent map has a dense set of unstable periodic orbits starting from rational initial conditions. This means, because of the trigonometric nature of the coordinate transformation (see Niven, 1956, Chapter 1), that the logistic map has also a dense set of unstable periodic orbits starting from a countable set of *irrational* initial conditions. For a measure-one set of initial conditions, the tent map's nonperiodic orbits are made up of iterates that are distributed uniformly over the unit interval, whereas each corresponding orbit of the logistic map is made up of a set of iterates that are distributed according to (2.68) as $n \to \infty$. In both cases, there are no regular regions of finite size in phase space, hence the term fully developed chaos.[2]

The ensemble average Liapunov exponent for the logistic map can be computed by using $f' = 4(1 - 2x)$ and (2.68) in equation (2.30), but it is simpler to observe that $\lambda = \ln 2$ because of invariance of the Liapunov exponent under differentiable coordinate changes.

2.6 Computation of nonperiodic chaotic orbits at fully developed chaos

At fully developed chaos (FDC), dense sets of unstable periodic orbits permeate the invariant set. We have as examples of fully developed chaos Bernoulli shifts with $D = 2, 3, 4, \ldots$, the binary tent map and the conjugate logistic map at $D = 4$. For all of these cases, the invariant set is the entire unit interval. For piecewise linear chaotic maps (Bernoulli shifts and tent maps), unstable nonperiodic (true chaotic) orbits start from irrational

---

[2] Our definition of the phrase 'fully developed chaos' does not agree with some definitions to be found elsewhere in the literature.

initial conditions. In the logistic map at FDC with $D = 4$, because of the trigonometric transformation from the tent map, rational as well as irrational initial conditions can generate nonperiodic orbits. Periodic orbits begin from a discrete subset of irrational initial conditions that can be computed to any prescribed degree of precision by backward iteration of the logistic map. The reason why the periodic orbits start from irrational initial conditions is because of the trigonometric coordinate transformation (see Niven, 1956, Chapter 1).

Our introduction to discrete maps began in this chapter with the Lorenz plot, which followed from an attempted numerical integration scheme for the Lorenz equations that was carried out by first replacing the solution of three coupled nonlinear differential equations by truncated power series expansions, yielding a three-dimensional iterated map, and then by iterating that map on any available digital computer by using the floating-point mode of arithmetic. Both of those operations, the integration scheme and the use of floating-point arithmetic, introduce uncontrolled and uncontrollable errors that are magnified exponentially fast by the action of a positive Liapunov exponent. Normally, the errors are extremely difficult to track, and one has no theoretical means, a priori, to judge the meaning of the numbers that come out of calculation. The resulting computed orbits are not correct orbits of the Lorenz model, but are examples of pseudo-orbits of the three-dimensional map that represents the integration method. A pseudo-orbit is defined to be any computed orbit of a dynamical system that is not an exact orbit of the system. A pseudo-orbit is the result, for example, of any attempt to compute where errors are introduced into the computation that are then magnified by the information flow property. Any attempt to compute a chaotic orbit of the logistic map at FDC, e.g., will produce a pseudo-orbit whenever floating-point arithmetic is used. It is normally assumed without proof that these sorts of mistakes do not matter, that the fact that the system is chaotic is somehow supposed to justify the use of bad arithmetic. However, it is far from clear, at least to the writer, that one does not lose something that might be important, or sometimes even get a completely wrong picture, by doing sloppy arithmetic, and the appeal to the belief that computer roundoff/truncation mimics the random perturbation of real, physical systems has not been justified by any adequate analysis, to date. In this text, we adopt the viewpoint that we want to discover all of the different classes of statistics that can be generated by the map, not merely those that are consistent with or are imposed by a computer's design. The question of what is, or is not, important for experiment and observation in nature is an entirely different question that surely cannot be settled by purely mathematical arguments: *nature* must be asked. One main point of this text is that the required finite-precision arithmetic can be done correctly. The payoff will come in Chapter 9, where, by following the digital approach to its natural conclusion, we systematically discover certain universal properties of an entire class of chaotic maps.

Pseudo-orbits, and therefore the statistics obtained from pseudo-orbits, are dependent upon the design of the computer's built-in method for making roundoff/truncation decisions. In Chapter 11, we discuss conditions that must be satisfied in order that pseudo-orbits, to within some degree of approximation, yield statistics that mimic the invariant density or, more generally, a particular invariant distribution. In fact, our method leads to a systematic way to test the extent to within which a given pseudo-orbit's statistics fall into one class or another. The statistics reflected by an invariant distribution are only one of an *infinity* of different possible statistical distributions that a chaotic system can generate whenever a computation is performed correctly, and there is no reason to believe in advance that the invariant density that follows from the Frobenius–Peron equation represents *preferred* statistical behavior for a given map in any application to nature. Therefore, we shall concentrate in what follows upon the computation of chaotic orbits and their statistics with finite, controlled precision, and with no prejudice in favor of any particular class of initial data.

The method of this book typically is the reduction of chaotic maps to automata whose orbits can be computed by the use of integer arithmetic, thereby avoiding the introduction of uncontrollable errors into orbit calculations. The price we pay for accuracy of description is that we must state explicitly the meaning of the algebraic symbol x. That makes it necessary to learn to think and to analyze in terms of digit strings. Fortunately, computers can be programmed to manipulate digit strings exactly.

So far as the arithmetic of chaos is concerned, we can make the following analogy: 'x' is an abstract algebraic symbol. When we choose a definite base of arithmetic then we have a digital representation for that symbol. In quantum mechanics or in functional analysis one studies abstract operators. Those operators provide a convenient way to do algebra and to discuss the theory in abstract terminology, but one must choose a *basis* and therefore a representation for an abstract operator in order to *calculate*. In chaos theory, 'x' is convenient for symbolic manipulations, either by hand or with the use of a symbolic programming language, but the information flow property cannot be handled on a computer without the introduction of uncontrollable errors unless one uses the language of digit strings, which a computer can be programmed to manipulate correctly, although (except while using a symbol manipulation language) no computer is constructed to interpret correctly what 'x' means (infinite precision is taken for granted when we write 'x'). The symbol x must be replaced by an algorithm for a digit string in order to get reliable arithmetic from the use of a computer whenever the system is chaotic.

We turn now to a method for avoiding the introduction of uncontrollable errors into the computation of chaotic orbits. The method can be implemented on a computer by using, e.g., a symbolic manipulation language where initial conditions are generated by algorithms. In particular, uniform approximations to chaotic orbits can be computed with controlled precision. This does not mean that chaotic orbits are necessarily easy to compute at long times: as we have already pointed out in our discussion of the information flow property, it is computer time that limits our ability to compute the future from the past starting from an exact initial condition. The limitation placed on predictability by considerations of computer time is the reason for unpredictability of solutions of deterministic chaotic systems of equations at long times.

The point to be made next is that, within the limits set by computation time, uniform approximations to nonperiodic orbits of chaotic maps can be computed with *controlled precision*. We discuss two examples: first the binary tent map, then the logistic map at a control parameter value that corresponds to a case of fully developed chaos. In the literature on chaos theory, one sometimes meets the phrase 'information loss'. When the arithmetic is done correctly, there is no information loss, and when there is information loss, then one has not computed the map correctly. Chaotic systems create information, and this information can be discovered digit by digit when one computes correctly, which means that floating-point arithmetic should be avoided.

Given the tent map (2.7) and an algorithm that generates an irrational initial condition, we can compute a nonperiodic orbit (2.8) to within any prescribed finite degree of accuracy by computing the bits $\varepsilon_i(n)$ in the nth iterate from the automaton

$$\varepsilon_i(n+1) = \begin{cases} \varepsilon_{i+1}(n), & \varepsilon_1(n) = 0, \\ 1 - \varepsilon_{i+1}(n), & \varepsilon_1(n) = 1. \end{cases} \tag{2.9}$$

By controlled precision, we mean that we can compute the sequences $z_1 = 0.\varepsilon_1(1)\cdots\varepsilon_N(1)\cdots$, $z_2 = 0.\varepsilon_1(2)\cdots\varepsilon_N(2)\cdots$, $z_n = 0.\varepsilon_1(n)\cdots\varepsilon_N(n)\cdots$ *exactly* for any finite-precision N, with the limits set by computation time. By an irrational initial condition z_0, we mean a nonperiodic binary sequence $0.\varepsilon_1(0)\varepsilon_2(0)\cdots\varepsilon_N(0)\cdots$ that is generated by an algorithm, so that all the bits $\varepsilon_i(0)$ are fixed by the algorithm. It is easy to write down such algorithms – the noninteger part of the square root of any integer like 2, 3, 5, 6, 7, 8, 10, etc. will do. For example, $\sqrt{2}-1=$ $0.0110101000001001\cdots$ is such an initial condition. Because of the information flow generated by $\lambda=\ln 2$ in the tent map, we know that if we want to know, e.g. the first 5 bits in z_1, then we have to compute the first 6 bits in z_0. Or, for example, if we want to know exactly what is the first bit in each of the first n iterates $z_1, z_2, \ldots, z_n$, then we must compute the first $n+1$-bits in z_0.

As a definite example, the trajectory that starts from $z_0=\sqrt{2}-1$ computed exactly to 4-bit accuracy has, as its first 5 iterates, $z_1=0.1101\cdots$, $z_2=0.0101\cdots$, $z_3=0.1010\cdots$, $z_4=0.1010\cdots$ and $z_5=0.1011\cdots$. It is clear that the resulting 4-bit orbit is *nonperiodic* and can be computed to any desired finite-precision N by our method. Furthermore, the computed orbit is a uniform approximation to the chaotic orbit that starts from the infinite-precision initial condition given by $\sqrt{2}-1$. This method of calculation was first suggested by J. Palmore as a simple example of parallel processing: you have to compute the initial condition along with the corresponding orbit as the map is iterated. In other words, the precision of the computation must be increased as you iterate the map in order to keep up with the information flow.

What happens with the logistic map with control parameter value $D=4$ (2.7c) is typical of maps that are not piecewise linear: if you start with a rational initial condition, then new digits (meaning longer strings for x_n) are created with every iteration of the map whenever you start with a rational initial condition. In the binary tent map, all the information is already contained in the initial condition and iterations of the map are only transformations of that information by the simple automaton (2.9). The automaton of the logistic map in base-2 is more complicated. In order to illustrate this, we compute a section of the logistic map's orbit starting from $x_0=\frac{1}{8}=0.001$, the simplest finite-length binary string for which the map does not iterate onto an unstable fixed point.

Multiplying binary strings together in $x_n=4x_n(1-x_n)$ and starting from $x_0=0.001$ ($D=100$ is a double binary-point shift to the right on $x_n(1-x_n)$) yields exactly $x_1=0.0111$, $x_2=0.111111$, $x_3=0.0000111111$, $x_4=0.00111011000111111$. In general, the nth iterate x_n has exactly $N(n)=2^n(N-2)+2$ bits if you start with an N-bit initial condition. Here, additional complexity in the form of longer and longer binary strings is produced by a simple initial condition. If you truncate x_n to N-bits, then after about $N\ln 2/\lambda=N$ more iterations, the first bit in x_{n+N} will (on the average) be wrong. This is the essence of deterministic chaos: we are limited, in the end, by the requirements of finite computer time in our effort to predict the future in deterministic-chaotic systems. In a system with a negative (or zero) Liapunov exponent, there is no corresponding limitation on our ability to predict the future from the past. The reader should notice that the limitation imposed on predictability due to information flow is just another way to describe sensitivity with respect to small changes in initial conditions if we think of the 'present state' x_n as the initial condition for x_{n+N} (the present state of the weather, which is itself determined by the earlier state of the weather, is the initial condition for the future state of the weather, for example). Again, the parallel-processing method can be used to compute the iterates of the logistic map or any other chaotic map to any desired finite precision. One computes the numbers

$$x_n = f^{(n)}(x_0) \tag{2.69}$$

to as many digits as one wants by replacing x_0 by an algorithm that generates a digit string. The more digits you want in x_n, the more you have to supply in x_0.

To make a contrast that emphasizes what a computer does when you program it in the fixed-precision mode, let us make a 6-bit computer (see Fig. 2.5): 12 bits are held in the multiplication register and the final result is truncated to 6 bits. Starting with $x_0 = \frac{1}{8}$ ($= 0.001$ in binary) and computing a trajectory of the map $f(x) = 4x(1 - x)$, we get $x_1 = 0.0111$, $x_2 = 0.111111$, $x_3 = 0.000011$, $x_4 = 0.001011$, $x_5 = 0.100100$, but $x_6 = 0.111111 = x_2$, so that the pseudo-orbit hops onto a 4-cycle after only two iterations, even though the true orbit is nonperiodic (periodic orbits are impossible for rational initial conditions). When floating-point arithmetic is used in single precision on a real binary computer, the resulting pseudo-orbit is one of very long period. The meaning of the statistics generated by periodic pseudo-orbits of logistic-like maps is discussed in Chapter 11.

What is the distribution of iterates that follows from $x_0 = \frac{1}{8}$, and from rational initial conditions in general, for exact (not pseudo-) orbits? The corresponding orbits are guaranteed to be nonperiodic, so does it follow that they are described by the invariant density $p(x)$ (2.68)? According to the measure-theoretic prediction, any one of 'almost all' initial conditions of the logistic map should, upon iteration as n goes to infinity, produce exactly the map's invariant density as the distribution of iterates. Other possible distributions of iterates for 'measure-zero' initial conditions can also occur, and these do not yield the map's invariant density. That the initial condition $x_0 = \frac{1}{8}$ does not belong to the 'measure-one' class of initial data for the logistic map follows from the analysis of Chapter 9.

2.7 Is the idea of randomness necessary in natural science?

Does randomness occur in deterministic chaos? Is the idea of randomness necessary in order to understand complex phenomena in nature? In order to discuss these questions, one must first understand the different uses of the word random and then decide which definition to use.

The word randomness, as applied to observations of nature, usually means the following or something that is more or less equivalent to the following: identical preparations of an initial state of a system combined with attempts to repeat an experiment do not produce the same results, but produce instead a sequence of results with a lot of 'scatter'. For example, the 'result' of one experiment can be a 'noisy' time series, and repeated identical laboratory preparations of the initial state, to within laboratory precision, do not yield the same time series, but a collection of different noisy time series. In this case, one attempts to organize the data by statistical methods. In observations of nature outside the laboratory,

Fig. 2.5 Computer graphics for orbits of the logistic map in binary arithmetic starting from the initial condition $x_0 = \frac{1}{8}$: a white pixel represents zero and a black pixel represents one. The initial condition $x_0 = 0.001$ along with several of the first iterates of the map are shown; the iterates represent the following: (a) Pseudo-orbit for a 6-bit machine that carries 12 bits in the multiplication register; the final number representing each iterate is truncated to 6 bits. The computed orbit is a 4-cycle because $x_2 = x_6$. (b) The exact orbit (for the same initial condition as in (a) is *nonperiodic* although, as we show in Chapter 9, the distribution of iterates for this initial condition *cannot* mimic the logistic map's invariant density.

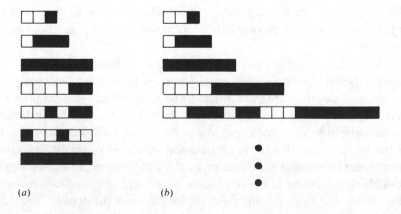

(a)   (b)

where one has no control over the preparation of initial conditions, one can partition a long, noisy time series into a collection of M different parts of the same length. Typically, there is no periodicity, and none of the sections is an exact repetition of any other section. In this case, the beginning of each section can be treated as the equivalent of a different initial condition.

True randomness implies breakdown of cause and effect: given exact knowledge of the state of a random system at one time, it is impossible to predict which of some set of alternatives will occur as the state of the system at the next instant in time. Such a breakdown of determinism occurs, e.g., for trajectories of particles that are described by quantum mechanics, where it is the probability amplitude and not the particle's momentum that obeys the deterministic time-evolution rule, the Schrödinger equation. We assume in this text that all length and time scales are such that quantum effects do not enter macroscopic physics. In this text, we use the word random in the sense of truly random, implying a breakdown of cause and effect.

A random number generator, in the context of computation, means a pseudorandom one: an *algorithm* is provided whereby a sequence of numbers is generated by *arithmetic*. The sequence is such that blocks of digits have a certain distribution, and a 'good' pseudorandom number generator is one where long runs of 'improbable' blocks of digits do not occur. Gambling devices such as slot machines are also pseurorandom; they are 'good' pseudorandom machines if it is very rare that a player can win a large sum of money on them.

Deterministic chaos is also pseudorandom because everything that can be computed follows from the rules of arithmetic: nothing that can be performed in finitely many arithmetic steps is truly random, although arithmetic sequences are commonly used to *mimic* randomness imperfectly. Sensitivity with respect to small changes in initial conditions is adequate, in principle, to account for apparent randomness in nature when finite resolution in the preparation of initial states in experiment is taken into account: consider a ball of initial conditions in phase space where the size of the ball reflects the limit of resolution of an experiment. From each initial condition inside the ball issues forth a different trajectory with its own statistics, so that the statistics of M *different* 'identical repeated experiments', starting from the same identically prepared initial state, will consist of the combined statistics generated by each of M different trajectories – this theme is continued in Chapters 7 and 9, based upon a formulation that is introduced in Chapter 4. It is therefore an interesting possibility that the idea of randomness in macroscopic physics may be completely unnecessary, that complex phenomena in nature may be, like the slot machine and the roulette wheel, merely pseudorandom. The interesting question for the future is whether we can discover the underlying pseudorandom number generators for complex natural phenomena in the form of *low*-dimensional chaotic maps. For the time being, we address only the following question: given a chaotic map, a map with the information flow property, how does one understand the statistics that are generated by its different possible orbits? The payoff for first understanding the simple maps comes later when we show how to use one of the simplest maps, the binary tent map, to compute the different possible statistics that can be generated by an entire class of more complicated maps that includes asymmetric, logistic-like maps where the motion can even take place on a fractal.

Finally, there is the question whether the phrase random choice has any meaning, especially when it comes down to the question: can one draw an initial condition randomly from the continuum? That an *arbitrary* choice is possible seems arguable, if by arbitrary one means that the choice was not made on the basis of careful thought. On the other hand, it is not likely that the human mind is capable of a *random* choice, in the sense of either true *or* pseudo-randomness. The brain is also a kind of computer, and although we do not yet understand very well how it computes, we should not use that ignorance as an excuse for ascribing special properties to the brain. Also, one may ask: to what extent is the continuum available to us in computation? Are all binary strings (all numbers) that can be formally defined also constructable by algorithms, or are some of them only illusions created by the mathematician in the sense that we can discuss them *formally*, but can never, by means of any available algorithm, get a handle on them for computation? This question is discussed in Chapter 11, following the path that was paved for us by Turing. In Chapter 9 we discuss the consequences of the random draw question and show how to generate distributions of iterates that agree with it, as well as distributions that differ considerably from it. In Chapter 10 we give evidence from turbulence that nature does not necessarily make a random draw from the continuum.

3
Conservative dynamical systems

3.1 Integrable conservative systems: symmetry, invariance, conservation laws, and motion on invariant tori in phase space

Systems that show instability with respect to small changes in initial data were neither studied nor taught in most physics departments until nearly two decades after Lorenz's discovery of the butterfly effect. The reason for this is in part that classical mechanics was regarded by many physicists as a dead subject and was often taught mainly as preparation for the study of quantum mechanics: dissipative systems were largely ignored and the study of Hamiltonian mechanics was restricted to a very special case, that of completely integrable (or just 'integrable') systems. Mechanics was typically taught as if there were no systems other than the integrable ones. One reason for this was the very strong influence of progress that had been made in quantum theory by studying systems that were completely solvable by using symmetry. Whenever you can solve a dynamics problem by symmetry, then the system is called integrable, or 'completely integrable'. This terminology has been used in quantum as well as in classical mechanics, and for the same reason: in both cases, the method proceeds by finding a (finite) complete set of commuting infinitesimal generators (complete set of commuting operators).[1]

Deterministic chaos cannot occur in completely integrable systems and vice versa. In the era of Lorenz's discovery, the search for 'hidden symmetries' was at the forefront of physics as a consequence of the success of the prediction of the Ω-particle by the use of the SU(3) group (the 'eightfold way') in quantum field theory. There are also classical mechanical

---

[1] A word of caution: the idea of independent generators of symmetries, well defined in classical mechanics, is not entirely free of ambiguity in quantum mechanics (see von Neumann, 1955, and especially Weigert and Thomas, 1991).

systems with a hidden symmetry: the Kepler problem is solvable not only in the traditional way, by the use of rotation symmetry in three dimensions, but, more deeply, by the use of the O(4) group (see Saletan and Cromer, 1974, and also Goldstein, 1980). Hence, in the time of Lorenz's discovery and the work by Hénon and Heiles, the most stimulating new development in quantum physics had led physicists even farther down the path of exclusive concentration upon integrable dynamical systems, because hidden symmetries were also sought for in classical mechanics as a guide to a better understanding of the hidden symmetries in quantum theory. The reader must take care not to confuse the term 'integrable', whose meaning has a very specific group-theoretic underpinning, with the notion of 'closed-form solution'. The Bernoulli shift, which we studied in the last chapter, is an example of a completely chaotic system that can be solved 'exactly' in closed, algebraic form, but the Bernoulli shift is also completely nonintegrable: its phase space is densely filled with unstable periodic orbits, and every orbit of the map shows sensitivity with respect to small changes in initial conditions. In contrast, the orbits of integrable systems are stable periodic or stable quasiperiodic and are confined to smooth tori in phase space.

The main purpose of this chapter is not to provide a complete discussion of conservative mechanics, because for that one needs an entire modern textbook (see Arnol'd, 1978; Lichtenberg and Lieberman, 1983; Tabor, 1989 and Zaslavsky *et al.*, 1991). The aim here is much more modest: to sketch, in part, the extremities of the different sorts of behavior that can occur in conservative systems and to provide an introductory explanation of the terms (completely) integrable, nearly integrable and nonintegrable by using the simplest known models that reflect such behavior. The former limit, that of complete integrability, is covered in all traditional classical mechanics texts, and the main point of this introductory section is to explain that all such problems are solvable by symmetry, leading to stable-periodic or stable-quasiperiodic motion on tori in phase space. *Therefore*, the possibility for chaotic motion (sensitivity with respect to small changes in initial conditions) is eliminated. We expect that the limit of fully developed chaos in a nonintegrable conservative system provides the foundation for classical statistical mechanics, including the description how an isolated conservative dynamical system can exhibit an apparently irreversible approach to a state of statistical equilibrium, for example, to a state described statistically by Gibbs' microcanonical ensemble. In the integrable case, there are no regions in phase space where chaotic motion occurs *because* the motion takes place on smooth tori. The linear circle map *universally* provides the paradigm for that limit. In the case of fully developed chaos, there are no regions in phase space where nonchaotic motion occurs – the motion is everywhere unstable-periodic or unstable-nonperiodic with positive Liapunov exponent, and the bakers' transformation and Arnol'd's cat map provide paradigms for this limit (for both of these maps, exact solutions can be written down, although these systems are completely nonintegrable). For nearly integrable systems, the simplest known paradigm is provided by the 'standard map' which is derived in section 3.3. In that, the most complex and most general case, there are both chaotic and nonchaotic regions in phase space and the map has a set of control parameter values where it reduces to the integrable case of the linear circle map in a certain limit. The second order of business is to understand why the condition that a system is completely solvable by symmetry confines the trajectories to stable periodic or quasiperiodic motion on tori in phase space. The first order of business is therefore to understand exactly what is the definition of complete integrability in terms of the *symmetries* of a physical system.

If there is enough symmetry, only stable periodic or stable quasiperiodic motion on smooth tori can occur. This is the theme whose essence we want to explore in what follows. We begin with a Hamiltonian system and a continuous transformation of coordinates,

$$q(\alpha) = \phi(q, p; \alpha),$$
$$p(\alpha) = \psi(q, p; \alpha),$$
(3.1)

in phase space that is generated by the single parameter α. q and p are canonically conjugate variables of a conservative system with Hamiltonian H,

$$\dot{q} = \frac{\partial H}{\partial p}, \qquad \dot{p} = -\frac{\partial H}{\partial q}$$
(3.1b)

(see also Chapter 1), and we want to guarantee that the new variables $q(\alpha)$ and $p(\alpha)$ are also canonically conjugate. The definition of canonical conjugacy is given below; first, we define the identity transformation as follows: when $\alpha = 0$, $q(0) = \phi(q, p; 0) = q$ and $p(0) = \psi(q, p; 0) = p$. For simplicity, we will develop the theory here for one degree of freedom and for a one-parameter group of transformations, but the theory is generalized to f degrees of freedom and r-parameter continuous groups below (see Appendix 3.A for details). Our aim in the beginning is to discuss the essence of the main idea with the least analytic complication, which is the reason for our temporary restriction to one degree of freedom.

For a scalar phase-function, $F(q, p) \to F + \delta F$ when $\alpha \to \delta\alpha$, where

$$\delta F = \delta q \frac{\delta F}{\delta q} + \delta p \frac{\delta F}{\delta p} \tag{3.2}$$

and

$$\delta q = \delta\alpha \frac{\partial\phi}{\partial\alpha}, \qquad \delta p = \delta\alpha \frac{\partial\psi}{\partial\alpha}. \tag{3.3}$$

If we restrict to canonical transformations, then Hamilton's equations and the Poisson bracket relations are preserved in form and the transform (3.1) is then called canonical (see also Appendix 3.A. The conditions $\{q, p\} = 1$, $\{q, q\} = 0$, $\{p, p\} = 0$ can be used to define a canonical set of variables (q, p) if the Poisson bracket of two scalar functions is defined by

$$\{A, B\} = \sum_{1}^{f} \left(\frac{\partial A}{\partial q_i} \frac{\partial B}{\partial p_i} - \frac{\partial A}{\partial p_i} \frac{\partial B}{\partial q_i} \right) \tag{3.3b}$$

for the case of f degrees of freedom (see Appendix 3.A, where we also show how the Poisson bracket relations $\{A, B\}$ that we define here for scalar functions $A(q, p)$, $B(q, p)$ of *canonical* variables follow from integrating the Poisson bracket relations $[V_k, V_l]$ for vector fields that were introduced in section 1.5). In terms of the new variables $q(\alpha)$, $p(\alpha)$, we can define a canonical transformation by imposing the conditions

$$\{q(\alpha), p(\alpha)\} = \frac{\partial q(\alpha)}{\partial q} \frac{\partial p(\alpha)}{\partial p} - \frac{\partial q(\alpha)}{\partial p} \frac{\partial p(\alpha)}{\partial q} = 1, \tag{3.4}$$

and

$$\{q(\alpha), q(\alpha)\} = \{p(\alpha), p(\alpha)\} = 0, \tag{3.4b}$$

where all brackets are now computed by using the 'old' variables (q, p). This guarantees that the Poisson bracket relations $\{F, G\}$ for arbitrary scalar functions F and G are *also* invariant under canonical transformations, because

$$\{F, G\}_\alpha = \frac{\partial F}{\partial q(\alpha)} \frac{\partial G}{\partial p(\alpha)} - \frac{\partial F}{\partial p(\alpha)} \frac{\partial G}{\partial q(\alpha)} = \frac{\partial(F, G)}{\partial(q(\alpha), p(\alpha))}, \tag{3.5}$$

where now we compute the brackets using the 'new' variables: if we use the chain rule for Jacobians

$$\{F, G\} = \frac{\partial(F, G)}{\partial(q, p)} = \frac{\partial(F, G)}{\partial(q(\alpha), p(\alpha))} \frac{\partial(q(\alpha), p(\alpha))}{\partial(q, p)}$$

$$= \{F, G\}_\alpha \{q(\alpha), p(\alpha)\}, \cdot \tag{3.6}$$

then we obtain $\{F, G\} = \{F, G\}_\alpha$ because $\{q(\alpha), p(\alpha)\} = 1$. Once a new set of canonical variables has been found, corresponding Poisson bracket relations hold in terms of the new variables as well. For an infinitesimal canonical transformation, the bracket condition for $q(\alpha)$ and $p(\alpha)$ yields

$$\left(1 + \frac{\partial \delta q}{\partial q}\right)\left(1 + \frac{\partial \delta p}{\partial p}\right) - \left(\frac{\partial}{\partial p} \delta q\right)\left(\frac{\partial}{\partial q} \delta p\right) = 1, \qquad (3.7)$$

or

$$\left(\frac{\partial}{\partial q} \delta q\right) + \left(\frac{\partial}{\partial p} \delta p\right) = 0 \qquad (3.8)$$

to first order in α. This is just the condition for a divergence-free vector field, and is satisfied in two dimensions by the introduction of a stream function G,

$$\delta q = \delta \alpha \frac{\delta G}{\partial p}, \qquad \delta p = -\delta \alpha \frac{\partial G}{\partial q}, \qquad (3.9)$$

which describes a streamline motion in phase space as the group parameter α is varied (here, we use the language of flows that was developed in Chapter 1).

If α is a canonical parameter for the group (α is then additive) then $\delta \alpha = d\alpha$ and the 'group velocity vector',

$$V_\alpha = \begin{pmatrix} \dfrac{dq(\alpha)}{d\alpha} \\[2mm] \dfrac{dp(\alpha)}{\partial \alpha} \end{pmatrix}, \qquad (3.10)$$

is the tangent vector to a streamline generated by the group motion, starting from the initial point $(q(0), p(0))$.

Since we have introduced the idea of a generator of infinitesimal transformations G, we can rewrite (3.2) as

$$\delta F = \delta \alpha \{F, G\} \qquad (3.11)$$

for the infinitesimal transformation of an arbitrary scalar phase function F. This leads to a very general connection between symmetry, invariance, and conservation laws, because the function G generates infinitesimal transformations corresponding to a definite symmetry operation on the dynamical system. If a scalar function F of the phase space variables is invariant under these transformations, then $\delta F = 0$, so that $\{F, G\} = 0$, and so the condition for the Hamiltonian H of the dynamical system to be invariant under the action of the one-parameter group is then

$$\{G, H\} = 0. \qquad (3.12)$$

On the other hand, it follows easily that

$$\frac{dG}{dt} = \dot{q}\,\frac{\partial G}{\partial q} + \dot{p}\,\frac{\partial G}{\partial p} = \{G, H\}, \tag{3.13}$$

so that invariance of H under transformations generated by G corresponds exactly to conservation of the dynamical quantity $G(q, p)$. That is, G is a constant of the motion whenever the dynamical system is invariant under the group generated by G (Noether's theorem). Note that since

$$\delta q(t) = \delta t\,\frac{\partial H}{\partial p}, \qquad \delta p(t) = -\delta t\,\frac{\partial H}{\partial q}, \tag{3.14}$$

the time evolution of the dynamical system is itself a canonical transformation that is generated by the Hamiltonian H. The classical connection between continuous groups and differential equations goes back to the Norwegian mathematician Sophus Lie, through the work of Engel, Scheffers, and Cartan.

The simplest example occurs if we set $\delta q = \delta\alpha$ and $\delta p = 0$, where $G = p$ follows from the equations for the infinitesimal transformation, so that the canonical momentum p generates translations in the direction of q. If $\{p, H\} = 0$, then H is independent of q and p is conserved. Conservation of the canonical momentum p therefore corresponds to the invariance of the Hamiltonian H under translations in the direction of the coordinate that is canonically conjugate to p.

One can attempt to construct, or at least formally introduce, a generating function $F_2(q, P)$ (here, we follow Goldstein's notation) that describes the global canonical transformation from one set of canonically conjugate variables (q, p) to another (Q, P):

$$p_i = \partial F_2/\partial P_i, \qquad Q_i = \partial F_2/\partial q_i. \tag{3.14b}$$

The details that lead to this formulation are given in Appendix 3.A. For this generating function, the identity operation ($Q_i = q_i$ and $P_i = p_i$) is generated by

$$F_2 = \sum_1^f q_i P_i, \tag{3.14c}$$

and the infinitesimal transformation $Q_i = q_i + \delta q_i$, $P_i = p_i + \delta p_i$ is generated, for an N-parameter continuous group, by

$$F_2 = \sum_1^f q_i P_i + \sum_1^N \delta\alpha_\sigma G_\sigma, \tag{3.14d}$$

where the N quantities G_σ are the infinitesimal generators of the group. This is just a generalization to N parameters of the case discussed in our introductory discussion: the form of (3.14d) follows from (3.9), (3.14d), and (3.14c), and the resulting differential equations of the group are a generalization of Hamilton's equations:

$$dq_i = \sum_1^N \delta\alpha_\sigma \frac{\partial G_\sigma}{\partial p_i}, \qquad dp_i = -\sum_1^N \delta\alpha_\sigma \frac{\partial G_\sigma}{\partial q_i}. \qquad (3.14e)$$

When the N generators close under the Poisson bracket operation

$$\{G_\alpha, G_\beta\} = \sum_1^N c_{\alpha\beta}^\gamma G_\gamma + d_{\alpha\beta} \qquad (3.14f)$$

then we have a Lie algebra (see Appendix 3.A). The constants $c_{\alpha\beta}^\gamma$ are called the structure constants of the algebra, and, for compact groups, the group elements are formed by exponentiating the linear operators

$$X_\sigma = \{ \quad , G_\sigma\}, \qquad (3.15)$$

of the algebra, yielding, for canonical group parameters (see Chapter 1), group elements that are noncommuting generalizations of the time evolution operator, as was discussed at the end of Chapter 1. In the time evolution operator, the time t is the canonical parameter of the one parameter group that is generated by H. The simplest example of this case is given by the components of the angular momentum $L_i = \varepsilon_{ijk} x_j p_k$ (sum over j and k) in Cartesian coordinates, which generate rotations through an angle θ_i in the plane perpendicular to L_i (see section 1.5).

It is straightforward to check that the operators X_σ obey commutation relations with the same structure constants that appear in the Poisson bracket relation (3.14f). Whenever the N dynamical quantities G_σ generate a Lie algebra, then corresponding vector fields do not commute, so that (for $N \geq f$) the f generators G_σ cannot simultaneously serve as f canonical momenta after a canonical transformation to another coordinate system because, as we have shown, (i) the canonical momenta obey $\{P_i, P_j\} = 0$, and (ii) all Poisson bracket relations are preserved under canonical transformations. If, however, we can find f functionally independent objects G_σ with *vanishing* Poisson brackets $\{G_\alpha, G_\beta\} = 0$ for α and $\beta = 1, 2, \ldots, f$, then these quantities may, in principle, serve as canonical momenta in a new coordinate system that is reached by a canonical transformation. According to a theorem by Lie and Engel, there is a canonical transformation to a new coordinate system where this is possible. A system of f functionally independent quantities G_σ where

$\{G_\alpha, G_\beta\} = 0$ for α and $\beta = 1, 2, \ldots, f$ is called an 'involution system' in the older literature (see, e.g., the books by Duff, 1962, Eisenhart, 1961, and Whittaker, 1965).

There is an intimate connection between the existence of integrals of the motion and the symmetry of the system. Consider f quantities G_σ that have vanishing Poisson brackets with each other. Each G_σ generates transformations that leave the dynamical system invariant, and each G_σ is also a constant of the motion, whenever $\{G_\sigma, H\} = 0$ for $\sigma = 1, 2, \ldots, f$. In other words, the f different transformations that are generated by our G_σ represent symmetry operations on the system. Because $\{G_\alpha, G_\beta\} = 0$, we can in principle construct a canonical transformation to a coordinate system where the f constants of the motion G_σ yield new generalized momenta P_k, satisfying $[P_i, P_k] = 0$, and that are all *constant*. In such a case, the new Hamiltonian $K(Q, P)$ yields

$$\frac{\mathrm{d}Q_k}{\mathrm{d}t} = \frac{\partial K}{\partial P_k} = \omega_k, \qquad \frac{\mathrm{d}P_k}{\mathrm{d}t} = -\frac{\partial K}{\partial Q_k} = 0, \tag{3.16}$$

and, because K depends *only* upon the P_k, each generalized frequency ω_k is constant. This puts the motion onto a torus: the f variables P_k can be taken as constant torus radii, and the corresponding f angular variables are given by $\phi_k = \omega_k t + \delta^k$ (see Fig. 3.1). The constants P_i are called isolating integrals, because each constant isolates one independent degree of freedom.

Now, we have reached the main point: an integrable ('completely integrable') f-degree of freedom Hamiltonian system is one where there exist f functionally independent, analytic first integrals (conserved quantities) $G_1, G_2, \ldots, G_f$ that are in involution: $\{G_i, G_j\} = 0$, $i, j = 1, \ldots, f$. For a Hamiltonian system in involution, we have reasoned above that one needs only f, not $2f$, functionally independent conserved quantities G_σ in order to integrate the system. The detailed explanation how this works, as Whittaker shows in his classic treatise on dynamics, is that the *remaining f* conserved quantities that are needed for a complete integration of the equations of motion follow automatically, once one has found the first f members of the set. (Whittaker's proof requires some knowledge of Pfaffian differential equations and is reproduced for the reader's benefit in Appendix 3.B, along with the necessary introduction to Pfaffian forms.) The *main* point is that the resulting motion takes place on an f-dimensional torus, and is either stable-periodic with rational frequency ratios, or stable-quasiperiodic with irrational frequency ratios. Invariant tori are generic because the stable equilibria of a Hamiltonian system are all elliptic points. And we have qualitatively illustrated the intimate connection between symmetry, invariance, conservation laws, and invariant tori.

Our experience with linear partial differential equations is that, whenever an equation is invariant under a continuous symmetry group, one can usually find a coordinate system in which the equation separates (decouples into independent ordinary differential equations). For example, Laplace's equation has rotation symmetry and is separable into three independent ordinary differential equations in spherical coordinates. In Hamiltonian dynamics, the relevant (nonlinear) partial differential equation is the Hamilton–Jacobi equation, which follows from the generating function F_2 (see Appendix 3.A). However, separation of variables was not assumed in the above discussion that led to tori and seems not to be required, even though motion on an f-torus corresponds to f decoupled, completely independent oscillations. In even rarer cases than that of separability of the Hamilton–Jacobi equation, the equations of motion separate into f independent degrees of freedom in some coordinate system because the

Fig. 3.1   (*a*) The motion in phase space of every completely integrable two-degree-of-freedom Hamiltonian system takes place on a 2-torus. This is an example of a weak sort of *topological* universality (see also the discussion of the dissipative circle map in Chapter 5). (*b*) Action-angle variables are indicated for a cross-section of a 2-torus.

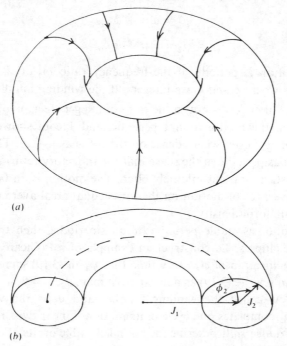

(*a*)

(*b*)

Hamiltonian itself decouples,

$$\bar{H} = \sum_{i=1}^{f} \bar{H}_i(\bar{q}_i, \bar{p}_i), \qquad (3.17)$$

with the result that Hamilton's equations

$$\dot{\bar{q}}_i = \frac{\partial \bar{H}_i}{\partial \bar{p}_i}, \qquad \dot{\bar{p}}_i = -\frac{\partial \bar{H}_i}{\partial \bar{q}_i} \qquad (3.18)$$

describe f completely independent dynamical systems. In this case the isolating integrals are the f independent Hamiltonians $\bar{H}_i$. Here, one can connect the appearance of f different elliptic points directly with an invariant torus. The bounded motion of a one-dimensional system between two turning points is periodic because the two-dimensional phase space has an elliptic point. Therefore, we obtain f separate frequencies for f different independent elliptic points. In particular, the motion of a mode with frequency ω_1 about an elliptic point takes place on a circle in the two-dimensional phase space, and the motion of two independent modes with frequencies ω_1 and ω_2 is either an open or a closed curve on a 2-torus in the four-dimensional phase space. With f degrees of freedom, one obtains a trajectory on an f-dimensional toroidal surface in the $2f$-dimensional phase space. For $f = 2$, for example,

$$\phi_1 = \omega_1 t + \phi_{10},$$
$$\phi_2 = \omega_2 t + \phi_{20}, \qquad (3.19)$$

and the motion is periodic if the frequency ratio ω_1/ω_2 is a rational number p/q, where p and q are integers. If the winding number ω_1/ω_2 is irrational, i.e. if ω_1/ω_2 is not in the ratio of integers (e.g., if $\omega_1/\omega_2 = \sqrt{2}$, e, or π) then the motion is not periodic, and Jacobi showed that the resulting curve is everywhere dense on the torus as $t \rightarrow \infty$. The resulting motion is quasiperiodic in the sense that the trajectory returns arbitrarily close to its starting point infinitely often. The motion is, in fact, ergodic: long-time averages of motion on the torus equal areal averages over the torus with uniform density.

If the motion is stable periodic or quasiperiodic, then the Poincaré sections are ellipses. To construct an example of a Poincaré section, we can plot the location of ϕ_2 every time that ϕ_1 in (3.19) increases by 2π. Now, we can rewrite (3.19) as a linear twist map: $\theta_{n+1} = \theta_n + 2\pi\Omega$, where $\Omega = \omega_2/\omega_1$ is called the winding number and θ_n is the value of ϕ_2 whenever ϕ_1 completes one cycle of its motion. Linear twist maps (linear circle maps) *universally* describe the bounded stable orbits of conservative dynamical systems, except for separatrix motion. The universality follows because bounded stable motion of an arbitrary conservative system is topologically equivalent to motion on a torus. If ω_1/ω_2 is rational, then we obtain for the Poincaré section a finite number of points as $t \to \infty$ (Fig. 3.2a). If ω_1/ω_2 is irrational, then as $t \to \infty$ the motion does not repeat, but Jacobi's theorem tells us that the Poincaré section will be densely covered by the points of the circle map as the number n of iterations goes to infinity (Fig. 3.2b).

'Ergodic' means that time averages (asymptotically as $t \to \infty$) can be replaced by phase space averages in some sense of convergence (point-wise, mean-square, or something else). The motion on the torus is ergodic, but it is not chaotic. One way to this is to show that the linear twist map has a vanishing Liapunov exponent. Therefore, there is no sensitivity to small changes in initial data. In particular, the motion on the 1-torus is ergodic, but a single harmonic oscillator with constant energy cannot approach thermal equilibrium. Ergodicity alone is completely inadequate as a basis for statistical behavior.

We have stated that quasiperiodic motion on the torus is not chaotic,

Fig. 3.2 Trajectories of the linear twist map for: (a) a rational winding number, (b) an irrational winding number.

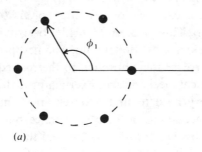

(a)

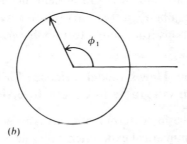

(b)

nor is it in any sense statistical-mechanical. To be specific, we can define statistical-mechanical behavior analytically as follows: a small uncertainty in initial data, represented by some volume element $\Delta\Omega(0)$, will become spread out over the available phase space of the system at large times, so that 'ignorance' that is initially small in magnitude becomes widely distributed; i.e. the system tends toward a state of statistical equilibrium. Clearly, there is an implicit requirement of action produced by pairs of positive and negative Liapunov exponents in order to achieve such a redistribution of initial phase volume. The way that the phase-volume spreading works was correctly anticipated by Gibbs without the aid of the necessary mathematical mechanism: pairs of positive and negative Liapunov exponents.

An important point is that the time development that is required for an approach to statistical equilibrium is apparently irreversible, and this seems superficially to contradict Poincaré's recurrence theorem. Poincaré's recurrence theorem for a conservative system states that, in a bounded system, the motion of a finite-volume element $\Delta\Omega$ must return infinitely often to the neighborhood of its original location as $t \to \infty$. It is easy to demonstrate the truth of the theorem: since the total available phase space is bounded, $\Delta\Omega(t)$ must intersect the region $\Delta\Omega(0)$ after a *finite* time, otherwise the phase space would have to be infinite. So must there occur second, third, ... intersections of $\Delta\Omega(t)$ with $\Delta\Omega(0)$ as $t \to \infty$ (because the system is conservative, there can be no attraction of $\Delta\Omega(t)$ to some region that excludes $\Delta\Omega(0)$). This theorem formed a basis for the arguments of Poincaré and Zermelo against Boltzmann, who first proposed a statistical explanation of the irreversible approach to thermal equilibrium of a gas on the basis of classical mechanics and certain probabilistic assumptions. Boltzmann tried very hard to find a purely mechanical (deterministic) explanation of the second law of thermodynamics, but (honorably) failed. When we study the bakers' transformation and the cat map, we shall see that no extra probabilistic assumptions are needed in order to understand the irreversible approach of a model conservative system (a fully chaotic area-preserving map) to statistical equilibrium: determinism is perfectly capable of doing the job. First, we survey some systems that show a transition via bifurcations from stable to unstable periodic orbits.

3.2 The Hénon–Heiles model: evidence for bifurcations from integrable to chaotic behavior

As qualitative evidence for deterministic chaos in conservative systems, we now review some numerical experiments that have been reported for a system introduced by Hénon and Heiles as a model in an astronomical context (the work was published in the same era as Lorenz's original work):

$$H = (p_x^2 + p_y^2)/2 + U = E = \text{constant}, \tag{3.20}$$

$$U = (x^2 + y^2 + 2x^2y - 2y^3/3)/2, \tag{3.21}$$

$$\dot{x} = \frac{\partial H}{\partial p_x} = p_x, \qquad \dot{y} = \frac{\partial H}{\partial p_y} = p_y. \tag{3.22}$$

$$\dot{p_x} = -\frac{\partial H}{\partial x} = -(x+2xy), \qquad \dot{p_y} = -\frac{\partial H}{\partial y} = -(y + x^2 - y^2).$$

The main idea in what follows is that a certain Poincaré map should be studied numerically, and it shows a sequence of bifurcations that lead eventually to the appearance of sensitivity with respect to small changes in initial conditions, in certain bounded subregions of phase space, as the energy E is increased.

Equilibria follow from $\dot{x} = \dot{y} = \dot{p_x} = \dot{p_y} = 0$, yielding the conditions

$$x(1 + 2y) = 0, \qquad y + x^2 - y^2 = 0, \tag{3.23}$$

so that either $x = 0$ or $y = -\frac{1}{2}$. If $x = 0$, then $y = 0$ or $y = 1$. If $y = -\frac{1}{2}$, then $x = \pm\sqrt{\frac{3}{4}}$. There are then four equilibria: $(0, 0)$, $(0, 1)$, $(-\sqrt{\frac{3}{4}}, -\frac{1}{2})$, and $(\sqrt{\frac{3}{4}}, -\frac{1}{2})$. The origin $(0, 0)$ is an elliptic point in the linear approximation, but a plot of U vs y yields a saddle in U at $(0, 1)$, so that near the origin the motion is bounded, whereas for $y > 1$ it is unbounded. We concentrate upon the region of bounded motion, corresponding to low enough energies E. The equipotentials (curves of constant potential U) are shown in Fig. (3.3), and we see that unbounded motion

Fig. 3.3 Equipotential curves of the Hénon–Heiles model.

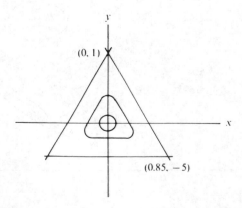

occurs if $E > U(0, 1) = \frac{1}{2}(0^2 + 1^2 - \frac{2}{3}) = \frac{1}{6}$, and bounded motion for $E < \frac{1}{6}$.

We concentrate here upon the region $0 \leq E \leq \frac{1}{6}$. Figs 3.4$a$–$c$ show the results of numerical computation for $E \cong \frac{1}{24}, \frac{1}{12}, \frac{1}{8}$, and $\frac{1}{6}$. The results are presented as a certain Poincaré section, or return map, whereby a point (y, p_y) is plotted whenever x passes through 0 with p_x positive. For $E = \frac{1}{24}$, the motion computed numerically is regular: it falls upon smooth curves which are sections of tori. The plot indicates that our (analytically unknown) return map has four elliptic and three hyperbolic points. The presence of only elliptic and hyperbolic points in the return map reflects the area-preserving property of a conservative map, which can be proven via Poincaré's integral invariants. For $E = \frac{1}{12}$, there is no qualitative difference with the previous result. For $E = \frac{1}{8}$, we see evidence of bifurcation; one can count at least four new elliptic points and some points are distributed irregularly, indicating we are looking at pseudo-orbits, because the computation was allowed to run beyond the point where the inversion of the integration scheme yielded the original initial data to within reasonable accuracy. In fact, sensitivity with respect to small changes in initial

Fig. 3.4   Poincaré sections for the Hénon–Heiles model for energies E of: (a) $\frac{1}{12}$, (b) $\frac{1}{8}$, and (c) $\frac{1}{6}$. The separatrices are not shown.

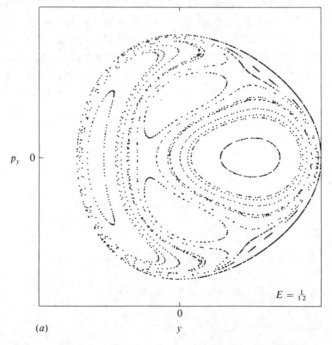

p_y 0

$E = \frac{1}{12}$

0

(a) y

Fig. 3.4—continued.

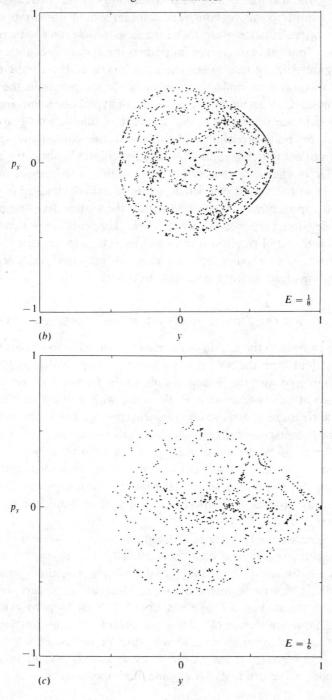

(b)

(c)

conditions can be verified numerically for the pseudo-orbit region. For
For $E = \frac{1}{6}$, the disorder generated by pseudo-orbits has increased along
with the number of bifurcations. The disorder would not disappear if a
correct orbit calculation were to be made: pseudo-orbits with positive
Liapunov exponents can emerge only when the system is chaotic.

The figures suggest that as the energy E increases, the elliptic regions
decrease in total area while the number of elliptic points in the return
map increases. Correspondingly, the number of hyperbolic points increases.
Our ability to see the extent of this sequence of bifurcations graphically
is limited by finite-precision arithmetic and the corresponding finite
resolution of computer graphics. Outside the elliptic regions, the motion
is irregular in the sense that orbits are unstable with respect to small
changes in initial conditions: small errors in initial conditions can be
magnified exponentially, which means that the system has one positive
and one negative Liapunov exponent. These exponents are not constants,
but vary with initial conditions from one trajectory to another.

We turn now to a well-known model return map that has a sequence
of bifurcations leading from integrable to chaotic motions.

3.3 Perturbed twist maps: nearly integrable conservative systems

If we return now to the nature of the chaos in the Hénon–Heiles system,
we must admit from the start that we face a more difficult problem than
that encountered with the Bernoulli shift. In the former problem, there is
a sequence of bifurcations as E is increased, with motion ranging from
integrable to more and more chaotic, whereby the total size of regular
regions in phase space decreases as chaos gains more and more ground.
The cat map, to be discussed later, provides us with a model of fully
developed conservative chaos, but in order to attempt to describe such
systems as the Hénon–Heiles model near the transition to chaos, it is
useful to start from a model that has an integrable limit, the perturbed
twist map.

If we begin with a two-degree-of-freedom integrable system in action-
angle variables (see Appendix 3.A), then $H(J_1, J_2) = E$, $\theta_i = \omega_i t + \theta_{i0}$, and,
as we have discussed in section 3.1, the motion is either periodic or
quasiperiodic on a two-dimensional torus. The radii of the tori are J_1 and
J_2, as is shown in Fig. 3.1. We can obtain a twist map by taking the
following Poincaré section of the motion: consider the intersection of the
orbit with the plane transverse to the torus of radius J_1, whenever θ_1
advances through 2π. If we denote J_2 by J_n and θ_2 and θ_n in the nth
intersection of the orbit with the plane, then the twist map is given by

the linear circle map

$$J_{n+1} = J_n,$$
$$\theta_{n+1} = \theta_n + 2\pi\alpha, \tag{3.24}$$

where α is the frequency ratio ω_1/ω_2. If α is rational then the θ_n-motion is periodic, consisting of a finite set of points. If α is irrational, then the θ_n motion is quasiperiodic and yields a dense set of points whose closure is a circle as $n \to \infty$. This is the universal nature of the return map for a two-degree-of-freedom integrable system. Liouville's theorem in this case guarantees an area-preserving map in the (J_n, θ_n) plane, corresponding to

$$\frac{\partial(J_{n+1}, \theta_{n+1})}{\partial(J_n, \theta_n)} = 1, \tag{3.25}$$

for (3.24) above.

In order to consider the perturbation of this integrable system, we follow Lichtenberg and Lieberman and assume for the perturbed twist map the form

$$J_{n+1} = J_n + \varepsilon f(J_{n+1}, \theta_n)$$
$$\theta_{n+1} = \theta_n + 2\pi\alpha(J_{n+1}) + \varepsilon g(J_{n+1}, \theta_n). \tag{3.26}$$

The extra terms should represent perturbation that makes the Hamiltonian nonintegrable, such as the nonquadratic part of the potential energy in the Hénon–Heiles problem (which has been proven to be nonintegrable). Our perturbed twist map must be area-preserving, so that

$$J = \frac{\partial(J_{n+1}, \theta_{n+1})}{\partial(J_n, \theta_n)} = \{J_{n+1}, \theta_{n+1}\} = 1 \tag{3.27}$$

but we need satisfy this condition only to $0(\varepsilon)$:

$$J = \{J_{n+1}, \theta_n + 2\pi\alpha + \varepsilon g\}$$
$$= \{J_{n+1}, \theta_n\} + \varepsilon\{J_{n+1}, g\}$$
$$= \{J_n + \varepsilon f, \theta_n\} + \varepsilon\{J_{n+1}, g\}$$
$$= 1 + \varepsilon\{f, \theta_n\} + \varepsilon\{J_{n+1}, g\}, \tag{3.28}$$

so that we must require

$$\{\theta_n, f\} + \{g, J_{n+1}\} = 0 \tag{3.29}$$

to $0(\varepsilon^2) = 0$. Since

$$\{f, \theta\} = \frac{\partial f}{\partial J_n} = \frac{\partial f}{\partial J_{n+1}} \frac{\partial J_{n+1}}{\partial J_n} \approx \frac{\partial f}{\partial J_{n+1}} + 0(\varepsilon), \qquad (3.30)$$

and since

$$\{g, J_{n+1}\} = \{g, J_n\} + \varepsilon\{g, f\} = \{g, J_n\} + 0(\varepsilon)$$

$$= \frac{\partial g}{\partial \theta_n} + 0(\varepsilon). \qquad (3.31)$$

then the area-preserving property requires that

$$\frac{\partial f}{\partial J_{n+1}} + \frac{\partial g}{\partial \theta_n} = 0. \qquad (3.32)$$

So we are free to model perturbed twist maps by different choices of f and g if we satisfy this constraint.

A map that has been much discussed in the literature is the radial twist map, which follows from assuming that

$$\frac{\partial f}{\partial J_{n+1}} = 0 \qquad \text{and} \qquad g = 0:$$

$$J_{n+1} = J_n + \varepsilon f(\theta_n), \qquad (3.33)$$

$$\theta_{n+1} = \theta_n + 2\pi\alpha(J_{n+1}).$$

We can now develop a perturbation theory as follows. We linearize about a period-one fixed point of the map $J_{n+1} = J_n = J_0$, so that $\alpha(J_0) = m$ is an integer. Then, we set $J_n = J_0 + \Delta J_n$ to obtain

$$\Delta J_{n+1} \cong \Delta J_n + \varepsilon f(\theta_n),$$

$$\theta_{n+1} = \theta_n + 2\pi m + 2\pi\alpha'(J_0)\,\Delta J_{n+1}. \qquad (3.34)$$

If then we set $I_n = 2\pi\alpha'\,\Delta J_n$, we find that

$$I_{n+1} = I_n + 2\pi\alpha'\varepsilon f(\theta_n),$$

$$\theta_{n+1} = \theta_n + 2\pi m + I_{n+1} \qquad (3.35)$$

$$= \theta_n + I_{n+1}, \qquad \text{mod } 2\pi$$

With $k = 2\pi\alpha'\varepsilon f_{\max}$ and $f^* = f/f_{\max}$ we obtain the radial twist map

$$I_{n+1} = I_n + kf^*(\theta_n)$$

$$\theta_{n+1} = \theta_n + I_{n+1}, \qquad \text{mod } 2\pi. \qquad (3.36)$$

The Chirikov–Taylor model, also called the standard map, follows from setting $f^*(\theta) = \sin\theta$. Analytic and numerical results for this model are discussed extensively in the text by Lichtenberg and Lieberman (1983). The reader is invited to iterate the standard map numerically and to observe that (for different choices of k), behavior qualitatively like that of Fig. 3.4a–d can be reproduced, as is illustrated in Fig. 3.5. We turn now to a brief description of the Kolmogorov Arnol'd Moser (KAM) theorem.

We begin with the perturbed twist map in the form

$$J_{n+1} = J_n + \varepsilon\bar{f}(J_n, \theta_n),$$
$$\theta_{n+1} = \theta_n + 2\pi\alpha(J_n) + \varepsilon\bar{g}(J_n, \theta_n),$$

(3.37)

which follows from (3.35) by iteration. When $\varepsilon = 0$, the 'bare' winding number α characterizes an invariant torus. When does this motion persist under weak perturbation: i.e., when is the motion with winding number α stable against perturbation? KAM gives a sufficient condition for the existence of perturbed invariant tori. The condition is a very nonintuitive one of 'sufficient incommensurability' of the winding number with respect to rational approximations. Consider the perturbed system, and consider the systematic approximation of an irrational winding number α by a sequence of rational approximations $p_1/q_1, p_2/q_2, \ldots, p_n/q_n, \ldots$. If, as $n \to \infty$, we find that

$$|\alpha - p_n/q_n| > K(\varepsilon)/q_n^{5/2}$$

(3.38)

as $\varepsilon \to 0$, then the orbit with irrational winding number α survives as a stable quasiperiodic orbit under the perturbation. $K(\varepsilon)$ is a positive constant dependent upon the perturbation strength ε. Useful sequences of rational approximations are generated by continued fractions. The details of this approach are discussed in Chapter 5, where we analyze the closely related dissipative circle map. Circle maps can be derived from standard maps by making the latter dissipative, and the theory of both systems has been discussed extensively by Arnol'd.

We can easily see that, according to the condition (3.38), almost all irrational tori should survive under weak perturbation, and therefore almost all of phase space should remain permeated by tori with quasi-periodic orbits. This can be understood by calculating the relative size of the intervals where the condition fails. For example, what is the fraction of the unit interval where the condition

$$|\alpha - p_n/q_n| \le K(\varepsilon)/q_n^{5/2}$$

(3.38b)

holds as $n \to \infty$ when $\varepsilon \ll 1$?

If we denote the length of that point set by $L(\varepsilon)$, we have

$$L(\varepsilon) < K(\varepsilon) \sum_{q=1}^{\infty} q^{-5/2} \approx k(\varepsilon), \qquad (3.39)$$

since the $q^{-5/2}$ series converges. Hence, the size of the gaps where the KAM condition (3.38) is violated vanishes as $\varepsilon \to 0$. What happens to the periodic and quasiperiodic orbits that do not survive? Of the periodic orbits, some remain stable while others become unstable-periodic: discussions are to be found under the heading of the Poincaré–Birkhoff theorem. We know from the numerical work on the Hénon–Heiles problem that new pairs of stable-periodic orbits emerge as older stable orbits are destroyed via pitchfork bifurcations (period-doubling bifurcations). Period doubling in conservative systems has been discussed by MacKay (see MacKay, 1984).

An example of an irrational winding number that is destroyed under perturbation is given by π, which has a rapidly convergent continued fraction expansion. On the other hand, the golden mean or any winding

Fig. 3.5 Typical numerical orbit for the standard map for parameter values of $k = 1.8$, $J_0 = 0.3$ and $\theta_0 = 5$.

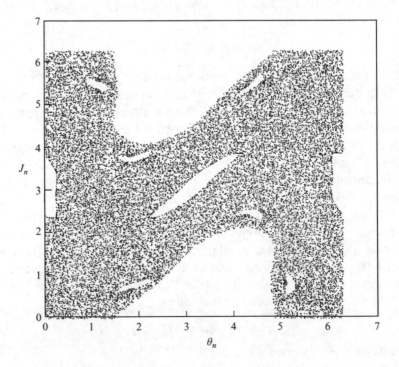

number whose continued fraction expansion is populated asymptotically by ones,

$$\alpha = \cfrac{1}{\cdots + \cfrac{1}{1 + \cfrac{1}{1 + \cfrac{1}{1 + \cdots}}}}, \tag{3.40}$$

will survive for small enough ε, since rational approximations to these winding numbers are the most slowly convergent of all algorithmically computable winding numbers (this class of winding numbers is sometimes mislabeled as 'the most irrational', but there are irrational numbers that cannot even be computed – see Chapter 11).

Our discussion of the KAM theory for twist maps follows the well-written article by Hénon and reflects Moser's proof of KAM. Arnol'd generalized the theorem to include Hamiltonian systems with an arbitrary finite number of degrees of freedom. Hamiltonian systems are discussed extensively in Lichtenberg and Lieberman (1983), and Tabor (1989), and results based upon the renormalization group method can be found in the paper by Kadanoff (1981).

3.4 Mixing and ergodicity: the approach to statistical equilibrium

According to Gibbs, who gave us modern statistical mechanics, the mixing together of two fluids is the proper analogy for the approach of a conservative dynamical system to statistical equilibrium. Consider a glass containing two fluids, namely water and a small droplet of red dye. Both fluids are approximately incompressible, so that the volume of each is separately conserved, and both are approximately nonviscous in the sense that we can ignore the effects of molecular diffusion of one fluid across the boundary of another; molecular diffusion alone is too slow to be of any use in mixing the two fluids together over time scales of practical interest.

Initially, the dye is localized within a small region A (see Fig. 3.6). Upon very slow stirring, the dye should become drawn into a long, thin filament which we denote by $A_t = U_t A$, where U_t is the time evolution operator representing the stirring. For $t \gg 0$, the stirring causes this filament to spread more or less uniformly throughout the glass of water, so that the entire fluid eventually appears uniformly pink in color. Here, the human eye performs the 'coarse-graining', which is the averaging required to give the appearance of the irreversible tendency toward statistical equilibrium. If one looks at the mixture on a small enough length scale, before molecular diffusion has had time to act, one should see a very long, thin, filament that threads throughout the water.

Gibbs discussed such an example as the likely explanation for the approach to statistical equilibrium in conservative systems, and we know now that Liapunov exponents provide a mechanism for the filamentation to occur. The droplet of dye is analogous to our initial uncertainty in initial conditions. The analog of stirring must be provided by the Hamiltonian of a completely chaotic system, and the glass of water corresponds to a bounded phase space. Conservation of volume of the dye (incompressibility) is the analog of Liouville's theorem. A uniformly red fluid corresponds to the state of maximum ignorance (statistical equilibrium, according to Boltzmann), and the Hamiltonian system should be completely nonintegrable, analogous to the bakers' transformation to be discussed in the next section, in order to accomplish all this.

The mixing condition can be formulated mathematically as follows: if B is any region in the fluid with volume V_B, then after a long time $t \gg 0$, the fraction of dye that threads throughout B should, with high probability, equal the fraction of the total space occupied by B:

$$\lim_{t \to \infty} \frac{V_{A_t \cap B}}{V_A} = \frac{V_B}{V_T},\tag{3.41}$$

where $A_t \cap B$ denotes the intersection of the two point sets A_t and B.

Fig. 3.6 Idealization of the sequence of events representing the nonturbulent mixing of a droplet of dye in a glass of otherwise still water by very slowly stirring. The droplet of dye is drawn out into a long, thin filament, as was first described by Gibbs: the water appears to have become uniformly coloured when the eye can no longer resolve the spacing between adjacent segments of the red filament.

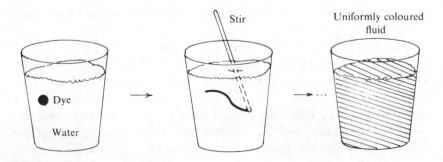

$A_t = U_t A$ and V_T is the total volume of fluid (size of the bounded phase space). So, if the region B occupies 4 percent of the total volume ($V_B/V_T = 0.04$), then with high probability, one should find about 4 percent of the dye threading throughout B as $t \to \infty$. If $\mu(A_t \cap B)$ denotes the measure of the intersection, i.e.

$$\mu(A_t \cap B) = V_{A_t \cap B}/V_T, \tag{3.42}$$

then mixing corresponds to the condition

$$\lim_{t \to \infty} \mu(A_t \cap B) = \mu(A)\mu(B), \tag{3.43}$$

where $\mu(A) = V_A/V_T$ and $\mu(B) = V_B/V_T$ represent the fractions of the total space occupied by fluids A and B initially.

It is interesting to note that O. Reynold's test for turbulence in a fluid was to inject a small amount of dye and then look for mixing: if the fluid motion mixes the dye, then the fluid is turbulent (see Fig. 10.2 for an example of mixing by chaotic but nonturbulent vorticity dynamics).

The mixing provided by the Bernoulli shift and the related bakers' transformation represent a *laminar* filamentation. Turbulent mixing requires vorticity, and chaotic vortex dynamics seems necessary in order to describe the energy-eddy cascade in fully developed turbulence (see Chapter 10 for the phenomenology of the eddy cascade). The examples of mixing illustrated below by the bakers' transformation do not have any chaotic swirls or eddies in phase space, hence are nonturbulent.

We have previously given an example of a dynamical system which is ergodic but nonmixing: quasiperiodic motion on a torus that arises from irrational frequency ratios. In that case, time averages equal space averages over the surface of the torus (ergodic motion), but the dynamical equations will not cause any filamentation of a small 'droplet' of initial data on the surface of the torus. The Liapunov exponents for that system are all zero, so that there is no mechanism to cause a droplet to become stretched.

Arnol'd and Avez (1986) show that a mixing system has no nontrivial invariant subspace: if we set $A = B$, then the mixing condition is given by

$$\lim_{t \to \infty} \mu(A_t \cap A) \to \mu(A)\mu(A). \tag{3.44}$$

On the other hand, $A = B = A_t$, so that

$$\mu(A_t \cap A) = \mu(A), \tag{3.45}$$

yielding $\mu(A) (1 - \mu(A)) = 0$. Hence, either A coincides with the entire bounded phase space ($\mu(A) = 1$) or else A is empty ($\mu(A) = 0$), which means that mixing implies ergodicity. We turn now to an example of a model conservative system that is mixing. In this model, we can see that the condition for the butterfly effect in the conservative system also yields Gibbs' filamentation: one needs two nonzero Liapunov exponents, one that is positive and one that is negative.

3.5 The bakers' transformation

The bakers' transformation

$$
\left.\begin{array}{c} x_{n+1} \\ y_{n+1} \end{array}\right\} = \begin{cases} 2x_n, & \\ y_n/2, & x_n \in [0, \tfrac{1}{2}), \\ 2x_n - 1, & \\ (y_n + 1)/2, & x_n \in (\tfrac{1}{2}, 1], \end{cases} \tag{3.46}
$$

is a model of the process whereby a baker kneads a blob of dough by stretching and folding. This process yields mixing via filamentation, as can be seen graphically by mapping finite regions (Fig. 3.7). As $n \to \infty$, the unit square is 'painted' by infinitely many thin, horizontal, colored filaments. When one can no longer resolve the distance between adjacent filaments, the system is said to be in a state of statistical equilibrium, so that the relaxation time depends upon the scale, or resolution of observation. Then, the entire unit square will appear to have been uniformly 'painted' by a small droplet of paint. In the figure, we can regard the shaded block of initial conditions that we iterated as a droplet of wet paint.

Since

$$
J(A) = \frac{\partial(x_{n+1}, y_{n+1})}{\partial(x_n, y_n)} = 2 \cdot \tfrac{1}{2} = 1, \tag{3.47}
$$

the total area of colored block ($0 \le x \le 1, 0 \le y \le \tfrac{1}{2}$) is preserved under

Fig. 3.7   Filamentation of a block of initial conditions in phase space by the action of the bakers' transformation.

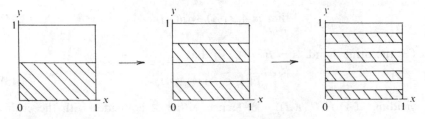

filamentation: even through the entire unit square will eventually appear to have been painted, the painted region actually occupies only half the total phase space.

The area-preserving filamentation which causes the mixing is due to two Liapunov exponents. Since the map is piecewise linear, small errors $(\delta x_0, \delta y_0)$ evolve according to the original map, so that with $|\delta x_n|$ and $|\delta y_n|$ both small compared with unity,

$$\delta X_n = \begin{pmatrix} 2 & 0 \\ 0 & \frac{1}{2} \end{pmatrix} \delta X_{n-1} = \begin{pmatrix} 2^n & 0 \\ 0 & \frac{1}{2^n} \end{pmatrix} \delta X_0, \tag{3.48}$$

where

$$\delta X_n = \begin{pmatrix} \delta x_n \\ \delta y_n \end{pmatrix}. \tag{3.49}$$

Since $\delta x_n = e^{n \ln 2} \delta x_0$ and $\delta y_n = e^{-n \ln 2} \delta y_0$, we obtain stretching of areas in the x-direction, with Liapunov exponent $\lambda_1 = \ln 2 > 0$, and contraction in the y-direction, with Liapunov exponent $\lambda_2 = -\ln 2 < 0$. The result, $\lambda_1 + \lambda_2 = 0$, corresponds to the area-preservation property $J(t) = 1$, since $J = e^{\lambda_1 + \lambda_2}$.

Does the filamentation caused by these Liapunov exponents produce a uniform invariant density, $p(x, y) = 1$, as $n \to \infty$, as one might expect? In order to investigate this question from a statistical mechanics viewpoint, it is useful to use the master equation, which is the equation of motion for $p_n(x, y)$, the probability density of the iterates, starting from some initial density. One can derive the equation of motion for $p_n(x, y)$ from Liouville's equation, which in this case is given by

$$p_n(x_n, y_n) \, dx_n \, dy_n = p_0(x_0, y_0) \, dx_0 \, dy_0$$
$$= p_{n-1}(x_{n-1} \cdot y_{n-1}) \, dx_{n-1}, dy_{n-1}. \tag{3.50}$$

Since all the relevant Jacobians equal unity, this means that

$$p_n(x_n, y_n) = p_{n-1}(x_{n-1}, y_{n-1}) = \cdots = p_0(x_0, y_0). \tag{3.51}$$

The inverse bakers' transformation is given by

$$\begin{aligned}
x_n &= x_{n+1}/2 \\
y_n &= 2y_n
\end{aligned} \Bigg\}, \qquad x_n \in [0, \tfrac{1}{2})$$

$$\begin{aligned}
x_n &= (x_{n+1} + 1)/2 \\
y_n &= 2y_{n+1} - 1
\end{aligned} \Bigg\}, \qquad x_n \in [\tfrac{1}{2}, 1], \tag{3.52}$$

which is the same as

$$
\left.\begin{array}{l} x_n \\ y_n \end{array}\right\} = \begin{cases} \left.\begin{array}{l} x_{n+1}/2 \\ 2y_{n+1} \end{array}\right\}, & y_n \in [0, \tfrac{1}{2}), \\[2ex] \left.\begin{array}{l} (x_{n+1} + 1)/2 \\ 2y_{n+1} - 1 \end{array}\right\}, & y_n \in [\tfrac{1}{2}, 1]. \end{cases} \tag{3.53}
$$

Replacing (x_{n-1}, y_{n-1}) by their expressions in terms of (x_n, y_n) on the right-hand side of the equation $p_n(x_n, y_n) = p_{n-1}(x_{n-1}, y_{n-1})$ and setting $(x_n, y_n) = (x, y)$ yields the master equation

$$
p_n(x, y) = \begin{cases} p_{n-1}\left(\dfrac{x}{2}, 2y\right), & y \in [0, \tfrac{1}{2}), \\[2ex] p_{n-1}\left(\dfrac{x+1}{2}, 2y - 1\right), & y \in [\tfrac{1}{2}, 1]. \end{cases} \tag{3.54}
$$

Is there an approach to statistical equilibrium, as is suggested by our interpretation of Fig. 3.7? The answer is yes, but it is in the 'weak' sense that asymptotic *averages* of well-behaved functions of the dynamical variables $f(x, y)$ can be computed as if there were convergence of $p_n(x, y)$ as $n \to \infty$ to a pointwise uniform invariant density $p(x, y) = 1$:

$$
\langle f(x, y) \rangle = \lim_{n \to \infty} \int_0^1 \int_0^1 dx\, dy\, p_n(x, y) f(x, y) = \int_0^1 \int_0^1 f(x, y)\, dx\, dy. \tag{3.55}
$$

We can also derive the master equation from the Frobenius–Peron equation:

$$
p_n(x, y) = \int_0^1 \int_0^1 dx'\, dy'\, p_{n-1}(x', y')\, \delta(x - f(x'))\, \delta(y - g(y')), \tag{3.56}
$$

where

$$
f(x) = \begin{cases} 2x, & 0 \le x < \tfrac{1}{2}, \\ 2x - 1, & \tfrac{1}{2} \le x \le 1, \end{cases} \tag{3.57}
$$

and

$$
g(y) = \begin{cases} y/2, & 0 \le x < \tfrac{1}{2}, \\ (y + 1)/2, & \tfrac{1}{2} \le x \le 1. \end{cases} \tag{3.58}
$$

Since

$$
\delta(x - f(x')) = \frac{1}{|f'(x')|}\, \delta(x' - f^{-1}(x)), \tag{3.59}
$$

and, using the fact that the location of x_n in the unit interval determines the y_n-branch that y_{n+1} came from, we have

$$p_n(x, y) = \frac{2}{2} \int_0^{1/2} dx' \int_0^{1/2} dy' p_{n-1}(x', y') \delta(x' - x/2) \delta(y' - 2y)$$

$$+ \frac{2}{2} \int_{1/2}^1 dx' \int_{1/2}^1 dy' p_{n-1}(x', y') \delta\left(x' - \frac{x+1}{2}\right)$$

$$\times \delta(y' - (2y - 1)). \tag{3.60}$$

So, if $0 \le y < \frac{1}{2}$, we have

$$p_n(x, y) = p_{n-1}\left(\frac{x}{2}, 2y\right), \tag{3.61}$$

whereas if $\frac{1}{2} \le y \le 1$ we obtain

$$p_n(x, y) = p_{n-1}\left(\frac{x+1}{2}, 2y - 1\right), \tag{3.61b}$$

the same results as before.

It is interesting to see graphically that the mixing process, i.e., the irreversible approach to statistical equilibrium, is consistent with Poincaré's recurrence theorem. Consider, e.g., a subset of the unit square, corresponding to some initial uncertainty in our knowledge of initial data. The point (x_0, y_0) lies within the block with even probability in this case. Then, under the action of the map, this initial uncertainty becomes filamented, as is shown in Fig. 3.8. So, for $n = 2$ there is already recurrence and clearly there will be infinitely many intersections of the filament with the location of the original block as $n \to \infty$. It follows that a nonperiodic orbit $(x_0, y_0) \to (x_1, y_1) \to \cdots \to (x_n, y_n) \to \cdots$ must return arbitrarily close to (x_0, y_0) infinitely many times as $n \to \infty$, in agreement with Poincaré's theorem.

In order to understand the unstable-periodic nature of the orbits of the bakers' transformation that start from rational initial conditions, we should study the bakers' map in binary arithmetic. With our initial conditions expressed as binary expansions,

$$x_0 = \sum_{j=1}^\infty \varepsilon_j/2^j = 0.\varepsilon_1\varepsilon_2\cdots, \tag{3.62}$$

$$y_0 = \sum_{j=1}^\infty \delta_j/2^j = 0.\delta_1\delta_2\cdots. \tag{3.62b}$$

The bakers' transformation,

$$x_n = \begin{cases} 2x_n, & = 0 \le x_n \le \frac{1}{2}, \\ 2x_{n-1}, & \frac{1}{2} \le x_n \le 1, \end{cases}$$

$$y_n = \begin{cases} y_n/2, & = 0 \le x_n < \frac{1}{2}, \\ (1 + y_n)/2, & \frac{1}{2} \le x_n \le 1, \end{cases} \tag{3.63}$$

can be understood as follows: $x_1 = 0.\varepsilon_2\varepsilon_3\cdots$ whether $\varepsilon_1 = 0$ or 1, $y_1 = 0.0\delta_1\delta_2\cdots$ if $\varepsilon_1 = 0$ and $y_1 = 0.1\delta_1\delta_2\cdots$ if $\varepsilon_1 = 1$. For $n = 2$, $y_2 = 0.0\varepsilon_1\delta_1\delta_2\cdots$ if $\varepsilon_2 = 0$ and $y_2 = 0.1\varepsilon_1\delta_1\delta_2\cdots$ if $\varepsilon_2 = 1$. So, $y_2 = 0.\varepsilon_2\varepsilon_1\delta_1\delta_2\cdots$. It is by now clear that the initial condition y_0 gets lost for large enough n, because the bits $\delta_1\delta_2\cdots$ are shifted from most significant into least-significant bits with each iteration (this is because the Liapunov exponent for the y-motion is $-\ln 2$), and in general we find that

$$x_n = 0.\varepsilon_{n+1}.\varepsilon_{n+2}\cdots,$$

$$y_n = 0.\varepsilon_n\varepsilon_{n-1}\cdots\varepsilon_1\delta_1\delta_2\cdots. \tag{3.64}$$

Fig. 3.8 Recurrence phenomena for the bakers' transformation: any initial state of the system is approximately repeated after a length of time that depends upon the accuracy to within which that state is known or prescribed; the finite precision to within which the state is prescribed corresponds geometrically to a block (or ball) of initial conditions in phase space (see also Fig. 3.6 for a qualitatively similar phenomenon).

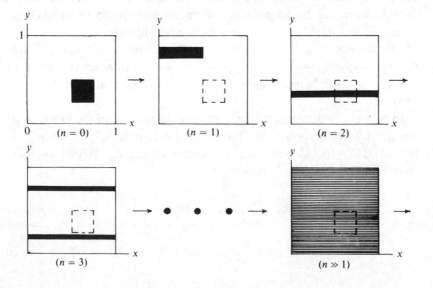

If we then write the *two-sided* binary sequence representing x_0 and y_0,

$$\cdots \delta_2 \delta_1 . \varepsilon_1 \varepsilon_2 \cdots, \tag{3.65}$$

then after n iterations the two-sided binary sequence,

$$\cdots \delta_2 \delta_1 \varepsilon_1 \varepsilon_2 \cdots \varepsilon_n . \varepsilon_{n+1} \varepsilon_{n+2} \cdots, \tag{3.66}$$

represents the bakers' transformation. x_n is given by the sequence of bits to the right of the binary point, whereas y_n is given by the reverse sequence of bits to the left of the binary point. So, the motion for large n is determined entirely by the distribution of bits in x_0, and not at all by y_0. If x_0 is rational, then the x-motion is unstable periodic, while the y-motion is unstable-almost-periodic. The instability corresponds to the fact that a small change in the last digits representing x_0 will produce an entirely different orbit, typically one with much larger period. If x_0 is irrational, both the x- and y-motions are truly chaotic and, for certain initial conditions, are also recurrent, whereby the neighborhood of any initial condition (x_0, y_0) will be visited arbitrarily often, but nonperiodically, by the system's orbit as n increases without bound.

The bakers' transformation models a completely nonintegrable system: there are no regions of finite size in phase space where the motion is regular because unstable periodic orbits are everywhere dense. The resulting filamentation of small errors in initial data produces an irreversible approach to statistical equilibrium that is consistent with Poincaré's recurrence theorem. To prove equivalent behavior for more realistic systems is very difficult, but we expect a dense set of unstable periodic orbits in the phase space of any system that shows an approach to statistical equilibrium.

3.6 Computation of chaotic orbits for an area-preserving map

We have seen how the Bernoulli shift can be generalized to an area-preserving map of the unit square, the bakers' transformation (3.67). In that case, nonperiodic orbits can be computed to any desired precision by using an algorithm to generate the nonperiodic binary string, bit by bit, of an irrational number such as $\sqrt{2} - 1$, $\pi - 3$, or $e - 2$, and then inserting the digit string into the two-sided Bernoulli shift (3.66). Instead of doing this in detail, we turn instead to a generalization of the Bernoulli shift that is an area-preserving map of the 2-torus. This model is interesting, because we can use the Palmore parallel-processing method (see section 2.6) to compute chaotic orbits of this map to any finite prescribed precision. First, we introduce the map and survey its general properties.

Arnol'd's cat map is defined by the transformation of the unit square,

$$x_{n+1} = 2x_n + y_n \bmod 1,$$
$$y_{n+1} = x_n + y_n \bmod 1,$$

(3.67)

and is area-preserving because

$$J = \frac{\partial(x_{n+1}, y_{n+1})}{\partial(x_n, y_n)} = 1.$$

(3.68)

Therefore, it is a discrete model of a conservative dynamical system. Because of the mod 1 condition on both x and y, the motion occurs on the surface of a two-dimensional torus: the points $x = 0$ and 1 are identical, as are the points $y = 0$ and 1.

The first thing to know is that the map has two nonzero Liapunov exponents, $\lambda_+ = \ln((3 + \sqrt{5})/2) = -\lambda_-$. This follows from writing

$$\begin{pmatrix} x_{n+1} \\ y_{n+1} \end{pmatrix} = \begin{pmatrix} 2 & 1 \\ 1 & 1 \end{pmatrix} \begin{pmatrix} x_n \\ y_n \end{pmatrix} = A \begin{pmatrix} x_n \\ y_n \end{pmatrix} \bmod 1,$$

(3.69)

where $(3 \pm \sqrt{5})/2$ are the eigenvalues of the matrix A. Hence, the map is chaotic (Arnol'd showed that the map has a uniform invariant density). Are there any regular regions in phase space? In order to analyze this question, notice that solutions have the form

$$\begin{pmatrix} x_n \\ y_n \end{pmatrix} = A^n \begin{pmatrix} x_0 \\ y_0 \end{pmatrix}, \qquad \bmod 1,$$

(3.70)

for any initial condition (x_0, y_0), and where it is easy to see that

$$A^n = \begin{pmatrix} F_{2n} & F_{2n-1} \\ F_{2n-1} & F_{2n-2} \end{pmatrix},$$

(3.71)

where F_n is a Fibonacci number: $F_{n+1} = F_n + F_{n-1}$ with $F_0 = F_1 = 1$. Now, all rational initial conditions can be written in the form

$$(x_0, y_0) = (P/N, Q/N),$$

(3.72)

where the unit square is partitioned into N^2 equal squares, each of side $1/N$, and P and Q are integers, $0 \le P, Q \le N$. Because there are N^2 cells, and A^n is an integer matrix (all entries are integers), the cell labeled by P and Q must be revisited after not more than N^2 iterations of the map, and, by uniqueness, the motion must then repeat. Hence, all rational initial conditions yield periodic orbits, and all of these orbits are unstable because $\lambda_+ > 0$. There are no 1-cycles or limit cycles because that would violate Liouville's theorem ($J = 1$). One can also prove that all cycles start from rational initial conditions and that irrational starting points yield unstable nonperiodic (chaotic) orbits. Here, we have another example of 'fully developed chaos'; there are no regular regions on the torus: unstable periodic orbits are everywhere dense. Like the Bernoulli shifts and tent maps, this dynamical system is uniformly hyperbolic (dense set of unstable periodic orbits with a constant positive Liapunov exponent). It is also an example of a mixing system, and Arnol'd (1983) has shown that the map's invariant density is uniform, meaning that a measure-one set of irrational initial conditions yields orbits whose distribution of iterates covers the torus uniformly as $n \to \infty$. Other irrational initial conditions produce nonuniform distributions of the map's iterates over the torus.

The main idea of the parallel-processing method of computing chaotic orbits with uniform precision is that the digit strings for the initial conditions must be computed along with those for the iterates of the map. Because the map is piecewise linear and $\lambda_+ > 0$, the information in the Nth bits of (x_0, y_0) is transformed at a finite rate, until in relatively short time it becomes information in the first bits of x_n and y_n. This is the butterfly effect for piecewise-linear maps of the square and for linear maps of the 2-torus (here, the map is linear but the phase space is nonlinear). Therefore, to know (x_n, y_n) to N-bit accuracy, we know that (x_0, y_0) must be computed to within roughly $n + N \ln 2/\lambda_+$ bit accuracy. This is easy to do. We need only use an algorithm for an irrational number (or a pair of algorithms), such as the algorithm for $\sqrt{2} - 1$ to write x_0 and y_0 in (3.70) as N-bit binary strings. The solution is then

$$x_n = F_{2n}x_0 + F_{2n-1}y_0, \qquad \text{mod } 1,$$
$$y_n = F_{2n-1}x_0 + F_{2n-2}y_0, \qquad \text{mod } 1,$$
(3.70b)

where the coefficients F_m are Fibonacci numbers. All the arithmetic can be done as binary-point shifts on x_0 and y_0 combined with the addition of finite-length strings. However, you must keep exactly so many bits in x_0 and y_0 as you need to get both x_n and y_n accurately to N-bits, where N is your decided-upon accuracy. That $F_m x_0$ consists only of binary-point shifts plus addition on x_0 (mod 1) is easy to see: with

$$F_m = \sum_{i=0}^{\mu} \varepsilon_i 2^i = \varepsilon_\mu \varepsilon_{\mu-1} \cdots \varepsilon_0.0,$$

where $\varepsilon_i = 0$ or 1, we have $F_3 = 11 = 10 + 1$, $F_4 = 101 = 100 + 1$,

$F_5 = 1000$, $F_6 = 1101 = 1000 + 100 + 1$, and so on. For example, given $x_0 = \sqrt{2} - 1 \simeq 0.0110$, then $F_4 x_0 = 0.10 + 0.01 = 0.11$ will be exact to 2 bits if there are no 'carries' in the neglected terms. If we want to know $F_3 x_0$ to 3 bits we have to use $\sqrt{2} - 1 = 0.01101$, so that $F_3 x_0 = 0.101 + 0.011 = 0.000$. For example, given $x_0 = \sqrt{2} - 1 = 0.01101010000001001 \cdots$ to 17 bits, we can compute

$$F_4 x_0 = 100 x_0 + x_0 \bmod 1 = 0.000100100001011 \cdots$$

to 14-bit accuracy. You should note that you have to have a method for handling 'carries'. For example, if we use $x_0 = 0.0110 \cdots$ to 4 bits to compute $F_4 x_0 = 0.10 + 0.01 = 0.11 \cdots$ to 2 bits, this is wrong, but $x_0 = 0.01101 \cdots$ to 5 bits yields the correct 3-bit result, $F_3 x_0 = 0.000$. So (x_n, y_n) can be computed from (3.70b) plus an algorithm that generates (x_0, y_0) as nonperiodic binary strings, to any prescribed precision N. The resulting computed orbits are uniform approximations to the exact chaotic orbits defined by (3.70b).

Appendix 3.A Generating functions for canonical transformations

It is useful to begin with the differential form,

$$L \, dt = \sum_1^f p_i \, dq_i - H \, dt, \qquad (3.A.1)$$

where $H(q, p, t)$ is the Hamiltonian, $L(q, \dot{q}, t)$ is the Lagrangian,

$$p_i = \frac{\partial L}{\partial \dot{q}_i} \qquad (3.A.2)$$

is the canonical momentum, and (q, p) is shorthand notation for the point $(q_1, \ldots, q_f, p_1, \ldots, p_f)$ in phase space. The Hamiltonian differential equations then follow from an action principle, whereby the functional S, called the action,

$$S = \int_{t_1}^{t_2} L \, dt \qquad (3.A.3)$$

is shifted by $2f$ differentiable but otherwise arbitrary variations

$$q_i \to q_i + \delta q_i, \qquad p_i \to p_i + \delta p_i, \qquad (3.A.4)$$

whose end points are fixed: δq_i and δp_i vanish at t_1 and t_2. It follows that

$$\delta S = \sum_{i=1}^{f} \int_{t_1}^{t_2} \left\{ \left(\dot{q}_i - \frac{\partial H}{\partial p_i} \right) \delta q_i + \left(-\dot{p}_i - \frac{\partial H}{\partial q_i} \right) \delta p_i \right\} dt, \quad (3.A.5)$$

so that for $2f$ independent variations the extremum gives Hamilton's equations:

$$\dot{q}_i = \frac{\partial H}{\partial p_i}, \qquad \dot{p}_i = -\frac{\partial H}{\partial q_i}, \quad (3.A.6)$$

q_i is the generalized coordinate and p_i is the canonical momentum conjugate to q_i; q_i and p_i are called canonically conjugate variables.

If next we consider a transformation to a new set of canonical variables (Q, P) where Hamilton's equations still hold with a new Hamiltonian $K(Q, P, t)$,

$$\dot{Q}_i = \frac{\partial K}{\partial P_i}, \qquad \dot{P}_i = -\frac{\partial K}{\partial Q_i}, \quad (3.A.7)$$

then the condition that both follow from the same action principle is that the differential form $L_1 \, dt - L_2 \, dt$ is exact:

$$L_1 \, dt - L_2 \, dt = \sum_{1}^{f} p_i \, dq_i - H \, dt - \sum_{1}^{f} P_i \, dQ_i + K \, dt = dF_1. \quad (3.A.8)$$

Here, we can study F_1 as a function of any $2f$ of the $4f$ variables (q, p, Q, P). The notation F_1, with subscript 1, is standard when we choose the $2f$ generalized coordinates (q, Q) as the independent variables, so that

$$p_i = \frac{\partial F_1}{\partial q_i}, \qquad P_i = -\frac{\partial F_1}{\partial Q_i}, \qquad \text{and} \qquad K = H + \frac{\partial F_1}{\partial t}. \quad (3.A.9)$$

Transformations that preserve the form of Hamilton's equations are called canonical transformations, and $F_1(q, Q, t)$ is an example of a generating function. Other generating functions can be reached by Legendre transforms. For example,

$$F_2(q, P) = F_1(q, Q) + \sum_{1}^{f} Q_i P_i, \qu(3.A.10)$$

so that

$$dF_2 = \sum_{1}^{f} (p_i \, dq_i + Q_i \, dP_i) - (H - K) \, dt, \quad (3.A.11)$$

which yields

$$p_i = \frac{\partial F_2}{\partial q_i}, \qquad Q_i = \frac{\partial F_2}{\partial P_i}, \qquad \text{and} \qquad K = H + \frac{\partial F_2}{\partial t}. \quad (3.A.12)$$

In fact, canonical transformations form a group: two successive Legendre transforms on a generating function can be written as a single Legendre transform (try it), the inverse transformations also can be constructed, and $F_2(q, P) = \sum_1^f q_i P_i$ generates the identity.

The dynamics problem would be solved if we could construct a canonical transformation to a new system (Q, P) where the f momenta P_i are constants, $P_i = \alpha_i = $ constant. Writing $F_2(q, P) = S(q, \alpha)$, we would then find that

$$p_i = \frac{\partial S}{\partial q_i}, \qquad Q_i = \frac{\partial S}{\partial P_i}, \tag{3.A.13}$$

where (Q, α) are the new canonically conjugate coordinates and momenta. For a time-independent canonical transformation (one where F_2 has no explicit time dependence), then

$$H\left(q, \frac{\partial S}{\partial q}\right) = K(\alpha), \tag{3.A.14}$$

where $K(\alpha)$ is the Hamiltonian in the new coordinate system. Because

$$\dot{P}_i = -\frac{\partial K}{\partial Q_i} = 0, \tag{3.A.15}$$

$K(\alpha) = $ constant, and the first-order nonlinear partial differential equation (A.15) is called the time-independent Hamilton–Jacobi equation (see Duff, 1962). One can also write down a time-dependent Hamilton–Jacobi partial differential equation. In practice, solutions are found only when one can separate variables in some coordinate system,

$$S(q_1, \ldots, q_f, \alpha_1, \ldots, \alpha_f) = \sum_1^f S_i(q_i, \alpha_i), \tag{3.A.16}$$

so that

$$H\left(q, \frac{\partial S}{\partial q}\right) = K(\alpha), \tag{3.A.17}$$

and the constant $K(\alpha_i)$ is then called an isolating integral, because it isolates one degree of freedom from the other $f - 1$ degrees of freedom. However, separability depends upon the *symmetry* of the Hamiltonian H, and separable systems are only a special case of integrable systems, systems which are entirely solved by symmetry (see Appendix 3.B).

This brings us to the main part of the story: the role of continuous symmetries in determining the solutions of conservative dynamical systems where all of the orbits are stable. Since $F_2 = \sum_1^f q_i P_i$ generates the identity transformation, an infinitesimal canonical transformation depending continuously and differentiably upon r parameters $a = (a_1, \ldots, a_r)$ is given by

$$F_2(q, P) \approx \sum_{i=1}^{f} q_i P_i + \sum_{\sigma=1}^{r} \delta \alpha^\sigma G_\sigma, \qquad (3.A.18)$$

so that we get the following generalization of Hamilton's equations:

$$dq_i = \sum_{\sigma=1}^{f} \delta \alpha^\sigma \frac{\partial G_\sigma}{\partial p_i}, \qquad dp_i = - \sum_{\sigma=1}^{f} \delta \alpha^\sigma \frac{\partial G_\sigma}{\partial q_i}. \qquad (3.A.19)$$

The simplest case is given by $G = H$ and $a = t$, which shows that Hamilton's equations are an infinitesimal transformation. It is because the global motion of the dynamical system is a canonical transformation, with infinitesimal generator determined by H, that one can attempt sensibly to solve the dynamics problem by the construction of a global generating function S. In the next paragraph, we use the results of section 1.5.

Consider the velocity fields

$$V_\sigma^i = \left(\frac{\partial G_\sigma}{\partial p_i}, -\frac{\partial G_\sigma}{\partial q_i} \right), \qquad (3.A.20)$$

which describe a certain phase space flow locally, the flow that is generated by G_σ. The corresponding infinitesimal operators (section 1.5) are given by

$$X_\sigma = V_\sigma^i \frac{\partial}{\partial x^i} = \frac{\partial G_\sigma}{\partial p_i} \frac{\partial}{\partial q_i} - \frac{\partial G_\sigma}{\partial q_i} \frac{\partial}{\partial p_i}, \qquad (3.A.21)$$

where $n = 2f$, $(x_1, \ldots, x^n) = (q_1, \ldots, q_f, p_1, \ldots, p_f)$ and we now use the summation convention (sum over repeated indices). If we consider the Poisson bracket of two vector fields V_1 and V_2 (see section 1.5),

$$[V_1, V_2]_k = \sum_{i=1}^{n} \left(V_2^i \frac{\partial V_1^k}{\partial x_i} - V_1^i \frac{\partial V_2^k}{\partial x_i} \right), \qquad (3.A.22)$$

then we can see that

$$[V_1, V_2]_k = \frac{\partial}{\partial q_k} \{G_1, G_2\} \qquad (3.A.23)$$

if $n = 1, \ldots, f$, and that

$$[V_1, V_2]_k = \frac{\partial}{\partial p_k} \{G_1, G_2\} \qquad (3.A.24)$$

if $n = f + 1, \ldots, 2f$, and where we have defined a *new* Poisson bracket

$\{A, B\}$ of two functions A and B as

$$\{A, B\} = \frac{\partial A}{\partial q_i} \frac{\partial B}{\partial p_i} - \frac{\partial A}{\partial p_i} \frac{\partial B}{\partial q_i}. \tag{3.A.25}$$

Therefore, we can also write

$$X_\sigma = \{\ , G_\sigma\}. \tag{3.A.26}$$

The integrability condition for an r-parameter group of transformations is the closure condition for the Lie algebra,

$$(X_\alpha, X_\beta) = c_{\alpha\beta}^\gamma X_\gamma, \tag{3.A.27}$$

where the objects $c_{\alpha\beta}^\gamma$ are the structure constants of the Lie group. By using Jacobi's identity, we obtain from this the conditions

$$\frac{\partial}{\partial x_i} \{G_\alpha, G_\beta\} = c_{\alpha\beta}^\gamma \frac{\partial G_\gamma}{\partial x_i}, \tag{3.A.28}$$

in terms of the new Poisson brackets, which we shall refer to from now on simply as the Poisson brackets. Integration then yields, from the closure condition for the Lie algebra, the closure condition

$$\{G_\alpha, G_\beta\} = c_{\alpha\beta}^\gamma G_\gamma + d_{\alpha\beta} \tag{3.A.29}$$

in terms of the Poisson brackets, where the $d_{\alpha\beta}$ are constants. These constants are important, because they allow us to interpret the 'fundamental' Poisson bracket relations

$$\{q_i, q_k\} = 0, \qquad \{p_i, p_k\} = 0, \qquad \{q_i, p_k\} = \delta_{ik} \tag{3.A.30}$$

group theoretically: the fundamental bracket relations can be taken as the definition of canonically conjugate variables, and mean that any set of $2f$ canonically conjugate variables (q, p) are infinitesimal generators for a complete set of *commuting* translations in the $2f$-dimensional phase space (it is common to call G_σ the infinitesimal generator of the corresponding transformation). In other words, the quantity $G_i = -q_i$ is the infinitesimal generator of translations along the p_i-axis, and p_i is the generator of translations along the q_i-axis. Since, for a function $F(q, p)$ with no explicit time dependence,

$$dF/dt = \{F, H\}, \tag{3.A.31}$$

it follows that H is the infinitesimal generator of the flow that represents the time development of the dynamical system.

Under an infinitesimal transformation of an r-parameter group, a scalar phase space function $F(q, p)$ changes by the amount

$$dF = \delta a^\sigma X_\sigma F = \delta a^\sigma \{F, G_\sigma\}, \tag{3.A.32}$$

so that functions that are invariant under transformations generated by G_σ satisfy the partial differential equation $X_\sigma F = \{F, G_\sigma\} = 0$. It follows from this, combined with the time evolution law $dF/dt = \{F, H\}$, that a conserved quantity leaves the Hamiltonian invariant and therefore generates a continuous symmetry operation on the dynamical system. From this follows the deep connection between symmetry, invariance, and conservation laws (see also discussions of Noether's theorem in the literature) that holds both in classical and in quantum mechanics.

As an example, consider the case where a Hamiltonian H is invariant under rotations in the xy-plane. Here, the angular momentum $L_z = xp_y - yp_x$ is the infinitesimal generator of rotations about the z-axis, and rotational invariance, which means that L_z is constant during the motion, yields the first-order partial differential equation

$$\{H, L_z\} = x \frac{\partial H}{\partial y} - y \frac{\partial H}{\partial x} + p_x \frac{\partial H}{\partial p_y} - p_y \frac{\partial H}{\partial p_x} = 0. \tag{3.A.33}$$

This means (if we integrate (3.A.33) by using the method of characteristics – see Duff (1962), for example) that H can depend upon the four variables (x, y, p_x, p_y) only in the form $H(p_x^2 + p_y^2, x^2 + y^2, xp_x + yp_y, xp_y - yp_x)$, that is, upon the quantities that are left invariant under a rotation about the z-axis.

Appendix 3.B Systems in involution

We now show that when f conserved quantities in involution (f functionally indendent mutually compatible constants of the motion) can be found, then the $2f$-dimensional system is integrated in the sense that the $2f$ canonically conjugate coordinates and momenta have their time dependence fixed, once and for all, by the conserved quantities. Before explaining how knowledge of only f quantities fixes $2f$ quantities, we first prove two results that will be used in the proof of the main result. The results require a brief excursion into the subject of Pfaffian differential equations.

Consider the differential form

$$\theta_d = \sum_1^n X_i \, dx_i, \tag{3.B.1}$$

where the coefficients X_i depend upon the n variables x_i, and also the form

$$\theta_\delta = \sum_1^n X_i \, \delta x_i, \tag{3.B.2}$$

where δx_i is any other variation in the x_i (as, for example, in a variational principle). To lowest order in $dx_i \, \delta x_j$ we obtain

$$\delta\theta_d - d\theta_\delta = \sum_{i,j=1}^n \left(\frac{\partial X_i}{\partial x_j} - \frac{\partial X_j}{\partial x_i} \right) dx_i \, \delta x_j. \tag{3.B.3}$$

In a transformation to a new set of variables where

$$\sum_1^n X_i \, \delta x_i = \sum_1^n Y_i \, dy_i, \tag{3.B.4}$$

the bilinear forms are also invariant:

$$\sum_1^n \left(\frac{\partial X_i}{\partial x_j} - \frac{\partial X_j}{\partial x_i} \right) dx_i \, \delta x_j = \sum_1^n \left(\frac{\partial Y_i}{\partial y_j} - \frac{\partial Y_j}{\partial y_i} \right) dy_i \, \delta y_j. \tag{3.B.5}$$

This completes the first part of the first lemma, namely that the linear and bilinear forms are invariantly connected with each other. Furthermore if we consider the n conditions whereby, for $j = 1, 2, \ldots, n$,

$$\sum_1^n \left(\frac{\partial X_i}{\partial x_j} - \frac{\partial X_j}{\partial x_i} \right) dx_i = 0, \tag{3.B.6}$$

then these n equations are (by Cramer's rule) mutually compatible because the $n\text{-}\times\text{-}n$ determinant of the coefficients of the dx_i vanishes. These n equations are called Pfaff's first system of equations, and are invariantly connected with the original differential form. That is the content of the rest of the lemma. In particular, if we take as independent variables the $n = 2f + 1$, q_i, p_i, and t, then Hamilton's equations

$$dq_i = \frac{\partial H}{\partial p_i} \, dt, \qquad dp_i = -\frac{\partial H}{\partial q_i} \, dt \tag{3.B.7}$$

are Pfaff's first system of equations for the form

$$L \, dt = \sum_1^f p_i \, dq_i - H \, dt, \tag{3.B.8}$$

which, as we know, is also the basis for setting up global canonical transformations (Appendix 3.A). This, along with the invariance of the bilinear form under coordinate changes, will be used at the very end of the proof of the main result.

Next, we prove a second lemma that will be used early in the proof of the main theorem: consider the involution system $\{\overline{G_\alpha}, \overline{G_\beta}\} = 0$, where $\alpha, \beta = 1, 2, \ldots, f$. If the conditions $u = 0$ and $v = 0$ follow from the f conditions $\overline{G_\sigma} = 0$, then it follows that $\{u, v\} = 0$. The proof is as follows: for conserved quantities $G_\sigma = a_\sigma$, we can also take as generators $\overline{G_\sigma} = G_\sigma - a_\sigma = 0$. If $u = 0$ is a consequence of the vanishing of all of the $\overline{G_\sigma}$, then the condition $u = 0$ is invariant under all infinitesimal transformations generated by the $\overline{G_\sigma}$, which means that $\{u, \overline{G_\sigma}\} = 0$ for $\sigma = 1, 2, \ldots, f$. We can turn this around: all of the $\overline{G_\sigma}$ are also invariant under the transformation that is generated by u. If $v = 0$ is another consequence of the f conditions $\overline{G_\sigma} = 0$, then it follows that $v = 0$ is also invariant under the transformation generated by u, which means that $\{u, v\} = 0$. In particular, if there are f conditions $u_i = 0$ that are consequences of the f conditions $\overline{G_\sigma} = 0$, where the $\overline{G_\sigma}$ are in involution, then the u_i are *also* in involution: $\{u_i, u_k\} = 0$ for $i, k = 1, 2, \ldots, f$. This completes the proof of the second lemma. We come now to the proof of the main theorem, which is due to Liouville.

Consider a $2f$-dimensional Hamiltonian system with f functionally independent, mutually compatible conserved quantities

$$G_\sigma = a_\sigma : \{G_\alpha, G_\beta\} = 0.$$

Knowledge of these f quantities is sufficient to fix the time-dependence of the $2f$ variables $q_1, \ldots, q_f; p_1, \ldots, p_f$. The way that this works is that the form

$$\sum_1^f p_i \, dq_i - H \, dt = dV \tag{3.B.9}$$

is exact, and from $V(q_1, \ldots, q_f, a_1, \ldots, a_f, t)$ follow the remaining f necessary constants of the motion

$$b_\sigma = \left(\frac{\partial V}{\partial a_\sigma}\right). \tag{3.B.10}$$

In order to prove the theorem, one begins with f conserved quantities G_σ that are in involution: $\{G_\alpha, G_\beta\} = 0$. These f equations $G_\sigma = a_\sigma$ must then be solved for the f canonical momenta $p_i = f_i(q_1, \ldots, q_f, a_1, \ldots, a_f, t)$, where the a_i are the constants of the motion. It follows from the last lemma proven above that the f conditions

$$p_i - f_i(q_1, \ldots, q_f, a_1, \ldots, a_f, t) = 0 \tag{3.B.11}$$

are also in involution, and from this it follows that

$$\{p_k, f_i\} + \{f_k, p_i\} = 0, \tag{3.B.12}$$

from which it follows that

$$\frac{\partial f_i}{\partial q_k} = \frac{\partial f_k}{\partial q_i}. \tag{3.B.13}$$

Next, we observe that

$$\dot{p}_i = \sum_1^f \frac{\partial f_i}{\partial q_k} \cdot \frac{\partial H}{\partial p_k} + \frac{\partial f_i}{\partial t} = \sum_1^f \frac{\partial f_k}{\partial q_i} \cdot \frac{\partial H}{\partial p_k} + \frac{\partial f_i}{\partial t}, \tag{3.B.14}$$

and so

$$\frac{\partial H}{\partial q_i} - \sum_1^f \frac{\partial f_k}{\partial q_i} \cdot \frac{\partial H}{\partial p_k} = \frac{\partial f_i}{\partial t}. \tag{3.B.15}$$

Note that the left-hand side is just the partial derivative of H_1 with respect to q_i, where by H_1 we mean H with p_i replaced by f_i:

$$-\frac{\partial H_1}{\partial q_i} = \frac{\partial f_i}{\partial t}. \tag{3.B.16}$$

This means that when we replace p_i by f_i everywhere in the fundamental differential form, we obtain

$$\sum_1^f p_i \, dq_i - H \, dt = \sum_1^f f_i \, dq_i - H_1 \, dt, \tag{3.B.17}$$

so that the differential form, considered as a function of the $2f + 1$ variables (q_i, α_i, t), is *exact*:

$$dV(q, a, t) = \sum_1^f f_i \, dq_i - H_1 \, dt. \tag{3.B.18}$$

If we next vary the constants a_σ, then we obtain

$$\sum_1^f p_i \, dq_i - H \, dt = -\sum_1^f \frac{\partial V}{\partial a_\sigma} \, dG_\sigma + dV. \tag{3.B.19}$$

By the lemma proven at the beginning of the appendix, Pfaff's first system of equations for the right-hand side of the equation, where the $2f + 1$ independent variables are $(q_1, \ldots, q_f, a_1, \ldots, a_f, t)$, is the transformed version of Hamilton's equations, which are Pfaff's first system for the left-hand side. Pfaff's first system for the form

$$-\sum_1^f \frac{\partial V}{\partial a_\sigma} \, dG_\sigma + dV = \sum_1^f \frac{\partial V}{\partial q_i} \, dq_i + \frac{\partial V}{\partial t} \, dt \tag{3.B.20}$$

is given by the f conditions $\mathrm{d}G_\sigma = 0$ and by the additional f conditions

$$\mathrm{d}\left(\frac{\partial V}{\partial a_\sigma}\right) = 0, \tag{3.B.22}$$

which means that there are now f *additional* constants of the motion given by

$$b_\sigma = \left(\frac{\partial V}{\partial a_\sigma}\right). \tag{3.B.23}$$

Hence, if the $2f$ conditions that a_σ and b_σ are constants can be solved algebraically for the $2f$ variables q_i and p_i, then the dynamics problem will have been completely solved. When this entire procedure can be carried out, then the dynamical system is called completely integrable, or integrable.

From the standpoint of Lie groups, the generators G_σ provide us with f linearly independent 'commuting' vector fields, namely, the f velocity fields of the group flow. Arnol'd (1978) proves rigorously by a group-theoretic argument that the motion of an integrable dynamical system takes place on a smooth f-torus that is embedded in the $2f$-dimensional phase space. We can understand qualitatively that, with f conserved quantities, the motion is confined to a closed f-dimensional surface in phase space. There are at least two topologically distinct possibilities for smooth closed surfaces: an f-dimensional sphere and an f-dimensional torus. The streamline motion on a sphere requires singularities (F. Klein, 'On Riemann's Problem . . .'), whereas our 'stream functions', the G_σ, must be analytic, so that conjugacy to motion on a sphere is ruled out.

We can also understand the restriction to toroidal motion as follows: the operators $U(\theta_1, \ldots, \theta_f) = U_1(\theta_1) \cdots U_f(\theta_f)$, with

$$U_i(\theta_i) = e^{\theta_i G_i}, \tag{3.B.24}$$

form the elements of a commutative group because $U_i U_k = U_k U_i$. A general group element can also be written in the form

$$U(\theta_1, \ldots, \theta_f) = e^{\boldsymbol{\theta} \cdot \mathbf{G}}, \tag{3.B.25}$$

and these operators leave invariant the manifold on which the motion of the dynamical system occurs. It follows from commutativity that $U(\alpha_1, \ldots, \alpha_f)U(\beta_1, \ldots, \beta_f) = U(\alpha_1 + \beta_1, \ldots, \alpha_f + \beta_f)$, so that each subgroup the group parameter θ_i is canonical (combines in additive fashion). Finally, we can see that each subgroup operator U_i has the effect of a rotation, so that θ_i has the character of a rotation angle: consider the action of the ith subgroup on the coordinate of a single particle in the system, say the x-coordinate of the kth particle:

$$x_k(\theta_i) = U_i(\theta_i)x_k. \tag{3.B.26}$$

For all values of θ_i, the coordinate $x_k(\theta_i)$ belongs to the manifold where the motion occurs, but the parameter θ_i is unbounded while the particle's motion, hence the coordinate x_k, is bounded. Therefore, $x_k(\theta_i)$ must be periodic in θ_i. So, the manifold that is left invariant by the commuting operators (3.B.25) is topologically guaranteed to be a torus (the rotations that describe a sphere do not commute).

In order to see an example of how symmetry can be used explicitly to solve an integrable problem completely, we recommend to the reader the group-theoretic solution of the Kepler problem which has been reproduced in Saletan and Cromer (1974), and in Goldstein (1980).

Both of these appendices might have been presented in more elegant style by the use of the modern language of differential forms (see Arnol'd, 1978), but the use of that formulation would likely have put the material beyond the reach of most of the readers of this book and therefore would have defeated the book's main purpose.

4
Fractals and fragmentation in phase space

4.1 Introduction to fractals

Smooth curves, or curves where the slope is at least piecewise well defined, have a definite length. When one measures the length of such a curve to higher and higher precision, by using a sequence of shorter and shorter rulers, for example, then one obtains corrections to the previous length measurement in the form of higher-order decimals, but the leading decimals are not changed if the former length measurement was at all accurate. These curves are the foundation for Euclidean and non-Euclidean geometry and we are familiar with them from standard courses in mathematics and physics. The elliptic orbit of the earth about the sun provides only one of many possible examples of the smooth orbits of integrable dynamics.

In stark contrast with mathematical smoothness, much of nature is made up of collections of fragments whose evolution and form seem at first glance to be beyond the possibility of description in the theory of differential equations, unless 'random noise' is included 'ad hoc' as an external force. A few examples are coastlines, mountain ranges, aggregates of soot particles, the porespace of sandstone, and the eddy cascade in turbulence.

That was, until recently, the prevailing philosophy in physics: that irregularity in nature, such as the development of fluid turbulence (irregular collections of eddies) could be explained only by the ad-hoc addition of random forces to otherwise deterministic differential equations. That randomness should somehow approximate the effect of the ignored 10^{23} degrees of freedom in a model described by nonlinear partial differential equations, or coupled nonlinear ordinary differential equations. Yet, methods based upon the addition of random forces to the

Navier–Stokes partial differential equations, e.g., have failed to shed new light upon the nature and evolution of fluid turbulence. Only in this generation have physicists begun to understand how coupled nonlinear differential equations with smooth coefficients can give rise both to statistics and to fragmented objects purely deterministically, without any need for the ad-hoc addition of 'random noise'. We have also understood that one does not need an Avogadro's number of degrees of freedom: low-dimensional deterministic systems of equations are enough to produce fragmentation with behavior that is 'effectively random'.

In contrast with the smooth curves with which we are very familiar, let us consider an example of fragmentation where more and more accurate measurements yield a sequence of increasingly different numbers for the length of the curve. The coastlines of Norway and Great Britain provide examples: the length that you obtain for the coastline depends strongly upon the length of the ruler that you use to measure it. Suppose, for example, that you could measure the Norgwegian coastline by laying a sequence of N_1 rulers end to end. If you should choose a ruler with length l_1, you would obtain a length $L_1 = N_1 l_1$ for the coastline, while with $l_2 = l_1/2$, the result for $L_2 = N_2 l_2$ will be about 2.4 times larger than L_1. Why? The reason is that Norway's coastline is highly fragmented, and you pick up more and more of this fragmentation in your measurement as you reduce the length of your ruler.

The important thing is that we can organize the fragmentation described above by introducing a scaling law which describes the most coarse features of the fragmentation, namely, that

$$N(l) \cong l^{-D} \tag{4.1}$$

is the number of rulers that would have to be placed end to end should one attempt to measure the coastline with rulers of length l. For Norway, when $4\,\text{km} < l < 370\,\text{km}$, D is about 1.28. When the fragmentation can be described by the scaling law (4.1) with noninteger exponent D, then we have an example of a fractal.

The mathematicians like to describe as fractal the curve or point set obtained in the limit $l \to 0$, and D is called the fractal dimension of the resulting uncountable set. In physics and in computation, the range of l is always finite, as is the smallest-length scale in the sequence, and so we call fractal any collection of fragments obeying (4.1) for some finite range of scales $l_{\text{min}} < l < l_{\text{max}}$, where l_{min} is large compared with an average interatomic spacing and l_{max} is small compared with the extent of the fragmented object (l_{max} should be small compared with the linear extent of Norway).

The interesting thing is not merely to understand that many different fragmented objects obey the scaling law (4.1) with some fractal dimension D for some range of length scales l; it is much more interesting to understand the dynamical origin of the fragmentation, hence the scaling law, from the standpoint of deterministic dynamics. In what follows, we first compute the dimensions of several relatively simple fractals. Then, we study two model dynamical systems (smooth discrete maps) that generate fractals.

4.2 Geometrically selfsimilar fractals

We begin with the definition of the middle-thirds Cantor set (cf. Fig. 4.1), which is the simplest mathematical fractal that one can study. In order to construct the set recursively, we begin with the unit interval and first remove the open middle third. As with all fractals that can be constructed, the fractal is constructed recursively and this is the first iteration in the recursion rule. The end points 0, $\frac{1}{3}$, $\frac{2}{3}$, and 1 belong to the Cantor set which is covered by the two closed intervals $[0, \frac{1}{3}]$ and $[\frac{2}{3}, 1]$, each with length $l = 3^{-1}$. In the second iteration of the recursive construction, the open middle-thirds of the two closed intervals $[0, \frac{1}{3}]$ and $[\frac{2}{3}, 1]$ are removed, so that (in addition to 0, $\frac{1}{3}$, $\frac{2}{3}$, and 1) the points $\frac{1}{9}$, $\frac{2}{9}$, $\frac{7}{9}$, and $\frac{8}{9}$ also belong to the Cantor set, which is now (by definition) covered by the four closed intervals $[0, \frac{1}{9}]$, $[\frac{2}{9}, \frac{2}{3}]$, $[\frac{2}{3}, \frac{7}{9}]$, and $[\frac{8}{9}, 1]$, each with length $l = 3^{-2}$. In the nth iteration of the recursive construction, the Cantor set is covered by $N_n = 2^n$ closed intervals, each with length $l_n = 3^{-n}$, so that combining $N = e^{n \ln 2}$

Fig. 4.1 The first three stages in the recursive construction that leads to the middle-thirds Cantor set. The mathematical Cantor set is itself nonconstructable because it would require the limit of infinitely many such iterations, and always lies beneath the 'bars', which represent closed sets.

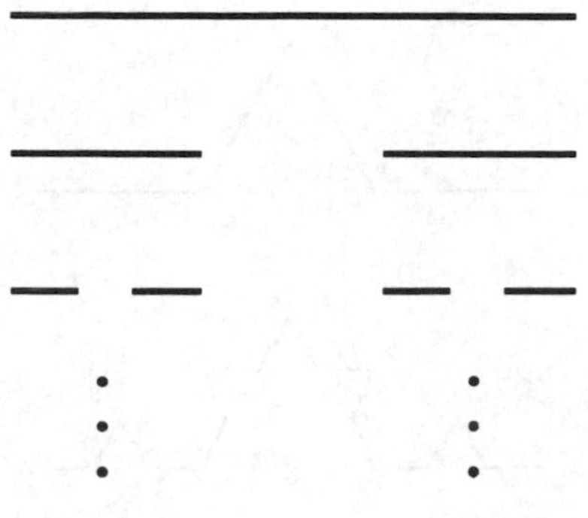

with $l_n = e^{-n\ln 3}$ in equation (4.1) yields

$$D = \frac{\ln 2}{\ln 3} \tag{4.2}$$

for the fractal dimension of the middle-thirds Cantor set. In the nth stage of construction, we have discovered $2^n + 2$ elements of the Cantor set, namely, the end points of all of the closed intervals. In the $n = \infty$ limit this point set is nonintuitive, hence is at first sight 'strange' in the sense it has 2^∞ points, the cardinality of the continuum. Therefore it is uncountable, yet it is 'intermittent' in the sense that there is so much space between the points that it occupies no space whatsoever in the unit interval: its length, or 'measure', is $L_n = l_n^{1-D} \to 0$ as $n \to \infty$, because $D < 1$. But there are other possibilities.

A Cantor set with finite measure and $D = 1$ follows from $N_n = 2^n$ with $l_n = (2 + y_n)^{-n}$, for $y_n > 0$, whereas $N_n = 2^n$ with $l_n = 3^{-n^2}$ yields a Cantor set with $D = 0$. The latter set is highly fragmented, so it is clear that the fractal dimension D alone is not an adequate measure of fragmentation. In fact, certain entirely different fractal objects can have the *same* fractal dimension: the fractal dimension is only one number, and, taken alone, is not enough to completely characterize a fractal.

An idealized model of a fractal coastline, the Koch curve, is made by the Cantor process of pulling up middle thirds to make tents (or roof-tops of houses), as is indicated in Fig. 4.2. There we need $N_n = 2^{2n}$ intervals, each of length $l_n = 3^{-n}$ to cover the Koch curve in the nth generation of its iterative construction, so that $D = 2\ln 2/\ln 3$. In this case, the Koch curve occupies no area but because $1 < D < 2$ it has infinite perimeter.

Fig. 4.2   The first three stages of recursive construction of the Koch curve.

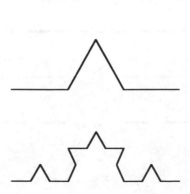

The Sierpinski carpet (see Fig. 4.3) is made by pulling out middle-ninth squares, so that the first generation consists of 2^3 squares, each with length $l = 3^{-1}$ and in the nth generation the Sierpinski carpet is covered by $N_n = 2^{3n}$ squares, each with side $l_n = 3^{-n}$ yielding the fractal dimension $D = 3 \ln 2/\ln 3 < 2$. Hence, the carpet occupies no area in the limit $n \to \infty$, but has infinite perimeter. We turn now to the study of dynamical systems that produce fractals by the fragmentation of phase space. The interesting thing is that we can get both statistical behavior *and* fragmentation from determinisic dynamics. The book by Barnsley (1988) provides the most detailed available text on this subject.

The fractals considered so far, and in the rest of this chapter, are constructed by iterating a single length scale ($l = \frac{1}{3}$, or more, generally, $l = 1/a$ with $a > 2$), so that the resulting fractals are geometrically selfsimilar: if one magnifies any infinite subset by an appropriate rescaling factor (by a^n, for an nth generation subset) then the result is a carbon

Fig. 4.3(a, b and c)   The first three stages of recursive construction of the uniform Sierpinski carpet. The carpet lies outside all of the black squares, which represent open sets.

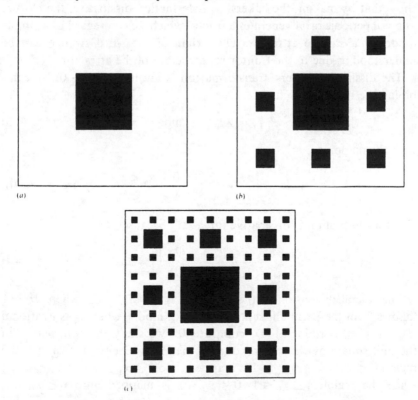

(a) $\qquad\qquad\qquad\qquad\qquad\qquad$ (b)

(c)

copy of the entire fractal. For fractals in nature – the Norwegian coastline is an example – there is only statistical selfsimilarity; strict geometric selfsimilarity does not hold. That is because real fractals are nonuniform: magnified subsets look more or less, but not exactly, like the whole coastline. In general, nonuniform Cantor sets can be constructed by an iterative procedure that is based upon at least *two* distinct length scales (see Chapter 7 for details).

4.3 The dissipative bakers' transformation: a model 'strange' attractor

The attractor for the Bernoulli shift with an irrational initial condition x_0 is the unit interval, with fractal dimension equal to 1. Any dynamical system with a well-defined continuous invariant density $p(x)$ has an attractor whose fractal dimension is unity. The reason why is that it can be covered by $N_n = e^{n\lambda}$ intervals, each with length $l = e^{-n\lambda}$, so that $N_n = 1^{-D}$ yields $D = 1$. The attractor for the bakers' transformation (Chapter 3) is the unit square, with fractal dimension equal to 2. We now show that by making the bakers' transformation dissipative, the closure of a nonperiodic orbit generates a fractal which we can regard as a simple model of a strange attractor. The origin of the word strange can be understood as due to the Cantor set structure of the attractor.

The dissipative bakers' transformation is given (with $a > 0$) by combining the Bernoulli shift

$$x_{n+1} = 2x_n, \qquad \text{mod } 1, \tag{4.3}$$

with the mapping

$$y_{n+1} = \begin{cases} ay_n, & 0 \le x_n < \tfrac{1}{2}, \\ ay_n + \tfrac{1}{2}, & \tfrac{1}{2} \le x_n \le 1. \end{cases} \tag{4.3b}$$

The transformation is dissipative for $a < \tfrac{1}{2}$, because

$$J = \frac{\partial(x_{n+1}, y_{n+1})}{\partial(x_n, y_n)} = \begin{vmatrix} 2 & 0 \\ 0 & a \end{vmatrix} = 2a. \tag{4.4}$$

The fractal dimension of the attractor is $D = D_x + D_y$, where $D_x = 1$ follows from the Bernoulli shift $x_{n+1} = 2x_n$, mod 1, when x_0 is irrational ($D_x = 0$ if x_0 is rational). To compute D_y, we study the contraction of the unit square by the bakers' transformation whenever $0 < 2a < 1$: the region $0 \le x_0 \le \tfrac{1}{2}, 0 \le y_0 \le 1$ is mapped onto $0 \le x_1 \le 1, 0 \le y_1 \le a < \tfrac{1}{2}$ while the region $\tfrac{1}{2} \le x_0 \le 1,\ 0 \le y_0 \le 1$ is mapped onto $0 \le x_0 \le 1$,

$\frac{1}{2} \le y_1 \le \frac{1}{2} + a$. Note that the total area shrinks by a factor $J = 2a$ per iteration, and the first few iterations of the unit square are illustrated in Fig. 4.4. In the first iteration, two rectangles of side $l_1 = a$ are generated and in the nth iteration the unit square is contracted onto $N_n = 2^n$ rectangles of side $l_n = a^n$, so that the y-component of the fractal dimension of the attractor is

$$D_y = \ln 2/|\ln a|, \qquad (4.5)$$

which is less than unity if $a < \frac{1}{2}$. Except for a measure zero point set (initial data that yield unstable periodic orbits), any initial condition in the unit square yields a nonperiodic orbit whose closure is a subset of our fractal attractor.

The map (4.3) has two Liapunov exponents, which can be obtained as follows: with

$$X_n = \begin{pmatrix} x_n \\ y_n \end{pmatrix}, \qquad (4.6)$$

we obtain for the error propagation equation

$$\delta X_{n+1} = \begin{pmatrix} 2 & 0 \\ 0 & a \end{pmatrix} \delta X_n = A \, \delta X_n \qquad (4.7)$$

so that the errors in x_n are stretched by a factor of 2, while errors in y_n are contracted by a factor of $a < \frac{1}{2}$. Furthermore, we can iterate (4.7) to obtain

$$\delta X_n = \begin{pmatrix} 2^n & 0 \\ 0 & a^n \end{pmatrix} \delta X_0, \qquad (4.8)$$

so that the Liapunov exponents are $\lambda_x = \ln 2$, $\lambda_y = \ln a < 0$ and are the reason for the filamentation.

Fig. 4.4 Filamentation due to the dissipative bakers' transformation: after infinitely many iterations of a finite blob of dough, there is none left to bake (in the third figure, representing the result of many iterations, the fractal set is beyond the resolution of the human eye).

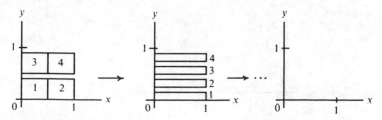

The chaotic orbits in the Lorenz model can be understood in terms of an attractor that has a certain fractal dimension, a 'strange attractor'. Unlike the above oversimplified model, but, like most fractals that occur in nature, the Lorenz attractor is nonuniform and cannot be characterized adequately by a single fractal dimension. Simple models of nonuniform fractals and their characterization are treated in Chapter 7.

4.4 The symmetric tent map: a model 'strange' repeller

Consider the symmetric tent map

$$f(x, a) = \begin{cases} ax, & x < \frac{1}{2} \\ a(1 - x), & x > \frac{1}{2} \end{cases} \tag{4.9}$$

for the case where $a > 2$ (Fig. 4.5). In this case, the unit interval is divided dynamically into an 'escape set', which occupies all of the space, and a fractal strange repeller, which is a uniform Cantor set (Fig. 4.6). With $a > 2$, initial conditions x_0 falling into the range $(a^{-1}, 1 - a^{-1})$ iterate to infinity as $n \to \infty$. This is because, with $x_0 \in (a^{-1}, 1 - a^{-1})$, $x_1 > 1$. Any

Fig. 4.5   A symmetric tent map that peaks above unity, hence generates bounded motion on a repelling, uniform Cantor set.

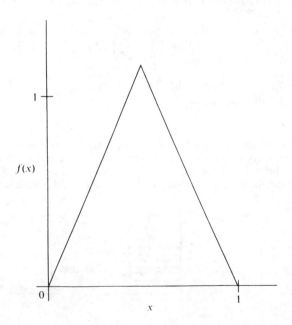

time that we obtain an iterate that exceeds unity, the succeeding $|x_n| \to \infty$ as $n \to \infty$. At the same time, the points left invariant by the tent map (4.7) lie beneath the two closed intervals $[0, a^{-1}]$, $[1 - a^{-1}, 1]$, each of length $l_1 = a^{-1}$. Hence, in one iteration, phase space is fragmented into two equal intervals. If we ask for the set of initial conditions yielding $x_2 > 1$, we obtain all points in the open intervals $(a^{-2}, a^{-1}(1 - a^{-1}))$ and $(1 - a^{-1} + a^{-2}, 1 - a^{-2})$, so that the invariant set (the strange repeller) must fall beneath the four closed intervals $[0, a^{-2}]$, $[a^{-1}(1 - a^{-1}), a^{-1}]$, $[1 - a^{1}, 1 - a^{-1} + a^{-2}]$ and $[1 - a^{-2}, 1]$, each with length $l_2 = a^{-2}$. The pattern is by now clear: the action of the Liapunov exponent $\lambda = \ln a > \ln 2$ generates a natural coarsegraining of the phase-space into $N_n = 2^n$ intervals, each of length $l_n = a^{-n} = e^{-n\ln a} = e^{-n\lambda}$ after n iterations of the map (4.9), and the invariant set is a uniform Cantor set whose fractal dimension is

$$D = \ln 2/\ln a, \tag{4.10}$$

which is less than unity because $a > 2$. $a = 3$ yields the best known uniform Cantor set, the middle-thirds Cantor set.

Because an attempt to define a pointwise invariant density would lead to an uncountable set of Dirac delta functions concentrated on the Cantor set, the pointwise invariant density is not a useful concept on a fractal. Coarsegrained stationary probabilities P_i and coarsegrained densities $\rho_i = P_i/l_i$ *are* useful, where $P_i = n_i/n$ is the fraction of iterates of the map falling into the ith of the 2^n bins after n iterations. One can ask, e.g., for the set of initial conditions $\{x_0\}$ that yields a more or less uniform

Fig. 4.6   The first few stages of construction of the uniform Cantor set that is the invariant set of the tent map in Fig. 4.5.

coarsegrained invariant distribution $\{P_i\}$ for large n. This question is answered in part in the next and last sections of this chapter. The main point is that we can use the map to generate the visitation frequencies P_i that can be interpreted as *empirical* probabilities for visiting different parts of the fractal.

The Cantor sets generated by tent maps are called strange repellers because points where $x_n > 1$ escape to minus infinity as $n \to \infty$, i.e., they are repelled by the Cantor set, which is the invariant set of the map f. It is the action of the Liapunov exponent that fragments the phase space into $N_n = 2^n$ pieces of length $l_n = e^{-n\lambda}$ for n iterations of the map. A fractal invariant set is always geometrically selfsimilar whenever the map's Liapunov exponent is constant. By geometrical selfsimilarity we mean that infinite subsets, upon magnification, are carbon copies of the original set.

There are several lessons to be learned from this example. One is that a fractal can be regarded as a hierarchy of generations of length scales that can be organized into a tree. This is important, because the scaling over a finite number of generations is, for the natural scientist or engineer, everything: for scientists and engineers, the $n = \infty$ limit is completely irrelevant. In the present example, the tree is binary, with $N_n = 2^n$ branches in the nth generation (Fig. 4.6), and all of the nth-generation scales $l_n = a^{-n}$ are equal to each other. In general, a fractal is nonuniform, meaning that the scales $\{l_i\}$ that are generated dynamically by the map are not all equal, and can be organized onto a tree with $N_n \simeq \mu^n$ branches in the nth generation where $\mu > 1$. When μ is nonintegral, some branches are missing from the tree, that is, the tree can be incomplete.

Another lesson is that fractals are generated by the action of Liapunov exponents in a dissipative, chaotic dynamical system: in the tent map, $l_n = e^{-n\lambda}$, when $\lambda = \ln 2/\ln a$ is the Liapunov exponent. Hence, the feature that is responsible for deterministic chaos, for generating statistical behavior, is also the cause of the fragmentation of phase space. Fractals can also occur at the transition to chaos (cf. Chapters 5 and 6) where the average Liapunov exponent vanishes.

4.5 The devil's staircase: arithmetic on the Cantor set

It is of interest to introduce the Cantor function and its graph, the devil's staircase, because several examples of the staircase occur in physics. One example appears in the theory of the circle map (Chapter 5) and another occurs in the theory of phase transitions. In what follows, we construct the devil's staircase for the uniform Cantor set with fractal dimension

$D = \ln 2/\ln a$ and restrict to the case where $a \geq 3$ is an integer and is the magnitude of the slope of the symmetric tent map (4.9). In this case, the function that describes the devil's staircase is the Cantor function, which can be understood, as we shall see, as a certain invariant probability distribution for the iterates of the symmetric tent map.

The Cantor set consists of all end points of intervals and limit points of end point sequences. Taken as initial data for the tent map (4.9), the end points iterate to zero in finite discrete time $n = N$ because all end points can be written as finite length a-ary strings,

$$x = \sum_{i=1}^{N} \varepsilon_i/a^i = 0.\varepsilon_1\varepsilon_2\cdots\varepsilon_N, \qquad (4.11)$$

where $\varepsilon_i = 0, 1, 2, \ldots,$ or $a - 1$.

Unstable periodic orbits on the fractal, in contrast, are given by rational elements of the Cantor set with $N = \infty$, whereas nonperiodic orbits correspond to irrational elements (nonperiodic a-ary strings $\varepsilon_1\varepsilon_2\cdots\varepsilon_N\cdots$ of infinite length). Both the periodic and nonperiodic orbits consist of limit points of infinite sequences of end points.

In the first stage of iterative construction of the Cantor set, we discovered that the end points a^{-1} and $1 - a^{-1}$ belong to the Cantor set. For $n = 2$, we discovered in addition the end points a^{-2}, $a^{-1} - a^{-2}$, $1 - a^{-1} + a^{-2}$, and $1 - a^{-2}$. These end-points, and all other end points, are elements of the Cantor set and can be written as finite-length a-ary strings using only two symbols, $\varepsilon_i = 0$ or $a - 1$. For example, $a^{-1} = 0.1 = 0.0\ a - 1\ a - 1\cdots$, while $1 - a^{-2} = 0.a - 1\ a - 1\ 00\cdots$, and so on. All of the elements of the Cantor set can be written using the binary symbols $\varepsilon_i = 0$ or $a - 1$, and all possible combinations of 0's and $a - 1$'s describe all the elements in the set. Hence, the Cantor set is in one-to-one correspondence with all possible binary strings, so that the cardinality of the set is 2^∞, which is the same as the cardinality of the continuum. This tells us that 'almost all' of the points in the Cantor set are irrational numbers.

We construct the Cantor function as follows. We interpret

$$P(x) = 0.\frac{\varepsilon_1}{a-1}\frac{\varepsilon_2}{a-1}\cdots\frac{\varepsilon_N}{a-1}\cdots \qquad (4.12)$$

as a binary number ($\varepsilon_i/(a-1) = 0$ or 1), where $x = 0.\varepsilon_1\varepsilon_2\cdots\varepsilon_N\cdots$ is an element of the Cantor set ($\varepsilon_i = 0$ or $a - 1$). It is easy to convince oneself that if x_1 and x_2 are the end points of a removed open interval, then $P(x_1) = P(x_2)$. For example, with $x_1 = a^{-1} = 0.0\ a - 1\ a - 1\cdots$ we get $P(x_1) = 0.01111 = 0.1 = \frac{1}{2}$. If $x_2 = 1 - a^{-1}$, then $P(x_1) = 0.1 = \frac{1}{2}$.

If $x_1 = 1/a = 0.01 = 0.00\ a-1\ a-1\cdots$, the $P(x_1) = 0.00111\cdots = 0.01\cdots = \frac{1}{4}$, whereas $x_2 = a^{-1} - a^{-2}$ also yields $P(x_2) = 0.01\cdots = \frac{1}{4}$. $x_1 = 1 - a^{-1} + a^{-2}$ yields $P(x_1) = 0.1011\cdots = 0.11\cdots = \frac{3}{4}$, whereas $x_2 = 1 - a^{-2}$ also yields $P(x_1) = 0.11\cdots = \frac{3}{4}$. If we extend $P(x) = $ constant on the closure $[x_1, x_2]$ of every removed open interval, then $P(x)$ has the following properties: P is continuous but nondecreasing and has a slope that vanishes almost everywhere. The first few staircase steps are plotted in Fig. 4.7.

The Cantor function can be understood as a certain probability distribution for the iterates of the tent map on the Cantor set. Consider any nonperiodic orbit of the symmetric tent map that generates, as its distribution of iterates, a uniform probability distribution on the Cantor set. Then we can interpret the Cantor function as follows: $P(x) = \frac{1}{2}$ for $x = a^{-1}$ means that half of the iterates of the map fall into the interval $[0, a^{-1}]$ and half in $[1 - a^{-1}, 1]$. $P(x) = \frac{1}{4}$ for $x = 1/a^{-2}$ means that $\frac{1}{4}$ of the iterates fall into the interval $[0, 1/a^{-2}]$, while $P(x) = \frac{3}{4}$ if $x = a^{-1} + a^{-2}$ means that $\frac{3}{4}$ of the iterates fall into three intervals $[0, a^{-2}]$, $[a^{-1} - a^{-2}, a^{-1}]$, and $[1 - a^{-1}, 1 - a^{-1} + a^{-2}]$. The remaining $\frac{1}{4}$ land in the interval $[1 - a^{-2}, 1]$.

For which *initial conditions* does $P(x)$ *correctly* describe the distribution of iterates of the tent map? We analyze this question in section 4.7. The reader should note that there is no probability density associated with $P(x)$ because, although the function is nondecreasing, its slope is zero almost everywhere in the unit interval. It is still possible to show that the Cantor function satisfies a master equation that is the generalization of

Fig. 4.7 Several steps of the devil's staircase are shown for the tent map of Fig. 4.5 for the parameter value $a = 3$.

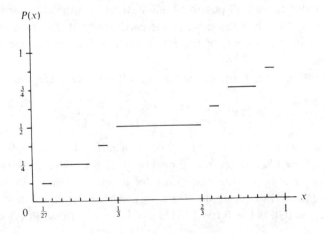

the Frobenius–Peron equation to the case where the probability distribution has no derivative and therefore no density (see Chapter 9).

Having introduced the idea of a hierarchy of coarsegrained descriptions of a fractal, or more generally of a map's invariant set, we turn next to a method for characterizing nonuniform distributions of iterates over the intervals that define a given generation of coarsegraining. It is important in the above discussion that the hierarchy of intervals (coverings of the Cantor set) were not arbitrarily chosen, but were generated directly by the tent map. The simplest way to generate the intervals is to use the tent map to iterate the unit interval backwards n times (Fig. 4.8).

4.6 Generalized dimensions and the coarsegraining of phase space

So far, we have introduced the fractal dimension D and have computed D for several geometrically selfsimilar fractals. The fractals that occur in

Fig. 4.8 Recursive construction of the uniform Cantor set that is the invariant set of the symmetric tent map by backward iteration of the map.

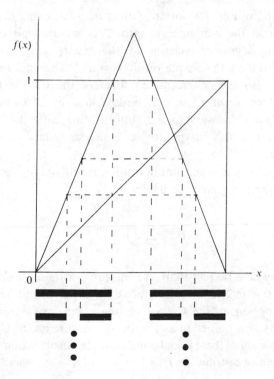

nature and in nontrivial model dynamical systems are nonuniform, and the knowledge of D alone is inadequate to characterize an invariant set with a certain distribution of iterates $\{P_i\}$. In fact, entirely different fractals can have the same fractal dimension. In what follows, we relabel the fractal dimension D as D_0. We have also discussed two examples of uniform probability distributions that are generated by maps: the invariant density of the binary tent map and the Cantor function for symmetric tent maps with integer slope magnitude $a > 2$.

In order to relax the two assumptions of uniformity, we generalize our previous considerations in the following way: we assume that fractal scaling holds statistically, with $N_n \approx l_n^{-D_0}$, but that the nth-generation intervals $l_1^{(n)}, l_2^{(n)}, \ldots, l_N^{(n)}$ are not necessarily all of equal size. The *characteristic size* of the intervals will be denoted by l_n. Also, we do not restrict our considerations to *uniform* probability distributions $\{P_i\}$, where P_i is the frequency with which the map visits the interval $l_i^{(n)}$. In this finite description, the phase space has been coarsegrained into N_n different intervals $\{l_i^{(n)}\}$ and $\{P_i\}$ is then the coarsegrained probability distribution that follows from iterating the map. Also important is that the intervals are determined by the dynamics, by the backward iteration of a one-dimensional map, as is explained in Chapters 7 and 9 and is illustrated in Fig. 4.8 for a symmetric tent map. If, for example, we coarsegrain the unit interval into 2^n equal intervals of size $l_n = 2^n$, then the coarsegrained version of the binary tent map's invariant density is given by the simple result $P_i = 2^{-n}$. The uniform distribution $P_i = 2^{-n}$ is also the coarsegrained form of the Cantor function $P(x)$ on the uniform Cantor set that is described by 2^n equal intervals of the form $l_n = a^{-n}$, where $a > 2$, but in this case the coarsegrained density $\rho_i = P_i/l_i$ has no pointwise limit, i.e., when one allows l_i to approach zero.

In order to take into account nonuniform probability distributions, the following generating function is useful:

$$\chi(q) = \sum_{i=1}^{N_n} P_i^q \simeq l_n^{(q-1)D_q}. \tag{4.13}$$

For the symmetric tent map, the motion occurs on a Cantor set with dimension $D = \ln 2/\ln a$, and all possible initial conditions of the map that yield motion on the Cantor set make up the Cantor set. We can think of the Cantor function as a uniform measure on the Cantor set. For a measure one set of these initial conditions, the distribution of iterates will yield the uniform distribution $P_i = 2^{-n} = l_n^D$ as $n \to \infty$, where $D = \ln 2/\ln a$.

Substituting the uniform distribution into (4.13) yields

$$\chi(q) = N_n \cdot P_i^q = l_n^{(q-1)D},\qquad(4.13b)$$

so that $D_q = D = \ln 2/\ln a$ for all q.

We show next that when we relax the assumption of a uniform probability distribution, then we obtain an entire spectrum of 'generalized dimensions' D_q, a set of scaling exponents that characterizes the *nonuniform* probability distribution as q varies from minus to plus infinity.

First, let $q = 0$. Then

$$\chi(0) = \sum_{i=1}^{N_m} P_i^0 = N_n,\qquad(4.13c)$$

which yields the relation guaranteeing *statistical* selfsimilarity,

$$N_n \sim l_n^{-D_0},\qquad(4.14)$$

where D_0 is what we previously called D, the 'box-counting' definition of the fractal dimension. That is, we should need N_n boxes, each of length l_n, in order to cover the set whose fractal dimension is D_0. As before, D_0 characterizes the fractal and not the probability distribution.

Next, let $q = 1$ in (4.13). Here, we must pass to the limit as follows: with

$$\sum_i P_i = 1,$$

we have

$$\ln \sum_i P_i^q = \ln\left(\sum_i P_i^q - 1 + 1\right) \approx \sum_i P_i^q - 1$$

for $q \sim 1$. Then, writing $q = 1 + \delta q$, we have

$$P_i^q = P_i^{1+\delta q} \simeq P_i\, e^{\delta q \ln P_i} \approx P_i(1 + \delta q \ln P_i - \cdots),$$

so that

$$\sum_i P_i^q - 1 \approx \sum_i P_i + \delta q \sum_i P_i \ln P_i - 1 \approx \delta q \sum_i P_i \ln P_i,$$

and so we obtain

$$D_1 \simeq \frac{\delta q}{\delta q}\sum_i p_i \ln p_i/\ln l_n$$

$$= \sum_{i=1}^{N_m} P_i \ln P_i/\ln l_n\qquad(4.14b)$$

as $\delta q \to 0$. Since

$$S(l) = -\sum_i P_i \ln P_i$$

is the Gibbs entropy in information theory and statistical mechanics,

$$e^{S(l)} \sim l_n^{-D_1} \tag{4.15}$$

defines the information dimension D_1.

For a uniform coarsegraining $P_i = 1/N_n$, we find that

$$S(l) = \sum_{i=1}^{N_n} \frac{1}{N_n} \ln N_n = \ln N_n, \tag{4.15b}$$

which is the Boltzmann entropy. Since $N_n \sim l^{-D_0}$, then $D_0 = D_1$, for a uniform distribution on a uniform invariant set (selfsimilar fractal, e.g.). In general, $D_1 \le D_0$ and $D_q \ge D_0$ for $q < 0$, $D_q \le D_0$ if $q > 0$.

Next, we show that D_2 is the correlation dimension. Define the pair correlation function

$$C(l_n) = \frac{1}{N_n^2} \sum_{i,j}^{N_n} \theta(l_n - |x_i - x_j|), \tag{4.16}$$

where $\theta(x) = 1$ if $x \ge 0$, 0 if $x < 0$. $C(l)$ is then the joint probability to find that any two iterates (x_i, x_j), $i, j = 1, \ldots, n$, have a separation smaller than l_n or, roughly speaking, fall into the same interval of size l_n. For the very special case where the iterates of the map are asymptotically (as $n \to \infty$), statistically independent (see Chapter 9), P_i^2 is the probability that any two iterates fall into the same interval i, so that $\sum_{i=1}^{N_i} P_i^2$ is also the probability that any two iterates are separated by a distance less than l. Hence $\sum_i P_i^2$ is the number of pairs of iterates with separation smaller than l, and so we can write

$$C(l_n) = \sum_{i=1}^{N_n} P_i^2,$$

and

$$D_2 = \lim_{l \to 0} \sum_{i=1}^{N_n} P_i^2 / \ln l_n, \tag{4.17b}$$

yielding

$$C(l) \sim l_n^{D_2} \tag{4.16b}$$

Finally, since $P_i < 1$, $P_{max} \approx l_n^{D_\infty}$ occurs for $q \to \infty$, whereas $P_{min} \approx l_n^{D_{-\infty}}$ for $q \to -\infty$. Hence, D_∞ and $D_{-\infty}$ describe the parts of a nonuniform fractal that are most and least often visited by the dynamical system.

We can also apply our definition of D_q to any continuous probability density by first using the density to construct a set of coarsegrained probabilities,

$$P_i = \int_{x_i}^{x_i + l_i} P(x) \, dx. \tag{4.17}$$

For example, if we use the invariant density

$$P(x) = \pi^{-1}(x(1-x))^{-1/2} \qquad (4.18)$$

of the logistic map at fully developed chaos and then partition the unit interval into $N_n = l^{-1}$ equal intervals, each of length l, obtaining

$$P_i \approx \frac{l}{\pi\sqrt{x_i(1-x_i)}} = a_i l, \qquad (4.17b)$$

where a_i is only slowly varying with x_i. This is valid for the $N_n - 2$ intervals that do not include the end points $x = 0$ and 1. For the end-point intervals we have $P_0 = P_{N_n}$, where

$$P_0 \simeq \frac{1}{\pi} \int_0^1 \frac{dx}{x^{1/2}} \cong \frac{2}{\pi}\sqrt{l}. \qquad (4.17c)$$

Our generating function therefore takes the following approximate form:

$$\chi(q) \cong C_1 l^{q/2} + N_n C_2 l^q$$
$$\simeq C_1 l^{q/2} + C_2 l^{q-1}, \qquad (4.19)$$

where C_1 and C_2 are both approximately constant.

In order to evaluate D_q in $\chi(q) \approx l^{(q-1)D_q}$, we ask for the dominant term in (4.19), and this yields for the generalized dimensions the result

$$D_q = \begin{cases} 1, & q < 2, \\ \dfrac{q/2}{q-1}, & q > 2. \end{cases} \qquad (4.20)$$

The interpretation of this result is as follows: notice that we can rewrite (4.19) in the suggestive (approximate) form

$$\chi(q) \approx 2P_{\max}^q + P_{\min}^q \cdot N_n \qquad (4.19c)$$

where $P_{\max}^q = l_n^{qD_\infty} \simeq C_1 l^{q/2}$ and $P_{\min}^q = l_n^{qD_{-\infty}} \simeq C_1 l^q$. Therefore, $D_\infty = \frac{1}{2}$ describes the most probable regions (the end-point intervals), while $D_{-\infty} = 1$ describes the region where the map spends the least time (the rest of the unit interval). The exponent $D_\infty = \frac{1}{2}$ represents the square-root singularity $P(x) \sim x^{-1/2}$ in the invariant density, cf. (4.17c), while the result $D_{-\infty} = D_0 = D_1 = 1$ reflects the fact that the rest of the phase space has finite length. Here, we have not used the intervals that are generated by backward iteration of the tent map, but in this case the negligence does not matter, because we concentrate only upon the end point singularities in $p(x)$.

The symmetric tent map with $a > 2$ generates a uniform fractal, because the iteration of a *single* length scale $l_1 = a^{-1}$ describes the coarsegraining of phase space. In this case, $D_q = D_0 = \ln 2/\ln a$, for all q, whenever initial conditions yielding the even probability distribution that is described by the Cantor function are chosen. When initial conditions are chosen that yield an uneven distribution of iterates, then a nontrivial D_q spectrum arises. The further development of the theory along these lines is continued in Chapter 7, where we discuss nonuniform fractals and nonuniform coarsegrained invariant probability distributions. The generation of non-uniform probability distributions by different classes of initial conditions of maps is the subject of Chapter 9. The main idea is that a nonuniform probability distribution on a fractal or nonfractal invariant set generates a D_q-spectrum. In order to illustrate this, assume without proof that it is possible to choose an initial condition for the symmetric tent map (slope a) so that the following distribution of iterates occurs: the distribution shows statistical independence, so that P_1 and P_2 are the probabilities for the two first-generation intervals, each with length $l_1 = a^{-1}$, to be visited and the probability to visit any of the 2^n nth-generation intervals $l_n = a^{-n}$ is $P_1^m P_2^{n-m}$, where $m = 0, 1, 2, \ldots, n$. In this case the generating function is binomial,

$$\chi(q) = (P_1^q + P_2^q)^n = a^{-n(q-1)D_q},\qquad(4.24)$$

and it follows that

$$(q-1)D_q = -(\ln(P_1^q + P_2^q))/\ln a.\qquad(4.22)$$

If we take $P_1 > P_2$, then $D_{-\infty} = -\ln P_2/\ln a$, $D_\infty = -\ln P_1/\ln a$, $D_0 = \ln 2/\ln a$, and the D_q spectrum is shown for $P_1 = \frac{3}{10} = 1 - P_2$ and $a = 2$ in Fig. 4.9. Here, the coarsegrained density $\rho_i = P_i/l_i$ is singular (has no pointwise limit in the $l_i = 0$ limit) unless $a = 2$ (nonfractal repeller) and an initial condition is chosen for the map so that $P_1/P_2 = 1$, in which case we recover $\rho_i = 1$ (invariant density of the binary tent map) and $D_q = 1$ for all values of q.

4.7 Computation of chaotic orbits on a fractal

We showed in section 2.6 how to compute, in finite precision, chaotic orbits that are nonperiodic over the unit interval. In section 3.6 we showed how to compute finite-precision chaotic orbits that are nonperiodic on the 2-torus. Here, we show how to use finite-precision arithmetic to compute uniform approximations to nonperiodic orbits on a repelling fractal. The orbits that we compute next for finite n have a closure, in the infinite-precision infinite-time limit that is a subset of the fractal. Just why one generally gets a subset and not necessarily the entire fractal is a question of choice of initial conditions: not every nonperiodic orbit visits the *entire* fractal as n goes to infinity, and those that do do not all visit different regions with equal frequency. Depending upon initial conditions, some parts of fractal are visited more frequently than other parts, producing *nonuniform* coarsegrained distributions $\{P_i\}$. This is not peculiar to motion on a fractal: the same is true for every other chaotic map on its own invariant set. Whether the map is symmetric or asymmetric turns out to be completely irrelevant, although the symmetry of the map can restrict the range of different Liapunov exponents that can occur for different initial conditions (Chapters 7 and 8).

Consider the symmetric tent map

$$x_{n+1} = \begin{cases} ax_n, & x_n < \frac{1}{2}, \\ a(1 - x_n), & x_n > \frac{1}{2}, \end{cases} \tag{4.9}$$

with $a > 2$ an integer. This map generates a repelling fractal, so that it is necessary to choose an initial condition that lies on the fractal in order to stay on the fractal for all n. If you choose an initial condition that is near a point on the fractal, then the map will stay near the fractal for a definite number of iterations until $x_n > 1$, for finite n, and then $x_n \to -\infty$ as $n \to \infty$.

Fig. 4.9 Spectrum of generalized dimensions for the binary tent map for a 'measure zero' class of initial conditions that generates a certain statistically independent distribution of iterates of the map, with $P_1 = \frac{3}{10} = 1 - P_2$ (see also Chapters 9 and 10).

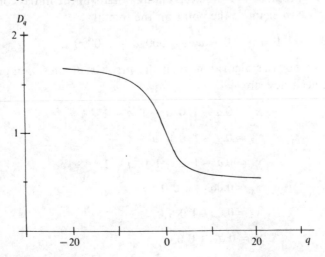

The map generates a uniform (geometrically selfsimilar) Cantor set with fractal dimension $D_0 = \ln 2/\ln a$, if $a > 2$. End points of the Cantor set (used as x_0 in (4.9)) yield $x_n = 0$ for some finite n, rational accumulation points of end points yield unstable periodic orbits on the fractal. The irrational accumulation points ('almost all' of the fractal, if we take the Cantor function as a uniform 'measure' on the Cantor set) yield unstable nonperiodic orbits. We want to show next how to compute uniform approximations to the nonperiodic orbits by using the parallel-processing method that we introduced in Chapter 2.

For the symmetric tent map, it is easy to invent algorithms for irrational numbers that belong to the Cantor set: in base-a arithmetic, any nonperiodic string made up of 0's and $a - 1$'s will do, so we are back to all possible binary strings. The underlying reason for this is that the coarsegraining of the fractal can be organized onto a *complete* binary tree: by backward iteration of the tent map, the unit interval $\to 2$ intervals of length $a^{-1} \to 4$ intervals of length $a^{-2} \to 8$ intervals of length $a^{-3} \to \cdots$ $\to 2^n$ intervals of length $a^{-n} \to \cdots$. So, given any binary string that is nonperiodic, e.g., $101001000100001\cdots$, we need only replace the symbol 1 by $a - 1$ and we are in business. Because we can write $1.000\cdots = 0.bbb\cdots$ (with $b = a - 1$) in base-a arithmetic, it is easy to write down an automaton for the tent map with slope a:

$$\varepsilon_i(n + 1) = \begin{cases} \varepsilon_{i+1}(n), & \varepsilon_1(n) = 0, \\ b - \varepsilon_{i+1}(n), & \varepsilon_1(n) = b. \end{cases} \tag{4.23}$$

By using this automaton, we can iterate the string for any initial condition forward to reconstruct the higher iterates of the map to any desired finite precision N, provided that we keep enough digits in the initial condition. If we take as an example the point on the fractal

$$x_0 = 0.a - 1\ 0\ a - 1\ 0\ 0\ a - 1\ 000\ a - 1\ 0000\ a - 1 \cdots, \tag{4.23b}$$

given by an obvious algorithm, then the first six iterates, computed to four-digit accuracy, are

$$x_1 = 0.a - 1\ 0\ a - 1\ a - 1\ \cdots$$

$$x_2 = 0.a - 1\ 0\ 0\ a - 1\ \cdots$$

$$x_3 = 0.a - 1\ a - 1\ 0\ a - 1\ \cdots$$

$$x_4 = 0.00a - 1\ 0\ 0\ \ \cdots$$

$$x_5 = 0.0a - 1\ 0\ 0\ 0\ \ \cdots$$

$$x_6 = 0.a - 1\ 0\ 0\ 0\ \ \cdots$$

$$\tag{4.24}$$

This four-digit orbit is nonperiodic even as $n \to \infty$ because the binary string in (4.23b) is nonperiodic, but it does not visit all regions of the fractal, even as n and N go to infinity. In order to compute x_6 to 4-digit accuracy, it was necessary to compute x_0 to 10-digit accuracy (with $\lambda = \ln a$, the information flow is one digit per iteration in base-a arithmetic). So, we have an example of a *uniform* approximation to an *exact* chaotic orbit on a repelling fractal. It makes good sense to refer to these finite-precision descriptions of exact chaotic orbits as 'coarsegrained chaotic orbits'.

Uniform approximations to true chaotic orbits can be calculated by doing careful arithmetic. The limitation on the ability to predict the future from the past is one of computation time, given an exact initial condition in the form of an algorithm that generates as many digits as you need. In observations of nature, we can never know initial conditions exactly and because systems that are observed in nature are never ideally closed, a perturbed orbit may not remain on a repelling fractal, but instead may remain near it and then leave it in finite time. Whether that 'finite time' is short or long on observation time scales is a question that cannot be answered without studying a particular dynamical system. Even then, it reduces to one of initial conditions, namely, how near to the repeller is the starting point, and how strong the perturbation. The nearer the starting point to the repelling invariant set (choose the first N digits of the starting point to coincide with the first N digits of a point on the invariant set), the longer the motion stays near the repeller, for small enough perturbations. This leads to discussions of the 'escape rate' for strange repellers and also to the necessity to be aware that observed orbits of natural dynamical systems that superficially appear to take place on an attractor may occur instead on or near a repeller, but with an escape time so large that, without further information about the system, one cannot tell the difference. The mere extraction of a bounded, finite time series from a deterministic chaotic dynamical system is inadequate to decide the question whether one has observed motion near an attractor or a repeller.

Finally, we encourage the reader to extend (4.24) to $n > 6$, and then to $N > 4$ and to compute examples of the coarsegrained probabilities P_i introduced in section 4.5. From these coarsegrained probabilities, the generalized dimensions D_q can be computed: each initial condition determines a hierarchy of coarsegrained invariant distributions, from which follows a particular D_q-spectrum. Different D_q-spectra are a way to characterize different probability distributions.

We have pointed out in section 4.5 that the Cantor function is a uniform measure on the Cantor set: as a probability distribution, it corresponds to initial conditions of the map whose iterates are distributed uniformly as the number of iterations of the map (in a gedanken computation) goes to infinity. In the language of section 4.5 we are talking about the even distribution $P_i = N_n^{-1} = 2^{-n}$ for every coarsegraining of the phase space into $N_n = 2^n$ intervals, each of width $l_n = a^{-n}$ (we use the symmetric tent map (4.9) here for the purpose of clarity). There are two important questions to be asked and answered: (i) which initial conditions yield the equal frequency distribution and (ii) how can one generate examples of those initial conditions?

It is easy to understand that, if we form the 'symbol sequence' $\varepsilon_1(0)\varepsilon_1(1)\varepsilon_1(2)\cdots\varepsilon_1(n)$ that is made up of the first digit of each iterate x_n of the tent map, then equal frequencies $P_1 = P_2 = \frac{1}{2}$ in the first stage $(l_1 = a^{-1})$ of coarsegraining of the fractal corresponds to having the same number of symbols $\varepsilon_1(n) = 0$ as $\varepsilon_1(n) = a - 1$ as $n \to \infty$. In this description of the dynamics, the tent map hops deterministically to the interval $[1, 1/a]$ when $\varepsilon_1(n) = 0$ and to $[1 - 1/a, 1]$ when $\varepsilon_1(n) = a - 1$.

In the next generation of coarsegraining, we must know the first two digits in each iteration of the map: $x_0 = 0.\varepsilon_1(0)\varepsilon_2(0), \ldots, \ldots, x_n = 0.\varepsilon_1(n)\varepsilon_2(n)\cdots$. The possibilities are (with $a - 1 = b$) $\varepsilon_1\varepsilon_2 = 00, 0b, bb$, or $b0$. If these four possibilities occur equally often as $n \to \infty$, then the second-generation probabilities are $P_1 = P_2 = P_3 = P_4 = \frac{1}{4}$ and we again have an even distribution of iterates over the 4 coarsegrained intervals $l_n = a^{-2}$ that cover the fractal. In this coarsegrained description of the motion, the map hops back and forth among the four intervals of size $1/a^2$.

By now, you should get the idea: the way that blocks of digits are distributed in the map's iterates tells you how often the map visits a given interval and therefore fixes the probabilities $\{P_i\}$. The distribution of blocks of digits in the map's iterates, in turn, is determined by the distribution of digits in the initial condition. We shall show in Chapter 9 that this is no less true for general nonlinear maps than it is for piecewise linear maps like the tent map.

How can one construct different initial conditions that yield the equal frequency distribution $P_i = N_n^{-1} = 2^{-n}$ for all generations of coarse-graining? Put differently, what are the initial conditions that generate the Cantor function as the tent map's probability distribution? In the crudest description of the dynamics on the fractal, there are two intervals $[0, 1/a]$ and $[1 - 1/a, 1]$ of length $l_1 = 1/a$, and we can assign to them the 'addresses' $\varepsilon_1 = 0$ and b respectively. In the next refinement of description, there are four intervals with widths $l_2 = 1/a^2$, and each can be assigned an address: $00, 0b, b0$ or bb because, for example, when $x_n = 0.bb\cdots$ then the map visits the interval $[1 - 1/a^2, 1]$. Similarly, each of the 2^n intervals

$l_n = a^{-n}$ in the nth generation can be assigned an n-digit address where the address is the same as the first n bits of the iterate whenever the map visits that particular interval. This information can all be arranged onto a complete binary tree, as is shown in Fig. 8.2. That the tree is 'complete' is the same as saying that, as $n \to \infty$, all possible binary strings occur. Therefore, the binary address labels are in one-to-one correspondence with all possible numbers in the unit interval, written as binary strings. Borel showed that 'almost all' numbers, written in binary (or in any other integral base of arithmetic) are what Lebesgue called 'normal'. A 'normal' binary string is one where 0 occurs as often as 1; 00, 01, 11, and 11 occur with equal frequency ($P_i = \frac{1}{4}$), and so on. There are no normal periodic strings – all normal strings are necessarily nonperiodic.[1] We use our rules derived in Chapter 2 for the binary tent map whereby the first N digits in x_n are determined by a corresponding $N + 1$ digit string in x_0. The reason for this is that, with the automaton (4.23) and the restriction to only two digits 0 and $b = a - 1$, we are again doing binary arithmetic on binary strings, exactly as with the binary tent map. Therefore, we can conclude that when 0 and b occur with equal frequency as the first bit of x_n, then 0 and b must occur with frequency $\frac{1}{2}$ in the initial condition, and 00, 0b, b0, and bb must each occur with frequency $\frac{1}{4}$. For 00, 0b, b0, and bb to occur with frequency $\frac{1}{4}$ as the first two bits of x_n requires that each of the eight different combinations of m 0's and 3-m b's, where $m = 0, 1, 2$, and 3, must each occur with the same frequency $\frac{1}{8}$ in the initial condition. It is clear by the continuation of this argument that the string that represents the initial condition must be a normal binary string. This means that if we think of the fractal as a collection of initial conditions for the tent map, then almost all initial conditions that can be defined would, *if they could be computed*, yield iterates that are distributed uniformly over any coarsegraining of the fractal. That is because normal strings occur with 'measure one' and (as we shall see later) have absolutely nothing to do with the symmetry of the tent map. These are the initial conditions that have the Cantor function as their probability distribution. In this case, $D_q = D_0 = \ln 2 / \ln a$ for all q. Other initial conditions yield nontrivial D_q-spectra. These other initial conditions have 'measure zero' (as measured by the Cantor function) on the fractal, but that does not mean that they can be ignored in computation or in applications to the description of nature: one must be careful in the application of the mathematics of the infinite to nature, for nature is discrete and is not built up either on the

---

[1] All normal numbers are irrational.

basis of the continuum or from the corresponding 'measure-one' conditions that follow from the mathematics of the continuum.

How can we find an example of an initial condition that generates the equal frequency distribution, a normal binary string? It is impossible to obtain such a number by the so-called mental act of a 'random draw' of numbers; instead, an algorithm is necessary. Here is an example of an algorithm that generates a normal string (it was provided in 1933 by an English graduate student, who also did not use the concept of a 'random draw' in its construction): write down the natural numbers in sequence, in any base of arithmetic (cf. Niven). In base-10, the algorithm yields $1234567891011\cdots$. In binary, the corresponding sequence is $110111001011101111000\cdots$, so that an example of a normal initial condition for the tent map is

$$x_0 = 0.bb0bbb00b0bbb0bbbb000\cdots. \tag{4.25}$$

For example, if we take $n = 5$ and we want to compute to four-digit precision, then we obtain

$$x_1 = 0.0bb0\cdots,$$

$$x_2 = 0.b000\cdots,$$

$$x_3 = 0.bbb0\cdots, \tag{4.26}$$

$$x_4 = 0.00bb\cdots,$$

$$x_5 = 0.0bb0\cdots.$$

Although the numerical details change with the map, the parallel-processing method combined with an algorithm for the digit string of an initial condition can be used to compute the iterates of a chaotic map. In other words, contrary to what is often stated without justification in the literature, information *flow* in chaotic maps does not have to mean information *loss* – otherwise, the computation of exact chaotic orbits with controlled precision would be impossible.

5
The way to chaos by instability of quasiperiodic orbits

5.1 From limit cycles to tori to chaos

We described earlier how, in the Lorenz model, a pitchfork bifurcation followed by the coalescence of two real, negative eigenvalues gives birth to a pair of stable spirals that become unstable at ρ_h, corresponding to a Hopf bifurcation. Normally, the Hopf bifurcation would give birth to two stable limit cycles (cf. the simpler example of a Hopf bifurcation in Chapter 1), but this is avoided in the Lorenz model because of an additional inverse Hopf bifurcation: as $\rho \to \rho_h^-$ two unstable limit cycles shrink down upon the two spirals, so that for $\rho > \rho_h$, there are no stable limit cycles. Computations or more careful analysis show that chaos begins in regions of phase space outside the unstable limit cycles at a finite value $\rho_c < \rho_h$. Therefore, the Lorenz model does not follow the detailed way to chaos proposed by Ruelle, Takens and Newhouse (RTN), which they predict to be a 'typical' route to chaos in systems of coupled nonlinear ordinary differential equations.

Historically, the RTN prediction followed several years after Lorenz's discovery of the butterfly effect and did not arouse much interest among physicists until after Feigenbaum's breakthrough. The reason for the earlier lack of attention is simple enough: Ruelle *et al.* did not propose a simple, transparent model of their effect that nonspecialists could analyze and understand. Mathematical predictions usually do not gain popularity in physics unless the results are presented in a language that physicists can follow, especially in the form of transparent models. Before any such simple model became widely known, the physicists Swinney and Gollub tried to test the RTN prediction in hydrodynamics experiments. Their results were not conclusive, although they seemed to rule out an earlier prediction by Landau, whereby turbulence should occur by a bifurcation sequence leading to infinitely many modes of oscillation. We turn now to a qualitative discussion of the prediction of Ruelle, Takens and Newhouse.

Typically, stable limit cycles grow from unstable spirals via a Hopf bifurcation: a zero-dimensional attractor dies and gives birth to a one-dimensional attractor. What else can happen that we are familiar with from nonchaotic mechanics? We know that quasiperiodic orbits can occur; we discussed them earlier for the special case of Hamiltonian systems. In the Hamiltonian case, the orbits occur on two dimensional tori in phase space and the quasiperiodic orbits show ergodic behavior as $t \to \infty$, in the sense that the closure of an orbit covers the entire 2-torus. In a driven dissipative system, the torus is an attractor which arises because of energy balance between driving and dissipative forces. According to RTN, the change from stable limit cycles (one-dimensional attractor) to stable quasiperiodic motion (two-dimensional attractor) occurs by way of a Hopf bifurcation. Next, according to RTN, the quasiperiodic orbits become unstable by yet another bifurcation, and *then* chaos can occur. The latter prediction is not hard to believe: nonperiodic orbits with a positive Liapunov exponent *are* unstable quasiperiodic orbits – we have seen examples of this already in our study of one-dimensional maps. For example, given a nonperiodic orbit for the binary tent map that is near an unstable periodic orbit, depending upon how close the initial conditions of the orbits lie to each other, you can iterate the map for a definite number of times before the extra digits in the irrational initial condition take over and show you that the orbit is not periodic after all. When the stable quasiperiodic orbits lie on a torus, the fractal that appears when they become unstable is sometimes described by the term 'torus breakup' or 'Cantori' (Cantor-set-tori).

So, while the original paper by RTN is hard for a nonspecialist to read, we now have models that illustrate what they were driving at. The RTN prediction (the 'quasiperiodic way to chaos') is: stable fixed point $\to$ stable limit cycle $\to$ stable quasiperiodic motion on a 2-torus $\to$ chaos (Fig. 5.1). But what about the transition to chaos via *periodic* orbits? This question was not analyzed by RTN, but has been analyzed in the literature based upon the Feigenbaum results for maps with quadratic maxima. The reason why it is important, as we shall show next, is that periodicity and quasiperiodicity coexist in periodically driven dissipative systems, and there may be no way to decide which path will be followed by a system in nature other than to observe and analyze it – mathematics alone cannot tell you.

5.2 Periodically driven systems and circle maps

We consider here the equation of the damped, periodically driven pendulum,

$$\alpha\ddot{\theta} + \beta\dot{\theta} + \gamma\sin\theta = A + K\cos\omega t, \qquad (5.1)$$

in which there are two competing 'bare' frequencies $\sqrt{\gamma/\alpha}$ and ω. This is the equation of the Josephson junction in the theory of superconductivity. With $x_1 = \theta$, $x_2 = \dot{\theta}$, and $x_3 = \omega t$ we have

$$\dot{x}_1 = x_2,$$

$$\dot{x}_2 = -\frac{1}{\alpha}[\beta x_2 + \gamma\sin x_1] + [A + K\cos x_3]/\alpha, \qquad (5.2)$$

$$\dot{x}_3 = \omega,$$

and a three-dimensional phase space, which is the minimum dimension required for chaos in solutions of ordinary differential equations.

Instead of (5.2), we can work in two dimensions by using the stroboscopic map (cf. Chapter 1),

$$\theta_{n+1} = G_1(\theta_n, \dot{\theta}_n),$$
$$\dot{\theta}_{n+1} = G_2(\theta_n, \dot{\theta}_n), \qquad (5.3)$$

where $n = 0, 1, 2, \ldots$ corresponds to times $0, 2T, 3T, \ldots$ and $T = 2\pi/\omega$ is the period of the external force. As we showed in Chapter 1, this choice of T guarantees that the functions G_1 and G_2 do not change with time. The Jacobian J of the map is simply the Jacobian

$$J(t + T, t) = \frac{\partial(\theta(t + T), \dot{\theta}(t + T))}{\partial(\theta(t), \dot{\theta}(t))} = e^{-\beta T/\alpha} \qquad (5.4)$$

of the first-order system that replaces (5.2), for a time shift $t \to \tau + T$.

Fig. 5.1 The Ruelle–Takens–Newhouse route to chaos by destruction of stable quasiperiodic motion on a torus as a control parameter is varied: a sink ($D_0 = 0$) becomes unstable and gives birth to a limit cycle ($D_0 = 1$), which eventually becomes unstable in favor of an attracting torus ($D_0 = 2$). At the next bifurcation, the typical attractor is predicted to be 'strange', which usually means fractal, among other things.

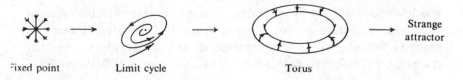

Fixed point   Limit cycle   Torus   Strange attractor

The result (5.4) follows from

$$\nabla \cdot V = \frac{\partial \dot{\theta}}{\partial \theta} + \frac{\partial \ddot{\theta}}{\partial \dot{\theta}} = 0 - \frac{1}{\alpha}(\beta) = -\beta/\alpha, \qquad (5.4b)$$

so that areas in the two-dimensional $(\theta_n, \dot{\theta}_n)$ phase space contract at the constant rate

$$b = e^{-2\pi\beta/\alpha\omega}, \qquad (5.5)$$

where $\beta > 0$ (case of ordinary friction). One can ask what is the dimension of the point set that the areas contract onto. If $\dot{\theta}_n$ should relax rapidly relative to θ_n, so that

$$\dot{\theta}_n \cong g(\theta_n), \qquad (5.6)$$

as $n \to \infty$, then we would be led directly to the study of the one-dimensional circle map

$$\theta_{n+1} = G_1(\theta_n, g(\theta_n)) = f(\theta_n), \qquad (5.7)$$

where $f(\theta_n + 2\pi) = f(\theta_n) + 2\pi$. To prove that this dimensional reduction actually occurs and that f is a smooth function is a nontrivial exercise, and no such general proof exists. All we can say is that it is a reasonable assumption within the nonchaotic regime to assume that the attractors have integer dimensions equal to 0 or 1. The study of this problem is mathematically similar to the study of the KAM problem for area-preserving maps: the question whether there exists a *smooth* map f of the circle onto itself is analogous to the question whether there exist smooth, invariant tori in KAM theory, and we know that so long as the map f is smooth and invertible, then no chaotic motion can occur. We can conclude from previous examples that noninvertibility is necessary for chaos in one dimension.

Consider next the sine circle map

$$f(\theta) = \theta + 2\pi\Omega - K \sin \theta, \qquad (5.8)$$

which reduces at zero − perturbation-strength ($K = 0$) to the simplest circle map, the linear circle map

$$\theta_{n+1} = \theta_n + 2\pi\Omega, \qquad (5.9)$$

which we discussed briefly in Chapter 3 in the context of stable orbits of Hamiltonian systems on tori in phase space. If Ω is rational, then the motion is periodic; irrational Ω corresponds to quasiperiodic motion that is ergodic but nonmixing. In what follows, we take $K \geq 0$. As we have seen in Chapter 3, this map can be used to represent one-dimensional motion on the surface of a two-dimensional torus for a two-degree-of-freedom integrable Hamiltonian system. There, Ω is the ratio of two frequencies. Since we began with a two-degree-of-freedom dissipative system and have assumed contraction onto a one-dimensional orbit asymptotically, it is not useful to think of Ω as the ratio of two competing frequencies in the original dissipative system. In practice, frequency-locking is observed in such systems: there is a tendency for the 'effective' or 'renormalized' frequency, also called the winding number W, to 'lock' at a rational value, reflecting the known tendency of two nonlinear oscillators to synchronize. This means that the RTN prediction, although mathematically general, is not necessarily the route to chaos that systems in nature will follow and we show below that stable periodic and stable quasiperiodic orbits coexist when the circle map is invertible.

In order to analyze the circle map, we introduce the winding number W, which can be defined as follows: beginning with (5.8), we ask for the most general form of map satisfying our circle map condition

$$f(\theta_n + 2\pi) = 2\pi + f(\theta_n), \tag{5.10}$$

and so we write

$$\theta_{n+1} = \theta_n + 2\pi\Omega + Kh(\theta_n), \tag{5.11}$$

where

$$h(\theta_n + 2\pi) = h(\theta_n) \bmod 2\pi. \tag{5.11b}$$

It follows that it is not Ω but the winding number

$$W = \lim_{n \to \infty} \frac{f(\theta_n) - \theta_0}{n} \tag{5.12}$$

that is the correct measure of periodicity or lack of it (if $\theta_{n+1} = \theta_n + 2\pi\Omega$, then $\theta_n = \theta_0 + n\Omega$, and we retrieve $W = \Omega$, as is expected). W can be understood as the average number of rotations through 2π per iteration, because $(f(\theta_{n-1}) - \theta_0)$ is the total distance travelled along the circle in n iterations. Arnol'd (1965) has shown that if W is irrational, and f is invertible, then the map $\theta_{n+1} = f(\theta_n)$ is equivalent to a rotation through some angle ϕ_n. In order to formulate this, we write

$$\phi_{n+1} = T\phi_n = \phi_n + 2\pi W, \tag{5.13}$$

which is a pure rotation with the same winding number W given by (5.12) for the original map $f(\theta)$. Next, we ask, when is it possible to find a transformation from θ to ϕ such that (5.13) holds for an arbitrary circle map f? That is, we want to construct analytically a transformation

$$\phi = H(\theta), \tag{5.14}$$

so that the mapping R defined by

$$R = HfH^{-1}, \tag{5.15}$$

where

$$\theta_{n+1} = R(\phi_n), \tag{5.16}$$

obeys

$$R(\phi) = \phi + 2\pi W. \tag{5.17}$$

According to Arnol'd (1965, 1983), such a transformation is possible whenever f is sufficiently smooth and invertible. In other words, there is an element of universality in the result: f can be a sine-circle map, or anything else that obeys (5.10), so long as the restrictions on f are obeyed. This equivalence holds for irrational winding numbers.

When W is rational, the motion 'locks' at every possible rational frequency P/Q for a finite range of Ω whenever K is finite. The winding number $W(K, \Omega)$ is continuous and monotonically increasing with Ω. This frequency-locking produces wedges of parametric resonance called Arnol'd tongues in the (Ω, K)-plane, and we derive a few of them in what follows for small K. We shall see that an Arnol'd tongue grows from every rational point on the Ω-axis as K increases from zero.

5.3 Arnol'd tongues and the devil's staircase

We now use a particular circle map to illustrate how the motion locks onto any given rational winding number in a finite width $\Delta\Omega$ of the Ω-axis and forms the resonant wedges called Arnol'd tongues in the (Ω, K)-plane. Examples of three such wedges are shown in Fig. 5.2. For

Fig. 5.2 Arnol'd tongues, representing regions of parametric resonance for three different rational winding numbers.

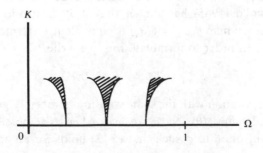

the cosine circle map

$$\theta_{n+1} = \theta_n + \Omega + K \cos 2\pi\theta_n, \tag{5.18}$$

if we set the winding number

$$W = \frac{\theta_n - \theta_0}{n} \tag{5.19}$$

equal to a definite rational number P/Q, then we can solve algebraically for the range of Ω over which this latter condition holds, for a fixed value K in the unit interval. Since $W = \Omega$ gives the average rotation through 2π when $K = 0$, we solve for Ω as an expansion in powers of K as $K \to 0$.

The main idea is to fix W and K and then to ask for the range of Ω (corresponding to different choices of θ_0) over which the periodic orbit exists. If $W = 0$, we have a fixed point

$$\theta_0 = \theta_0 + \Omega + K \cos 2\pi\theta_0 \tag{5.20}$$

or

$$\Omega + K \cos 2\pi\theta_0 = 0, \tag{5.21}$$

and this yields the wedge boundaries

$$\Omega(\varepsilon) = \pm|K| \tag{5.22}$$

because $-1 \le \cos 2\pi\theta_0 \le 1$ (see Fig. 5.3a).

Fig. 5.3 Arnol'd tongues for the cosine–circle map, representing three rational winding numbers: (a) $W = 0$, (b) $W = 1$, and (c) $W = \frac{1}{2}$.

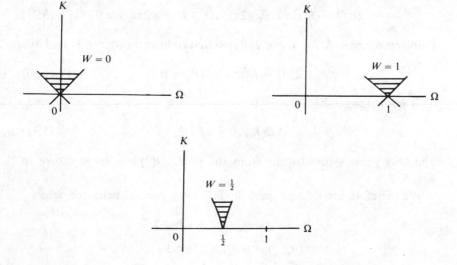

When $W = 1$, we must solve

$$\theta_1 = \theta_0 + \Omega + K \cos 2\pi\theta_0 \tag{5.23}$$

with

$$\theta_1 = \theta_0 + 1. \tag{5.24}$$

Setting $\Omega = 1 + \Delta\Omega$, we again obtain

$$\Delta\Omega(K) = \pm|K| \tag{5.25}$$

for the wedge boundaries (Fig. 5.3b).

The value $W = \frac{1}{2}$ corresponds to $\theta_2 = \theta_0 + \frac{1}{2}$. With

$$\theta_2 = \theta_0 + \Omega + K \cos 2\pi\theta_0 + \Omega + K \cos 2\pi(\theta_0 + \Omega + K \cos 2\pi\theta_0),$$
$$\tag{5.26}$$

we again can set $\Omega = \frac{1}{2} + \Delta\Omega$ for a given θ_0 to obtain

$$2\Delta\Omega + K \cos 2\pi\theta_0 + K \cos 2\pi(\theta_0 + \Omega + K \cos 2\pi\theta_0) = 0. \tag{5.27}$$

Since $\cos 2\pi(\theta_2 + \frac{1}{2}) = -\cos 2\pi\theta_0$ and $\sin 2\pi(\theta_0 + \frac{1}{2}) = -\sin 2\pi\theta_0$, we can use

$$\cos 2\pi(\theta_0 + \tfrac{1}{2} + \Delta\Omega + K \cos 2\pi\theta_0)$$
$$= \cos 2\pi(\theta_0 + \tfrac{1}{2}) \cos 2\pi(\Delta\Omega + K \cos 2\pi\theta_0)$$
$$- \sin 2\pi(\theta_0 + \tfrac{1}{2}) \sin 2\pi(\Delta\Omega + K \cos 2\pi\theta_0) \tag{5.28}$$

to obtain

$$2\Delta\Omega - (K \sin 2\pi\theta_0)(2\pi)(\Delta\Omega + K \cos 2\pi\theta_0) \approx 0 \tag{5.29}$$

to lowest order in $\Delta\Omega + K \cos 2\pi\theta_0$, so that to lowest order in K we have

$$2\Delta\Omega - K^2\pi \sin 4\pi\theta_0 \approx 0. \tag{5.30}$$

This yields tongue boundaries as

$$\Delta\Omega(K) \simeq \pm\pi K^2/2, \tag{5.31}$$

and they grow symmetrically from the point $\Omega(0) = \frac{1}{2}$, as is shown in Fig. 5.3c.

We therefore are led to expect the following general behavior: when

$$\theta_{n+1} = \theta_n + \Omega + Kh(\theta_n),$$
$$h(\theta_{n+1}) = h(\theta_n), \qquad \text{mod } 1, \tag{5.32}$$

under suitable differentiability conditions on h, and for rational winding numbers,

$$W = P/Q, \tag{5.33}$$

we should solve for cycles by writing

$$\theta_Q = \theta_0 + QW = \theta_0 + P \tag{5.34}$$

or

$$\theta_Q = \theta_0 + Q\Omega + F(\theta_0, \Omega, K) = \theta_0 + P \tag{5.34b}$$

where F is obtained from functional composition, with $\Omega = P/\Omega + \Delta\Omega$, in order to obtain the shape of the resonant wedge for a cycle as $K \to 0$. By this technique, we can understand how the wedges grow from any given rational winding number on the Ω-axis.

In order to understand qualitatively how the winding numbers of quasiperiodic orbits can define a Cantor set of finite measure, yielding a complementary devil's staircase of resonant wedges, let us make the following simplifying assumption. Suppose that for fixed $|K| \ll 1$ all Q-cycles have roughly the same tongue widths $\sim K^Q$, where there are at most $Q - 1$ distinct cycles with winding numbers $W = P/Q, P = 1, 2, \ldots,$ $Q - 1$. We can then form a Cantor set by systematically removing the tongues of width $Q \cong K^n$, $n = 1, 2, 3, \ldots,$ one by one. Next, consider the intersection of the remaining set with a line corresponding to $K = $ constant in the (Ω, K) plane. With $Q - 1$ distinct wedges of roughly maximum width K^Q, the total length m_T removed from the interval $0 \le \Omega \le 1$ would be of the order of

$$m_T \sim 2K + \sum_{Q=2}^{\infty} (Q - 1)K^Q + \frac{K^2}{(1 - K)^2} \sim K, \tag{5.35}$$

where various numerical factors have been ignored. This procedure defines a Cantor set, so that the remaining tongues are removed from a Cantor set, producing yet another Cantor set. The argument given here is not complete, but is designed to reflect the essence of the truth.

That $m_T \to 0$ as $K \to 0$ corresponds to the contraction of the frequencies of periodic orbits onto a zero-measure set in $0 \le \Omega \le 1$ when $K = 0$ (the winding numbers of quasiperiodic orbits live between adjacent Arnol'd tongues). For $K > 0$, the allowed irrational winding numbers in $0 \le \Omega \le 1$ must occupy a Cantor set with finite measure $\sim 1 - m_T \sim 1 - K$. The plot of winding number vs Ω, for small K, yields the complementary point-set called the devil's staircase (see Fig. 5.4). The devil's staircase that arises here is different in detail but topologically similar to the one constructed for the symmetric tent map in Chapter 4.

The devilish aspect of the staircase follows from the fact that every rational W has a finite width $\Delta\Omega$ growing from every rational value of Ω, and is a non-decreasing function of Ω. An explicit numerical representation of such a function, called a Cantor function, was constructed for the middle-thirds Cantor set in Chapter 4. That staircase has measure one. Any measure-one staircase is called complete. Since $m_T < 1$ when $K > 0$, we have an incomplete devil's staircase. Bohr, Bak and Jensen (1984, 1985) were led by a numerical analysis to suggest that, as $K \to 1^-$, $m_T \to 1^-$, and have obtained numerical evidence for the scaling law

$$1 - m_T \sim (1 - K)^\beta \qquad \text{as } K \to 1^-. \tag{5.36}$$

They predict a fractal dimension of $D_0 \sim 0.87$ for the complementary zero-measure Cantor set, to which irrational winding numbers are confined at $K = 1$. To obtain an analytical description of criticality as $K \to 1^-$, one must use an adaptation to circle maps of Feigenbaum's renormalization group treatment of period doubling for maps with quadratic maxima. What is assumed without proof in what follows in the next section is that the quasiperiodic orbits lose their stability at $K = 1$ and become true chaotic orbits with positive Liapunov exponents when $K > 1$.

Fig. 5.4   A few steps in the devil's staircase for the nonlinear circle map are indicated. The dots represent the devilishly many missing steps. The staircase is selfsimilar: the study of a small region ('window') of the staircase reveals a structure that, upon magnification, is statistically similar to that of the entire staircase.

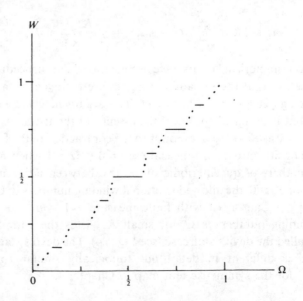

The Arnol'd tongues pose a conceptual problem geometrically, because the rational frequencies are dense on the Ω-axis and a tongue of finite width grows from every rational value of Ω. Therefore, why do the tongues not overlap for every finite value of K (analysis shows that instead overlap occurs at $K_c = 1 + \delta K$ with $\delta K > 0$)? The proof that resonance overlap does not occur when K is infinitesimally small is equivalent to the proof that rational numbers are countable. We leave it to the reader to prove that as an exercise.

Because noninvertibility of the map is essential for chaos in one dimension, the map

$$f(\theta) = \theta + \Omega - K \sin 2\pi\theta / 2\pi \tag{5.37}$$

has the possibility of chaotic behavior only if $K > 1$. Because

$$f'(\theta) = 1 - K \cos 2\pi\theta, \tag{5.38}$$

$f(\theta)$ is monotonically increasing for $K < 1$. Since $f(1) = f(0) \bmod 1$, $f(0)$ is single-valued, hence invertible (Fig. 5.5a).

Fig. 5.5 (a) Invertible sine–circle map for $K = \frac{1}{2}$, where all closed orbits are stable. (b) Sine–circle map at the transition to chaos, $K = 1$. (c) Noninvertible sine–circle map, here drawn for $K = \frac{3}{2}$, where the quasiperiodic orbits are unstable.

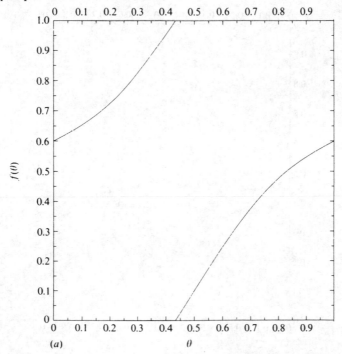

(a)

Fig. 5.5—continued.

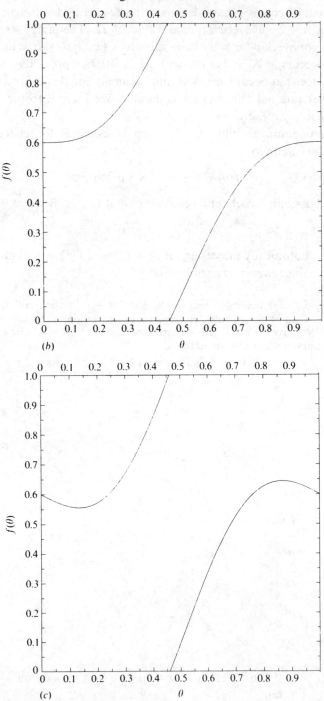

(b)

(c)

At $K = 1$, $f'(\theta) = 0$ at $\theta = 0$ ($=1$ mod 1), as is indicated in Fig. 5.5b. When $K > 1$, $f(\theta)$ has a maximum and minimum (mod 1) at

$$\theta = 1 - \frac{1}{2\pi} \cos^{-1} \frac{1}{K} \qquad (5.39)$$

and

$$\theta = \frac{1}{2\pi} \cos^{-1} \frac{1}{K} \qquad (5.40)$$

respectively (see Fig. 5c). Therefore, chaos requires $K > 1$.

5.4 Scaling laws and renormalization group equations

So far, we have concluded the following for $0 < K < 1$: when W is rational, the motion corresponds to stable cycles (stability follows from a negative Liapunov exponent, $\lambda < 0$) in a finite range of Ω in $0 \leq \Omega \leq 1$, yielding the Arnol'd tongues that describe the frequency-locking regime, and stable quasiperiodic orbits. Here, the map is invertible and chaos cannot occur. Stable quasi-periodic motion corresponding to irrational W occurs on a Cantor set of finite measure in Ω when $K < 1$. If the quasiperiodic orbits have winding numbers that occupy a Cantor set at $K = 1$, then we expect that these orbits become chaotic at $K > 1$. We now follow a renormalization group method of analyzing the transition to chaos due to instability of quasiperiodic orbits. The following method of analysis, due to Shenker (1982), is based upon Kadanoff's treatment of the KAM problem for low-degree-of-freedom conservative systems.

Because one cannot determine directly from the winding number

$$W = \lim_{n \to \infty} \frac{f'(x_0) - x_0}{n}, \qquad (5.41)$$

which value of Ω, for a given initial condition θ_0, generates a given irrational value of W, we shall use a continued fraction expansion of W to describe the approach to irrationality of W via a sequence of stable cycles of increasingly higher order. Because its continued fraction development leads to a particularly simple form for the renormalization-group equations, the so-called golden-mean winding number[1]

---

[1] According to Arnol'd (1965), stable quasi-periodic orbits with irrational winding numbers that have the slowest convergence rate by continued fraction approximants will yield a uniform invariant density. The golden mean belongs to this category.

$$\bar{W} = \frac{\sqrt{5} - 1}{2} \tag{5.42}$$

has been historically singled out for study. The continued fraction of $\bar{W}$ representation follows from the properties of the Fibonacci sequence, whose initial conditions are $F_0 = F_1 = 1$ and the rest of the sequence is generated recursively:

$$F_{n+1} = F_n + F_{n-1}, \qquad n \geq 1. \tag{5.43}$$

If we define a rational winding number by the Fibonacci ratio

$$W_n = F_n/F_{n+1}, \tag{5.44}$$

then the recursion relation for F_{n+1} yields the finite continued fraction representation

$$W_n = 1/(1 + 1/1 + 1/1 + \cdots + 1). \tag{5.44b}$$

If now we ask for the limit

$$\bar{W} = \lim_{n \to \infty} W_n \tag{5.45}$$

then we obtain on the one hand the infinite continued fraction (corresponding to irrational $\bar{W}$)

$$\bar{W} = 1/(1 + 1/1 + 1/1 + \cdots) \tag{5.46}$$

and, on the other, the algebraic result

$$\bar{W} = 1/(1 + \bar{W}) \tag{5.47}$$

yielding $\bar{W}(1 + \bar{W}) = 1$, whose positive solution is given by

$$\bar{W} = \frac{\sqrt{5} - 1}{2}. \tag{5.48}$$

If we choose a point $\theta_{0,m}$ belonging to the F_{m+1} cycle with winding number $W_m = F_m/F_{m+1}$, then we obtain

$$W_m = \frac{f^{F_{m+1}}(\theta_{0,m}) - \theta_{0,m}}{F_{m+1}} \tag{5.41b}$$

where $f^{F_{m+1}}$ denotes $f(\theta_{F_{m+1}})$, the F_{m+1}th iterate of f, and if we choose coordinates so that $\theta = 0$ belongs to the same cycle, then

$$W_m = \frac{F_m}{F_{m+1}} = \frac{f^{F_{m+1}}(0)}{F_{m+1}} \tag{5.41c}$$

follows because $f^{F_{m+1}}(0) = F_m$. So, if we fix $K \in [0,1]$, we can calculate the particular rational bare frequency Ω_m that yields a particular F_{m+1}-cycle.

It is easy to prove that the sequence $\{W_n\}$ converges geometrically. We begin with $W_n = W_\infty - c\delta^{-n}$, with $\delta^{-1} = -\bar{W}^{-2}$. First, note that

$$W_{n+1} - W_n = -\frac{F_n F_{n+1}}{F_{n+1} F_{n+2}} (W_n - W_{n-1}). \tag{5.49}$$

Next, we see that

$$\frac{W_{n+1} - W_n}{W_n - W_{n-1}} = -\frac{F_n F_{n+1}}{F_{n+1} F_{n+2}} \to -\bar{W}^{-2} \qquad \text{as } n \to \infty, \tag{5.50}$$

so that

$$\delta^{-1} = \lim_{n \to \infty} \frac{W_{n+1} - W_n}{W_n - W_{n-1}} = -\bar{W}^{-2} \tag{5.51}$$

follows.

Arnol'd has pointed out that irrational winding numbers that fall victim to rapidly convergent approximation by rationals exhibit continuous but nonuniform invariant densities, corresponding to 'ghosts' of nonexistent cycles. That is because the low-order irrational W_0, W_1, etc. are accurate approximations to the irrational winding number W for relatively large iteration numbers n. Among the irrationals that can be generated by continued fractions, those ending in strings of 1's have the slowest convergence rates, so that for these we expect to obtain $p(x) = 1$.

One can also show that

$$\lim_{n \to \infty} \frac{\Omega_{n+1} - \Omega_n}{\Omega_n - \Omega_{n-1}} = -\bar{W}^{-2}, \tag{5.52}$$

when $|K| < 1$: this result is qualitatively suggested by the geometry of the devil's staircase (plot a few steps of W vs Ω as an exercise in order to understand it). Hence, we have $\Omega^n(K) \sim \Omega_\infty(K) - c\delta^{-n}$ with $\delta^{-1} = -\bar{W}^2$ for a given $|K| < 1$. This is one of the scaling laws in the subcritical case. Shenker assumed the scaling relation

$$\delta^{-1} \simeq -\bar{W}^y \tag{5.53}$$

and found numerically that $y = 2.1644 \cdots$ for a sine–circle map when $|K| = 1$, or $\delta \simeq -2.8336 \cdots$.

The second scaling relation, $d_n/d_{n+1} \to \sim -\bar{W}^{-1}$, as $n \to \infty$ for $|K| < 1$, also follows from the properties of the Fibonacci sequence. We begin with the definition

$$d_n = f^{F_n}(0) - F_{n-1} \qquad (5.54)$$

and assume that f is fixed at the frequency Ω_n corresponding to winding number $W_n = F_n/F_{n+1}$ (i.e. an F_{n+1}-cycle) where also $f^{F_{n+1}}(0) - F_n = 0$. If we write down the d_n for the case where f is the pure rotation with winding number W_n,

$$f(\theta) = \theta + W_n, \qquad (5.55)$$

it follows that d_n is the distance to the point on the cycle nearest to $\theta = 0$ (mod 1): for example if $f(x) = x + W$ then $f(0) = W_n$ and $F^{F_n}(0) = F_n W_n = F_n^2/F_{n+1}$. It follows in this case that

$$d_n = F_n^2/F_{n+1} - F_{n-1} = F_n(F_n/F_{n+1} - F_{n-1}/F_n)$$
$$= F_n(W_n - W_{n-1}), \qquad (5.56)$$

and we can understand from a few special cases how this works. If $n = 2$, $W_2 = F/F = \frac{1}{2}$, which is a 2-cycle and we obtain $d_2 = -\frac{1}{2}$, half a rotation away from $x = 0$ mod 1. If $n = 3$, $W_3 = F_3/F_4 = \frac{1}{3}$, which is $\frac{1}{3}$ of a full rotation (mod 1) beyond $x = 0$. The main assumption that is made is that $\Omega^n(K) \to \bar{\Omega}(K)$ as $n \to \infty$ for fixed K where $\bar{\Omega}(K)$ yields $\bar{W} = (\sqrt{5} - 1)/2$ as winding number for the map f. The scaling follows from the geometrical rate of convergence of the winding numbers to $\bar{W}$:

$$\alpha = d_n/d_{n+1} = (F_n/F_{n+1})(F_n/F_{n+1} - F_{n-1}/F_n)/(F_{n+1}/F_{n+2} - F_n/F_{n+1})$$
$$= (F_n/F_{n+1})(W_n - W_{n-1})/(W_{n+1} - W_n)$$
$$\to \bar{W}(-\bar{W}^{-2}) = -\bar{W}^{-1} \qquad \text{as } n \to \infty. \qquad (5.57)$$

At $|K| = 1$ (criticality), Shenker used a numerical method to find instead that $d_n/d_{n+1} \leftarrow -1.2857\cdots = \bar{W}^{-x}$ with $x \approx 0.5267\cdots$, but there is an analytic way to calculate such results, the so-called renormalization group method.

We can now use the recursion properties of the Fibonacci sequence along with the circle map condition $f(\theta + 1) = 1 + f(\theta)$ to set up the recursion formulae that are required for the renormalization group analysis of the instability at $K = 1$. From $f(\theta + 1) = 1 + f(\theta)$ we have also that

$$f(\theta + 2) = f(\theta + 1) + 1 = f(\theta) + 2$$
$$\vdots$$
$$f(\theta + F_n) = f(0) + F_n, \qquad (5.58)$$

where F_n is any integer, in particular, F_n is a Fibonacci number. On the circle, it follows also that

$$f(\theta - F_n) = f(\theta) - F_n. \tag{5.59}$$

Define $f^{(n)}(\theta) = f^{F_{n+1}}(\theta) - F_n$ where $f^{(n)}$ refers to the F_{n+1}-cycle with winding number $W_n = F_n/F_{n+1}$. Then

$$f^{(n+1)}(\theta) = f^{F_{n+2}}(\theta) - F_{n+1} = f^{F_{n+1}+F_n}(\theta) - F_n - F_{n-1}$$

$$= \underbrace{f \cdots f}_{F_{n+1} \text{ times}} \cdot \underbrace{f \cdots f}_{F_n \text{ times}} (\theta) - F_n - F_{n-1} \tag{5.60}$$

holds for an F_{n+2}-cycle with winding number $W_{n+1} = F_{n+1}/F_{n+2}$. We can then rewrite the latter expression as follows:

$$f^{(n+1)}(\theta) = f \underbrace{(f \cdots f}_{F_{n+1} - 1 \text{ times}} \cdot \underbrace{f \cdots f}_{F_n \text{ times}} (\theta) - F_{n-1}) - F_n$$

$$= f(f \underbrace{(f \cdots f}_{F_{n+1} - 2 \text{ times}} \cdot \underbrace{f \cdots f}_{F_n \text{ times}} (\theta) - F_{n-1})) - F_n$$

$$\vdots$$

$$= \underbrace{f \cdots f}_{F_{n+1} \text{ times}} \cdot \underbrace{(f \cdots f}_{F_{n+1} \text{ times}} (\theta) - F_{n-1}) - F_n$$

$$= f^{F_{n+1}}(f^{F_n}(\theta) - F_{n-1}) - F_n$$

$$= f^{F_{n+1}}(f^{(n-1)}(\theta)) - F_n$$

$$= f^{(n)}(f^{(n-1)}(\theta)). \tag{5.61}$$

By a similar procedure, we can show that

$$f^{(n+1)}(\theta) = f^{(n-1)}(f^{(n)}(\theta)). \tag{5.62}$$

To go further, we define

$$f_n(x) = \alpha^n f^{(n)}(\alpha^{-n}x), \tag{5.63}$$

a form that is motivated by Shenker's scaling assumption

$$f^{(n)}(x) \simeq \alpha^{-n}\bar{f}(\alpha^n x), \tag{5.64}$$

where $\alpha = \bar{W}^{-x}$ is the scaling length and $\bar{f}$ is universal. This approach was motivated by Feigenbaum's treatment of scaling near criticality for one-dimensional maps with quadratic maxima. An analytic development of the scaling theory can be found in Ostlund, Rand, Sethna and Siggia (1983). The objective is to derive a circle map version of the renormalization group equation analogous to the one for the Feigenbaum period-doubling sequence.

We begin with $f^{(n)}(\alpha^{-n}x) = \alpha^{-n}f_n(x)$ and rewrite it as $f^{(n)}(x) = \alpha^{-n}f(\alpha^n x)$, along with $f^{(n+1)}(x) = f^{(n-1)}(f^{(n)}(x))$. Then

$$f^{(n+1)}(x) = \alpha^{-n-1}f_{n+1}(\alpha^{n+1}x) \tag{5.65}$$

and

$$f^{(n-1)}(z) = \alpha^{-n+1}f_{n-1}(\alpha^{n-1}z). \tag{5.66}$$

We set $z = f^{(n)}(x) = \alpha^{-n}f_n(\alpha^n x)$ in order to obtain

$$\alpha^{-n-1}f_{n+1}(\alpha^{n+1}x) = \alpha^{-n+1}f_{n-1}(\alpha^{n-1}\alpha^n f_n(\alpha^n x)) \tag{5.67}$$

with $\alpha^{n+1}x \to x$ or $x \to \alpha^{-n-1}x$, we have

$$f_{n+1}(x) = \alpha^2 f_{n-1}(\alpha^{-1}f_n(\alpha^{-1}x)). \tag{5.68}$$

We can also use $f^{(n+1)}(x) = f^{(n)}(f^{(n-1)})(x))$ to obtain the recursion formula

$$f_{n+1}(x) = \alpha f_n(\alpha f_{n-1}(\alpha^{-2}x)). \tag{5.69}$$

These are analogs of the renormalization group equations of critical phenomena and the assumption of scaling corresponds to the assumption that the $n = \infty$ limit of these equations can be solved. The $n = \infty$ limits are

$$\bar{f}(x) = \alpha^2 \bar{f}(\alpha^{-1}\bar{f}(\alpha^{-1}x)) \tag{5.70}$$

and

$$\bar{f}(x) = \alpha \bar{f}(\alpha \bar{f}(\alpha^{-2}x)). \tag{5.71}$$

Notice that $\bar{f}(x) = -1 + x$ is an exact solution, e.g.

$$\begin{aligned}
-1+x &= \alpha^2(-1 + \alpha^{-1}(-1 + \alpha^{-1}x)) \\
&= \alpha^2(-1 - \alpha^{-1} + \alpha^{-2}x) \\
&= -\alpha^2 - \alpha + x
\end{aligned} \tag{5.72}$$

$$\therefore -1 = -\alpha^2 - \alpha$$

$$\alpha^2 + \alpha - 1 = 0, \qquad \alpha = \frac{-1 \pm \sqrt{5}}{2}.$$

Hence, we obtain golden-mean universality with the classical exponents $x = 2$, $y = 1$ when $|K| < 1$. The behavior is universal in the sense that it holds for all invertible circle maps f that are sufficiently well behaved.

The solution for the case $|K| = 1$ has been worked out by Feigenbaum and coworkers (Feigenbaum, Kadanoff and Shenker, 1982).

Universality is restricted here to maps with the same winding number: Shenker found approximately the same critical exponents $x \simeq 0.527$ and $y \simeq 2.164 \cdots$ by using the continued fraction expansion approximations to $(\sqrt{5} - 1)/2$, for the map

$$x_{n+1} = x_n + \Omega' - K(\sin 2\pi x_n + 0.2 \sin (6\pi x_n))/2\pi. \qquad (5.73)$$

However, the original map with winding number

$$W = \cfrac{1}{2 + \cfrac{1}{2 + \cfrac{1}{2 + \cdots}}} \qquad (5.74)$$

yields different critical exponents,

$$x \simeq 0.523 \cdots \quad \text{and} \quad y \simeq 2.174. \qquad (5.75)$$

When $K = 1$ the quasiperiodic orbits lose their stability and become chaotic for $K > 1$. Stable periodic orbits do not immediately disappear, but pass through a period-doubling sequence to chaos as K increases beyond unity. This case also has been treated in the literature.

However, for $K > 1$, the circle map is only a mathematical laboratory for studying chaos in one-dimensional maps: the Josephson junction problem requires a two-dimensional map for its description whenever $K > 1$.

5.5 The Farey tree

One of the main themes of this book is the organization of successive generations of the coarsegrained description of fractals onto trees where the addresses of different branches are provided by the dynamics. For the circle map, the Farey tree does the job. The tree addresses are the rational frequencies of resonant wedges, the intervals are successive distances between wedges taken in the Farey tree order, and a particular path on the tree corresponds to the golden-mean approximants. Let us see just how this works.

We start with the fact that every rational number $P/Q \in [0, 1]$ is the frequency of a definite Arnol'd tongue. Consider the wedges with $\Omega \in [0, 1]$ for a fixed value of K. With $\Omega = P/Q$, we know that the width of the wedges is decreasing as the denominator Q is increased. For any two frequencies $\Omega_1 = P/Q < \Omega_2 = P'/Q'$, the next frequency Ω_3 with the smallest denominator and such that $\Omega_1 < \Omega_3 < \Omega_2$ is given by the Farey fraction $\Omega_3 = (P + P')/(Q + Q')$. This defines the construction of the Farey tree, which is shown as Fig. 5.6.

The generalization of the box counting condition $N(l)l^D \sim 1$ to include nonuniform intervals is

$$\sum_{i=1}^{N_n} l_i^D \sim 1. \tag{5.76}$$

The Cantor set of interest here is the one defined by the irrational winding numbers (the point set between the infinity of resonant wedges). If we denote by $\bar{s}$ the distance between the wedges that are labeled by P/Q and P'/Q', by s_1 the distance from the boundary of the P/Q-wedge to that of the $(P + P')/(Q + Q')$-wedge, and by s_2 the distance between wedges with frequencies $(P + P')/(Q + Q')$ and P'/Q', then $l_1 = s_1/\bar{s}$ and $l_2 = s_2/\bar{s}$ defines the first generation of coarsegraining that describes the fractal that lies between Ω_1, Ω_3, and Ω_2, so that the crudest approximation to D is given by $l_1^D + l_2^D \sim 1$ (Fig. 5.7). Successive stages of refinement (and better approximations to D) are given by the intervals corresponding to higher generations on the Farey tree. According to Bak and coworkers, $D \approx 0.87$ when $K = 1$, but we must find $D = 1$ if $0 < K < 1$ for $N_n \gg 1$ in (5.76).

That the branches of the Farey tree are self-similar in a certain sense to the whole tree can be seen if we rewrite each address as a continued fraction expansion: $\frac{1}{2} = 1/(1 + 1)$, $\frac{2}{3} = 1/(1 + 1/(1 + 1))$, and so on. The

Fig. 5.6 The first few branches in the recursive construction of the Farey tree.

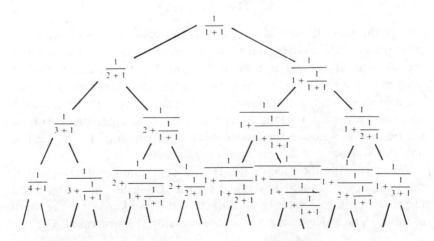

point is that you can get the $(n + 1)$th generation addresses from those of the nth generation by a replacement of the last integer 1 in each of the latter by 2 and then by $\frac{1}{2} = 1/(1 + 1)$. For example, in the fourth generation $\frac{2}{5} = 1/(2 + 1/(1 + 1))$ gives birth to $1/(2 + 1/(2 + 1)) = \frac{3}{7}$ and $1/(2 + 1/(1 + 1/(1 + 1))) = \frac{3}{8}$ in the fifth generation. The golden mean approximants correspond to the path indicated by arrows in Fig. 5.8. The

Fig. 5.7 Diagram indicating the crudest approximation to the Hausdorf dimension that follows from the Farey tree construction.

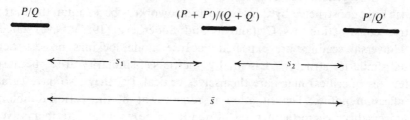

Fig. 5.8 The golden mean approximants are indicated on the Farey tree by the thick path with arrows.

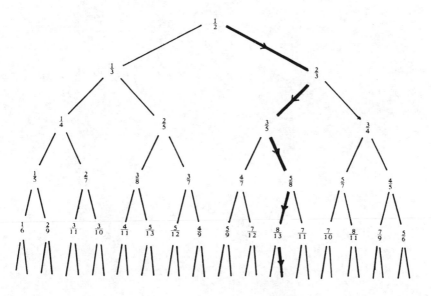

golden mean is only one of countably many computable irrational winding numbers that can be studied. In particular, it is not (as is sometimes mistakenly claimed) the 'most irrational' winding number. It is, however, an example of the class of computable irrational numbers that has the slowest convergence rate when approximated by a sequence of rational numbers given as continued fractions; every number that has a 'tail' of ones in the continued fraction expansion will have a similarly slow convergence rate.

A more general renormalization group equation than equation (5.71) has been constructed by Cvitanović and coworkers based upon the Farey tree construction. As Cvitanović and Søderberg (1985) have stated: 'Universal scalings are expected because mode lockings possess rich self-similarity structure. Essentially, metric self-similarity arises because iterates of critical maps are themselves critical, i.e., they also have cubic inflection points. The situation is reminiscent of the period-doubling universality, where the iterates of quadratic polynomials are themselves locally quadratic, and where this local selfsimilar structure leads to the universal scaling laws.'

6
The way to chaos by period doubling

6.1 Universality at transitions to chaos

We use the word 'universal' in the sense of 'including, pertaining to, affecting all members of a class or group'. The important idea is therefore that of a 'universality class' and it is necessary to have a criterion for membership.

The modern use of these phrases in physics stems from the renormalization group theory of phase transitions in classical statistical mechanics, where all models with the same symmetry and dimension yield the same critical exponents independently of other details of the Hamiltonian, so long as corrections to scaling are ignored. The predictions of Lorenz and Ruelle and Takens had been made, but had generated no large following; just as broken symmetries dominated the physics of the 1960s, the 1970s were the heyday of critical phenomena and Wilson's renormalization group method. So, when Feigenbaum (1978, 1980) argued near the end of that decade that the 'period-doubling' transition to chaos yields universal critical exponents, many physicists became excited to learn what was meant by 'deterministic chaos' and then had to learn about the 'strange' mathematical objects called Cantor sets, because the period-doubling limit defines a particular Cantor set.

Feigenbaum developed his renormalization group theory of period doubling by starting with an iterative map of the interval, the logistic map $x \to f(x, D) = Dx(1 - x)$. He showed that all smooth maps of the interval with a quadratic maximum (Fig. 6.1) should yield the same critical exponents, so that the order of the maximum is one factor that determines the universality class.

That the logistic map is chaotic was long known. Von Neumann (1951) had derived the invariant density describing fully developed chaos ($D = 4$)

173

in the 1940s, before the phrase 'deterministic chaos' was coined. Many other properties of the map away from criticality were known, and the map had even been used by theoretical biologists as a model of population growth (see May, 1976). Certain universal properties of the map's symbol sequences had been discovered earlier by Metropolis, Stein, and Stein (1973). In Germany, Grossmann and his collaborators independently discovered scaling at the period-doubling critical point, but did not arrive at universality (Grossman and Thomae, 1977).

Without working on deterministic chaos, Mandelbrot taught us how to generate many different fractals recursively, and thereby taught us about the geometry of Cantor sets. The attractors and repellers that occur in chaotic systems were at first called 'strange', but, in all examples that we understand how to calculate, they are Cantor sets of one sort or another. And, in analogy with Mandelbrot's recursive geometrical construction of fractals, we can generate fractals recursively by backward iteration of certain chaotic maps. So, we now have a geometry to describe fragmentation that occurs in nature, and we also understand that fragmentation is generated by deterministic chaos and at the borderline of chaos.

Fig. 6.1   Maps with a quadratic maximum define a universality class for the transition to chaos by period-doubling of orbits. Here, we have plotted $f(x) = 3.7x(1 - x)$.

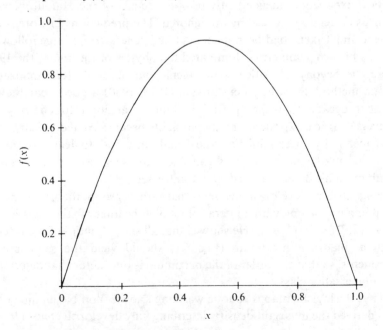

A real challenge in physics is to construct model dynamical systems that generate the fractals that actually occur in nature. At a transition to chaos, different maps may show the same scaling behavior and thereby belong to the same universality class. The universal critical exponents describe the scaling near the transition.

We have seen in Chapter 2 that, when $D = 4$, the logistic map is conjugate to the binary tent map. In the binary tent map, every rational initial condition yields a periodic orbit, so that periods of all orders occur. These unstable-periodic orbits map onto unstable periodic orbits of the logistic map at FDC. The instability of all orbits of order 2^n for the logistic map occurs as $D \to D_\infty$, where D_∞ is the period-doubling critical point. The instabilities of all periodic orbits of orders different from 2^n develop when $D_\infty < D \le 4$ and follow an ordering discovered by Sarkovskii. For example, the 3-cycle consisting of the points $\frac{2}{9}$, $\frac{4}{9}$, and $\frac{8}{9}$ for the binary tent maps onto an unstable 3-cycle of the logistic map at $D = 4$. That 3-cycle does not exist when $D \le D_\infty$.

An invariant set I of a map f is any set left invariant by the map $f: I \to I$. Particular attractors and repellers are examples of invariant sets. When $D = 4$, the entire unit interval is the invariant set of the logistic map.[1] When $D < D_\infty$, a stable cycle of period 2^n is an invariant set. There are two regions in D where the invariant set of the logistic map is fractal (is a Cantor set). One is at D_∞, the period-doubling critical point; the other is at $D > 4$ which we treat in Chapters 7 and 8. For $D < D_\infty$ the invariant set contains an attractor; when $D > 4$ the invariant set is a repeller.

We have sketched the application of Feigenbaum's method to the instability of quasiperiodic orbits for circle maps. In what follows, we follow the transition to chaos by the instability of all orbits of period 2^n as $n \to \infty$.

6.2 Instability of periodic orbits by period doubling

Consider a one-dimensional map of the interval

$$x_{n+1} = f(x_n, D), \tag{6.1}$$

where D is a control parameter. We are interested in the maps with quadratic maxima where, as D increases, a stable fixed point gives birth to a stable 2-cycle (pitchfork bifurcation), which then gives birth to a

---

[1] In this case, the unit interval divides into non-overlapping invariant subsets: unstable-periodic orbits and unstable-nonperiodic orbits.

stable 4-cycle, and so on until at $D = D_\infty$ all cycles of order 2^n are unstable and the invariant set of the map consists of 2^∞ points. The bifurcation diagram in this range is therefore a complete binary tree (Fig. 6.2). The logistic map belongs to this category.

The condition for a fixed point is that

$$\bar{x} = f(\bar{x}, D), \tag{6.2}$$

corresponding to a 1-cycle. In order to decide the stability of the fixed point, set $\delta x = x - \bar{x}$ and study the approximate linear map

$$\delta x_{n+1} \approx f'(\bar{x}, D)\, \delta x_n \tag{6.3}$$

whose solution is

$$\delta x_n \approx (f'(\bar{x}, D))^{n-1}\, \delta x_0. \tag{6.3b}$$

The fixed point is stable if $|f'(\bar{x}, D)| < 1$, unstable if $|f'(\bar{x}, D| > 1$ and the value D_1 where $|f'(\bar{x}, D_1)| = 1$ signals a bifurcation.

For the logistic map, the fixed points are $\bar{x} = 0$ and $\bar{x} = 1 - D^{-1}$. Because $f'(x, D) = D(1 - 2x)$, $\bar{x} = 0$ is stable when $|D| < 1$. The second fixed point belongs to $[0, 1]$ if $D \geq 1$,[2] and its stability range is then $D \leq 3$,

---

[2] The reader is invited to show that, for $0 < D \leq 4$, $x_{n+1} \in [0, 1]$ if $x_n \in [0, 1]$.

Fig. 6.2 For maps with a quadratic maximum, the bifurcation diagram describing the path to chaos is a binary tree (period-doubling path to chaos).

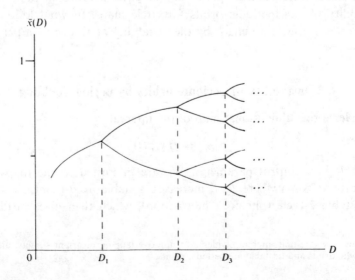

as D is increased. It is easy to see graphically (Fig. 6.2) that the map has a pitchfork bifurcation at $D_1 = 3$, and that the invariant set is a 2-cycle if D is a little greater than 3. That is, for $D > 3$, all iterations $x_n \to (\bar{x}_1, \bar{x}_2)$, where

$$\bar{x}_2 = f(\bar{x}_1, D),$$
$$\bar{x}_1 = f(\bar{x}_2, D),$$
(6.4)

or, better said,

$$\bar{x}_i = f^{(2)}(\bar{x}_i, D) = f(f(\bar{x}_i, D)) \tag{6.5}$$

where $f^{(2)}$ is the second iterate of f. The second iterate has two fixed points, $\bar{x}_1$ and $\bar{x}_2$ (Fig. 6.3). In order to determine the stability range of these two fixed points, we set $\delta x_n = x_n - \bar{x}_i$ and study the linear map

$$\delta x_{n+2} \approx f^{(2)\prime}(\bar{x}_1, D)\, \delta x_n, \tag{6.6}$$

where

$$f^{(2)\prime}(\bar{x}_1, D) = f'(f(\bar{x}_1, D))f'(\bar{x}_1, D)$$
$$= f'(\bar{x}_1, D)f'(\bar{x}_2, D). \tag{6.7}$$

Fig. 6.3   At $D = 3$, the logistic map has a pitchfork bifurcation to a 2-cycle (period-doubling).

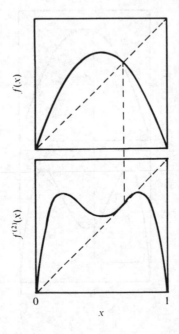

The stability range $3 < D < D_2$ is therefore determined by the condition

$$|f'(\bar{x}_1, D_2)|\,|f'(\bar{x}_2, D_2)| = 1, \qquad (6.8)$$

which can be solved to yield D_2. It is easy to see graphically (Fig. 6.4) that the 2-cycle becomes unstable and gives birth to a 4-cycle when $D > D_2$. Again, the instability at D_2 is of the pitchfork brand.

The above procedure continues, with stability of a cycle of order 2^n determined by the range $D_{n-1} < D < D_n$, where

$$\prod_{i=1}^{2^n} |f'(\bar{x}_i, D)| < 1 \qquad (6.9)$$

and $(\bar{x}_1, \bar{x}_2, \ldots, \bar{x}_{2^n})$ is the invariant set. Feigenbaum discovered asymptotic scaling laws for the binary tree (Fig. 6.2) as $D_n \to D_\infty$, where all cycles of order 2^n become unstable, yet the motion is bounded: $x_n \in [0, 1]$ if $x_{n-1} \in [0, 1]$. Therefore, when $D = D_\infty$, the map wanders deterministically on a set with 2^∞ points. The main idea is that the scaling behavior at D_∞ is the same for all maps with a quadratic maximum. That is, the critical exponents are universal.

Fig. 6.4   At $D > D_2$, the logistic map's 2-cycle is unstable but has given birth by a pitchfork instability to a new, stable 4-cycle (period-doubling).

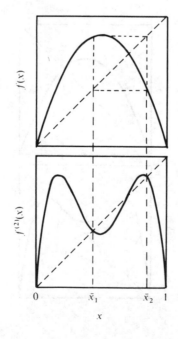

6.3 Universal scaling for noninvertible quadratic maps of the interval

Consider the branching diagram shown in Fig. (6.5), where $x_\infty(D) = x(D)$ is plotted against the control parameter D. The figure represents the period-doubling sequence of instabilities for maps with quadratic maxima and is a complete binary tree. Universality for this case was discovered by Feigenbaum, who studied the logistic map's bifurcation sequence with the aid of a pocket calculator. His discoveries include the scaling laws for large n: (i) the ratio $(D_n - D_{n-1})/(D_{n+1} - D_n)$ approaches a constant δ, so that $D_n \simeq D_x - c\delta^{-n}$, where c is a constant, and (ii) superstable cycles obey $d_n/d_{n+1} \simeq -\alpha$. A superstable 2^n-cycle is one that includes the point $x = \frac{1}{2}$, and $d_n = f^{2^{n-1}}(\frac{1}{2}, D'_n) - \frac{1}{2}$ is the distance to the point on the 2^n cycle nearest to $x = \frac{1}{2}$ (see Fig. (6.5)). It follows also that $D'_n \simeq D_\infty - c'\delta^{-n}$, where D'_n is the control parameter's value where the superstable cycle occurs. δ is analogous to a critical exponent in the theory of phase transitions, and, by an adaptation of the renormalization group idea to the study of quadratic maps, Feigenbaum showed that

$$\alpha = 2.502\,907\,875\,1\cdots,$$

$$\delta = 4.669\,201\,609\,1\cdots, \tag{6.10}$$

$$D_\infty = 3.569\,945\,6\cdots.$$

Fig. 6.5 Superstable cycles for the logistic map; $d_n = f^{2^{n-1}}(\frac{1}{2}, D'_n) - \frac{1}{2}$ is the distance to the point on the 2^n-cycle nearest to the point $x = \frac{1}{2}$.

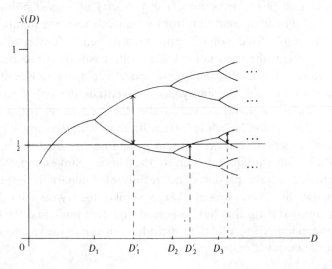

The seed from which the renormalization group theory of universal critical exponents grew was Kadanoff's qualitative discussion of the scaling of blocks of spins in the Ising model (Kadanoff, 1966). The motivation at that time was the lack of an explanation why entirely distinct physical systems could have approximately the same critical exponents. Kadanoff's block spin method was a way to derive the scaling laws without being able to predict a single critical exponent. Later, through the renormalization group method, it was explained why, e.g., thin superfluid ^{4}He films and an XY model of a planar magnet could have the same critical behavior: the critical exponents depend only upon the symmetry and dimension of the system. For example, the two-dimensional planar rotator model of a magnet has the same critical exponents as a two-dimensional superfluid because both have rotation symmetry in a plane, whereas the two-dimensional Ising model, with distinct critical exponents, has a discrete symmetry (reflection symmetry).

Setting $T = T_c$ in statistical mechanics corresponds to setting a coupling constant equal to a critical value in the renormalization group. The analog of a critical value of a coupling constant for the case of a quadratic map is the control parameter D_∞ at the period-doubling limit, where all cycles of period 2^n with $n = 0, 1, 2, \ldots$, are unstable. The instability of cycles of arbitrarily long period is superficially analogous to the role played by divergent fluctuations in critical phenomena, where the correlation length diverges as $\xi \sim |T - T_c|^{-\nu}$, with critical exponent ν as T approaches the critical temperature.

In critical phenomena, universality arises because the renormalization group method yields iterative equations whose stable fixed-point solutions describe certain phase transitions. The property of a fixed point is that the detail of the initial condition from which you start the iterations gets lost as the stable fixed point is approached: hence, 'universal' critical exponents arise if the fixed point has a finite basin of attraction. In the case of critical phenomena, the 'phase space' is a space of Hamiltonians (more accurately said, thermodynamic potentials derived from partial trace operations), so that the details of the Hamiltonian get lost as $T \to T_c$: the basin of attraction includes all Hamiltonians with the same dimension and symmetry group. The attempts to generalize this method to include systems far from equilibrium such as the Navier–Stokes equations and fluid turbulence have produced no results of comparable importance. Feigenbaum, however, showed that a similar idea works for discrete one-dimensional maps that have a period-doubling limit, and the method was later generalized by others to include both conservative and dissipative two-dimensional maps.

So, the aim is to find an iterative description of period doubling whose limit yields the same fixed-point solution for all quadratic maps of the interval, irrespective of other details of the maps. Because the composition rule for period doubling is $2^n + 2^n = 2^{n+1}$, in analogy with the method of Chapter 5, we can begin with the functional composition rule

$$f^{(2^n)} \cdot f^{(2^n)} = f^{(2^{n+1})} \tag{6.11}$$

and the corresponding scaling ansatz is

$$f^{(2^n)}(x) \cong (-\alpha)^n \bar{f}((-\alpha)^n x), \tag{6.12}$$

where $\bar{f}$ is the fixed-point solution of (6.11). If we combine the rule (6.11) with the scaling ansatz (6.12), we obtain

$$f^{(2^n)} \cdot f^{(2^n)}(x) = (-\alpha)^{-n} \bar{f}((-\alpha)^n (-\alpha)^{-n} \bar{f}((-\alpha)^n x))$$
$$= (-\alpha)^{-n} \bar{f}(\bar{f}((-\alpha)^n x)). \tag{6.13}$$

Since it follows also that

$$f^{(2^{n+1})}(x) \simeq (-\alpha)^{-n-1} \bar{f}((-\alpha)^{n+1} x), \tag{6.12b}$$

we can combine this result with (6.11) and (6.13) to obtain

$$(-\alpha)^{n-1} \bar{f}((-\alpha)^{n+1} x) \simeq (-\alpha)^n \bar{f}(\bar{f}(-\alpha)^n x), \tag{6.14}$$

and also the fixed-point composition rule

$$\bar{f}(x) = -\alpha \bar{f}(\bar{f}(x/\alpha)) \tag{6.14b}$$

in the space of functions of noninvertible quadratic maps of the interval. This equation was first derived by Feigenbaum and Cvitanović.

Given a solution $\bar{f}(x)$, it is easy to verify that $g(x, \mu) = \mu \bar{f}(x/\mu)$ is also a solution for arbitrary μ. We can set the scale by choosing $\mu \bar{f}(0) = 1$.[3] Therefore, we can look for solutions of $g(x) = -\alpha g(g(x/\alpha))$ in the form

$$g(x) = 1 + bx^2 + \cdots, \tag{6.15}$$

where $b < 0$ is necessary. This expansion, solved to lowest order in x^2, yields the approximate results $1 = -\alpha(1 + b)$ and $b = -\alpha/2$. The condition $b < 0$ yields then $\alpha = 1 + \sqrt{3} \simeq 2.723 \cdots$ as a rough first approximation to the known result $\alpha = 2.5029 \cdots$.

The binary tree of Fig. (6.5), hence the invariant set of the map, has 2^∞ points when $D = D_\infty$ (for any $D < D_\infty$ the invariant sets are one stable

---

[3] With this terminology, it is necessary to work with the logistic map in the form $f(x) = 1 - ax^2$, rather than in the (transformed) form $Dx(1 - x)$. Then $a_x = 1.401 \cdots$ corresponds to $D_x = 3.569 \cdots$.

cycle of period 2^n and all unstable cycles of periods $2^{n-1}, 2^{n-2}, \ldots, 1$). Is the invariant set a Cantor set, a finite part of the interval, or some other uncountable set? Both the middle-thirds Cantor set and the continuum (with numbers expressed as binary strings) can be organized onto a complete binary tree of 2^∞ points. The invariant set turns out to be a nonuniform zero-measure Cantor set (an example of a multifractal), but we defer the discussion of this Cantor set until the end of Chapter 7.

In order to illustrate how the numbers α, a_∞ and the critical exponent δ can be computed by a renormalization group method, we use the logistic map in the form $f(x, a) = 1 - ax^2$, and follow a more elementary line of reasoning than the one that begins with (6.14b). Let us denote by $\tau(a_n) = 2^n$ the period of the 2^n-cycle, so that $\tau(a_{n+1}) = 2\tau(a_n)$. In the spirit of critical phenomena in statistical mechanics, we define a critical exponent v by assuming that

$$\tau(a_n) \sim (a_\infty - a_n)^{-v} \qquad (6.16)$$

as $n \to \infty$. Also in the spirit of critical phenomena is the consideration of the 'coupling constant' recursion formula (for $n \to \infty$),

$$a_{n+1} = R(a_n) \cong a_\infty + R'(a_\infty)(a_\infty - a_n) + \cdots, \qquad (6.17)$$

whose derivation we shall illustrate below by an example. If we combine $\tau(a_{n+1}) = 2\tau(a_n)$ with (6.17) then we obtain

$$|R'(a_\infty)|^{-v}(a_\infty - a_n)^{-v} \approx 2(a_\infty - a_n)^{-v}, \qquad (6.18)$$

and consequently

$$v = -\ln 2/\ln |R'(a_\infty)|. \qquad (6.19)$$

Because the bifurcation rate is geometric, $\delta(a_{n+1} - a_n) \approx a_n - a_{n-1}$, we can write

$$a_\infty - a_n \sim \delta^{-n}, \qquad (6.20)$$

and comparison with (6.17) yields

$$\delta = 1/|R'(a_\infty)|. \qquad (6.21)$$

In order to compute approximations to a_∞, v, and δ, we must estimate the magnitude of the slope $|R'(a_\infty)|$ at the critical point.

Let us denote by $S_n(a_n)$ the slope of the 2^nth iterate of $f(x, a)$ evaluated at any point on the stable 2^n-cycle:

$$S_n(a) = \prod_{i=1}^{2^n} f'(x_i, a). \qquad (6.22)$$

Because $|S_n(a)| = 1$ at a bifurcation point, we can then write

$$S_n(a_n) = S_{n+1}(a_{n+1}) \qquad (6.23)$$

and the critical parameter a_∞ is a fixed-point solution of this equation as $n \to \infty$ (one must solve (6.23) to find the recursion formula $a_{n+1} = R_n(a_n)$, where $R_n \to R$, the fixed-point function). Clearly,

$$\delta = 1/(da_{n+1}/da_n)_{a_n = a_x} \qquad (6.21b)$$

In order to obtain accurate approximations to δ, we must consider longer and longer cycles. In order to illustrate how the method works, it is enough to look at the bifurcation from a stable fixed point to a stable 2-cycle. The 1-cycles are given by solving $\bar{x} = 1 - a\bar{x}^2$ and the one of interest to us is

$$\bar{x} = \frac{-1 + \sqrt{1 + 4a}}{2a}. \qquad (6.24)$$

In this case, $S_1(a) = -2a\bar{x} = 1 - \sqrt{1 + 4a}$. The 2-cycle is defined by $\bar{x}_2 = f(\bar{x}_1) = f(f(\bar{x}_2))$ and consists of the points

$$\bar{x}_{1,2} = \frac{1 \pm \sqrt{4a - 3}}{2a}. \qquad (6.25)$$

yielding $S_2(a) = 4(1 - a)$. In the lowest-order approximation to the period-doubling limit, we naively set $S_1(a_\infty) \approx S_2(a_\infty)$ to obtain

$$a_\infty = (7 + \sqrt{17})/8 = 1.39\cdots, \qquad (6.26)$$

$$\delta = 1/(da_{n+1}/da_n)_{a_x} \cong 2\sqrt{1 + 4a_\infty} = 5.12\cdots \qquad (6.27)$$

and

$$v \approx \ln 2/\ln 5.12 \approx 0.42\cdots. \qquad (6.28)$$

Superficially, these numbers can be compared with $a_\infty = 1.401\cdots$ and $\delta = 4.669\cdots$, but because we considered the limit $n \to 1$ rather than the limit $n \to \infty$, the agreement between the estimates (6.26)–(6.28) the exact results should be considered accidental rather than significant (no such 'fast-and-easy' convergence follows for the circle map at its critical point, for example). The purpose here is not numerical accuracy, but to illustrate how the renormalization group method works.

Because elliptic points in conservative systems lose their stability by a pitchfork bifurcation, one should expect to find period doubling as a way to chaos in area-preserving maps (see MacKay in Cvitanović (1984)).

The renormalization group method in statistical mechanics was once described by K. G. Wilson as a 'search for analyticity' (Wilson, 1974). Phase transitions are driven by divergent fluctuations in one variable or another, causing power series expansions of the thermodynamic potentials to diverge at the critical point. In other words, if one starts with an expansion at $T > T_c$, then an infinite resummation of the series is necessary in order to continue analytically to $T < T_c$. The renormalization group, an iterative method, was adapted to statistical mechanics in order to handle that problem in a smooth, analytic way. At the transition to chaos, the invariant set of a map is typically a Cantor set, one of the most singular objects in mathematics. One cannot, for example, define a probability *density* on a Cantor set, because it would require the mathematical nonsense of uncountably many Dirac delta functions. The $f(\alpha)$-spectrum, which we discuss in Chapter 7, can be thought of as an 'analytic' (smooth) way to describe Cantor sets, although the $f(\alpha)$ spectrum is not restricted to Cantor sets.

The transition to chaos is an order–disorder transition. We can understand this in the following way. In Chapters 7 and 8 it is observed that the symbol sequences of chaotic orbits can be thought of as lattice configurations and an idea of lattice disorder immediately arises: different lattice configurations can give rise to the same Liapunov exponent. If there are $N(\lambda)$ configurations with the same Liapunov exponent λ, then the Boltzmann entropy of this disorder is $S(\lambda) = \ln N(\lambda)$. The fractal dimension of the set of initial conditions that generates these lattice configurations is the Boltzmann entropy per iteration divided by the Liapunov exponent. We have already seen an example of this: the fractal dimension of the uniform Cantor set generated by a symmetric tent map with slope $a > 2$ is $D_0 = \ln 2/\ln a = \ln 2/\lambda$. $\ln 2$ is the maximum possible entropy per iteration, reflecting the very special class of initial conditions where half of the iterates fall into $[0, 1/a]$ and half into $[1 - 1/a, 1]$ as $n \to \infty$, and so on. The entropy $\ln 2$ per iteration represents maximum disorder, corresponding to a certain class of symbol sequences of the map; this point of view is developed in Chapter 8. The stable periodic orbits at $D \leq D_\infty$ represent perfect order: they are a zero-entropy system, with only one stable periodic orbit for each (negative) Liapunov exponent. A difference with critical phenomena is that even simple maps such as the logistic map can have a sequence of critical points where chaos disappears and then reappears, the so-called windows of periodicity that occur for $D_\infty < D < 4$. In ^{4}He, in contrast, the system is superfluid at $T < T_\lambda$ and normal for $T > T_\lambda$ – superfluidity does not reappear at higher temperatures. The reader is referred to the monograph by Collet and Eckmann

(1980) for a description of the infinity of bifurcations that occurs in the range $D_\infty < D < 4$, where, depending upon D, either chaotic or stable periodic motion can occur.

That period doubling is relevant for understanding the transition to turbulence in fluid mechanics was shown by Libchaber (Libchaber *et al.*, 1984). Later, in Chapter 10, we show how maps that generate fully developed chaos can be used to reproduce the statistics inferred from one-dimensional cuts of fully developed turbulence. For that, as well as to analyze the period doubling limit's invariant set, we need the ideas of multifractal sets and multifractal probability distributions.

7

Introduction to multifractals

7.1 Incomplete but optimal information: the natural coarsegraining of phase space

How can one characterize an invariant set of a chaotic dynamical system and the motion on that set in a way that makes basic theoretical quantities of interest available for comparison with experiment or computation? Because a formulation of the theory based upon finite resolution is needed in both cases, we generalize the results of Chapter 4 and introduce a formalism that is based upon a hierarchy of more and more refined coarsegrained descriptions of both the invariant set and the motion. Within this finite description, it is sometimes useful to introduce the $f(\alpha)$-spectrum. The $f(\alpha)$-spectrum, in the end, depends upon two things: the invariant set and the frequency with which the dynamical system visits different parts of the invariant set. A coarsegrained description of both properties is demanded by virtue of the fact that invariant sets of chaotic dynamical systems have, mathematically, the cardinality of the continuum.

We have seen by the example of the Lorenz model that strange attractors can lead at least approximately to one-dimensional chaotic maps (Chapter 2). In that case, one starts with a certain time series $\{z(t)\}$ that follows from plotting maxima of the z coordinate against the time, yielding a sequence of numbers $z(t_n) = z_n$ at discrete times t_n, where $n = 1, 2, 3, \ldots$. In the Lorenz model, the phase space flow is three-dimensional and the points on the orbit that include the discrete time series $\{z_n\}$ do not fall within a single plane, nor is there any known simple analytic pattern among the various times $\{t_n\}$. None of that matters; the central point for us is that when z_{n+1} is plotted against z_n then the points fall approximately onto a single curve (Fig. 2.3) with a single maximum, suggesting an underlying one-dimensional map

$$z_{n+1} = f(z_n; \beta, \sigma, \rho), \tag{7.1}$$

with control parameters (β, σ, ρ), as the approximate pseudorandom number generator for the maxima of $z(t)$ in the Lorenz model. There is an underlying pseudorandom number generator for the maxima of $z(t)$ *because* the motion is deterministic. The resulting map is, as we know from our analysis in Chapter 2, only *approximately* one-dimensional. Furthermore, we know that the Poincaré map of a three-dimensional set of differential equations is invertible (solutions of the differential equations are invertible so long as the Jacobian is nonzero), so it is clear that a one-dimensional chaotic map, which is by necessity *noninvertible*, cannot contain *complete* information about the dynamics of the full, three-dimensional Lorenz model. That does not matter: what matters is that the map derived from the Lorenz plot is *approximately* one-dimensional, that it has a *single* maximum, and that it contains certain useful *partial* information about the Lorenz model. In other words, it permits us to gain admittedly *incomplete* information about a very difficult dynamics problem.

One general prescription for the analysis of chaotic systems is therefore as follows: one begins with a time series, experimental or computational in origin, and attempts to extract from that series a low-dimensional discrete map. There is no single, fixed prescription for how to do this. One way is to construct a Lorenz plot from the maxima of a time series, but this may be fruitful only if the underlying map is approximately one- or two-dimensional. Even if the points don't fall approximately onto a single curve, the plot may still reveal order that was not at all evident merely by looking at the time series. A second method of attack is described in Chapter 10, using the $f(\alpha)$-spectrum, and most of this chapter is devoted to laying the foundation for that method.

Time series from experiment can be understood from the standpoint of discrete maps if noisy, macroscopic phenomena in nature are not random but are pseudorandom. Mainly, it is a question whether one can obtain a useful description by using a low-dimensional map. In a truly random system, there would be a failure of cause and effect in the sense that, given the state of a random system at one time, there is in principle and in practice no way to predict the state of the system at the next instant in time. Faced with such a system, one could at best try to discover or prescribe probabilities for the various alternative states. Deterministic chaos, in contrast, is pseudorandom: given the state of the dynamical system at one time, the state at the next instant follows from the laws of

arithmetic. The *apparently* random (pseudorandom) behavior of deterministic chaotic systems is accounted for by the fact that deterministic
trajectories can generate probability distributions (distributions of iterates
of the map, e.g.) whenever there is a positive Liapunov exponent. This
feature permits us in principle to account for the apparent irreproducibility of *experimental* results where statistics follow from repeated attempts
to reproduce the same irregular time series by attempting to prepare the
dynamical system in exactly the same initial state. The attempted state
preparation fails to reproduce identical time series in repeated experiments
on deterministic chaotic systems owing to finite resolution, because the
butterfly effect generates a different chaotic trajectory each time that the
dynamical system starts from a nearby initial condition. Different initial
conditions will arise in any attempt to repeat the preparation of a definite
experiment. Usually, it is not the initial conditions of the dynamical system
that are controlled in state preparation, but some other set of parameters.
In that case, an entire class of initial conditions can be consistent with
the externally constrained variables, even initial conditions that are not
near one another in phase space. Classes of initial conditions that generate
the same statistics are considered in Chapter 9. The main point is that
we can account deterministically for apparent randomness in nature if we
can deduce a map or even a class of related maps from experimental data.

It is convenient to introduce the $f(\alpha)$-spectrum in the context of
one-dimensional maps of the unit interval, although $f(\alpha)$ is not confined
to one-dimensional chaotic maps, nor to deterministic dynamics. The
reason for our choice is to illustrate as clearly and as simply as possible
the content of $f(\alpha)$ for the case of deterministic chaos. Furthermore,
complete generality in presentation is not necessary: if one understands
the construction of $f(\alpha)$ for a map of the interval with a single maximum,
then one can generalize the method directly to handle other classes of
maps. We shall discover that the three important questions are: (i) What
is the order of the tree that the dynamical system generates and onto
which the dynamics can be organized? (ii) Is the tree complete or
incomplete? (iii) What are the sizes of the first few generations of intervals?
Our method proceeds systematically by the organization of the hierarchy
of finite (coarsegrained) descriptions of the map's invariant set onto a
tree. The tree, its order, its degree of completeness, and all of the other
important details are generated by the dynamical system, as we illustrate
below for the class of systems where each map generates a complete binary
tree.

Let us review the reason why it is necessary to insist upon providing
a finite description of the dynamics, one that avoids taking the limit of

zero length scale in the theory. *Finite resolution* is a fact of life in both computation and in experiment, whereas a chaotic dynamical system, mathematically, has an invariant set that consists of 2^∞ points (the middle-thirds Cantor set, e.g., is the invariant set of the ternary tent map, and the entire unit interval is the invariant set of the binary tent map: both point sets have the cardinality 2^∞ of the continuum). In contrast, fractal scaling in *nature*, at least for macroscopic geometrical objects, occurs *only* over a finite range of scales $l_{max} > l_n > l_{min}$, where the upper limit is smaller than the geometric extent of the object and the lower limit is large compared with an average interparticle spacing. In other words, fractals *in nature* can be regarded as made up of a finite number of finite segments where the number and sizes of the segments that one can observe depends upon the resolution of the physical measurement. The mathematical analog of this is to work with a *hierarchy* of coarsegrained descriptions of the dynamics in phase space. For us, there is then the question: what determines this hierarchy of finite descriptions of the dynamics? Another way to put the question is: how can we use a map to construct a fractal recursively?

The invariant set of a chaotic map can be *arbitrarily* coarsegrained by the free choice of N_n intervals $\{l_i\}$, but there is also an *optimal partitioning*, a most efficient covering of the invariant set, that is determined by the map itself. In other words, there is a way to eliminate freedom of choice by the use of an algorithm and to benefit from it in efficiency and clarity of description. For example, we can always cover the invariant set of the ternary tent map by an *even* partitioning of 2^n intervals, where $l_n = 2^{-n}$, but we can learn directly from the ternary tent map that the most efficient covering for a given generation n is provided by 2^n intervals where $l_n = 3^{-n}$. How does the tent map give us this information? By backward iteration: the n-times backward iteration of the ternary tent map generates exactly 2^n intervals, each of width 3^{-n}, and the location of these intervals is such that the invariant set of the tent map, the middle-thirds Cantor set, always lies *exactly* beneath the intervals so constructed. No other choice of partitioning does this job as well as the tent map does it for us in backward iteration. These intervals are, in fact, the *signature* of the ternary tent map. Unfortunately, they are not a unique signature: they are also the signature of a ternary Bernoulli shift. However, they come as close as possible to providing us with a *finite* signature of the map, and they provide the optimal coarsegraining of the invariant set of the ternary tent map. They provide, in every sense of the word, the best *finite* description of the map's invariant set.

Now, we can generalize: consider any map of the interval, symmetric or asymmetric, that peaks at or above unity (Fig. 7.1) and contracts intervals in backward iteration. The map's invariant set is a repeller in forward iteration, but it is an attractor for backward iterations of the map. If we iterate the map n times backward, then N_n intervals $\{l_i^{(n)}\}$ are generated, and the invariant set of the map always lies beneath this covering, no matter how large n is taken to be. This is the natural partitioning of phase space by the map. If we specialize to maps that have precisely a two-valued inverse (Fig. 7.1), then $N_n = 2^n$ is the number of intervals that cover the invariant set in the nth generation of coarsegraining. We can think of the intervals as providing the finite description of the invariant set, which we can pin down to as high a finite accuracy as we please by taking n to be large enough.

This brings us to a practical important point: One will typically not be able to retrieve an exact map from the analysis of time series, but one may be able to extract a hierarchy of intervals $(l_i^{(n)}\}$ and probabilities P_i for visiting those intervals, for some range of the index $n = 1, 2, \ldots, n_{max}$. One can then try, to within experimental error, to infer a qualifying *class* of maps, as is illustrated in application in Chapter 10. From this standpoint, to say that one knows 'the attractor' (or repeller, or other invariant set) from experiment means that one knows at least the first few generations of intervals that are generated by the dynamical system. The

Fig. 7.1   We consider the class of maps that peak at or above unity and contract intervals in backward iteration.

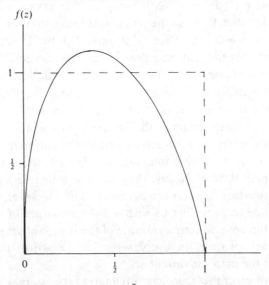

converse, from this standpoint, is that if one does not know at least the first-generation partitioning that is generated naturally by the dynamical system then one has no systematic information about the dynamical system's invariant set: *A noisy time series, taken alone with no further analysis, does not define an attractor (or repeller)*. To summarize our point of view, while a one-dimensional map provides a useful incomplete *theoretical* description of a higher-dimensional chaotic dynamical system, a finite collection of the intervals generated by the map provides a *more incomplete*, but practical, even optimal, description of the map. By stages, we are giving up the idea that we can know everything in infinite detail, which we cannot, in favor of concentrating upon quantities that can be either computed or measured. That is the reason for our insistence upon sticking to a finite-precision approach to chaotic dynamics.

While the hierarchy of intervals generated by backward iteration characterizes the map, the frequency with which the map visits the different intervals characterizes a set of initial conditions within the invariant set of the map. Given a noninvertible map of the interval

$$z_{n+1} = f(z_n), \tag{7.2}$$

the choice of an initial condition z_0 leads to a particular trajectory, a set of iterates $\{z_n\}$ of the map, and if we denote the natural partitioning by $\{l_i^{(n)}\}$, then we denote by $P_i^{(n)}$ the frequency with which the ith interval $l_i^{(n)}$ is visited by the particular trajectory. Of course, in order to compute the frequencies $P_i^{(n)}$ for a given generation n we must know a large number of the iterates of the map to at least an accuracy given by the *smallest* of the N_n intervals, $(l_i^{(n)})_{min}$. This can be accomplished by the use of the parallel-processing method (Chapters 2 and 4), whereby the initial condition is represented by an algorithm for an irrational number, but another way to carry this out in practice is developed in Chapter 9. In what follows, we make the assumption that our coarsegrained distributions $\{P_i^{(n)}\}$ are stationary, which amounts to assuming the infinite time limit in the iteration of the map. A closer look at this particular assumption is taken in Chapter 9, where we consider the effects of 'deterministic noise'.

In what follows, we assume three properties for our chosen class of maps: they have a single maximum, peaking at or above unity, the maps contract intervals in backward iteration, and the hierarchy of intervals so generated lives on a *complete* binary tree. This class of map includes both symmetric and asymmetric tent maps, symmetric and asymmetric logistic maps, and Bernoulli shifts. The main point is that the n-times backward iteration of the map generates a *complete binary tree*, with $N_n = 2^n$ intervals

$\{l_i^{(n)}\}$, not all of which need be equal, depending upon the asymmetry of the map (Fig. 7.2). These intervals provide the optimal partitioning because any other partitioning provides a less efficient covering of the map's invariant set. If the map peaks above unity, then there is an escape set with fractal dimension of one, and the invariant set is a zero-measure Cantor set with fractal dimension D_0 less than one. When the map peaks at unity then the invariant set is the entire unit interval, because the sum of the 2^n intervals always equals unity (Fig. 7.3). However, if the map is asymmetric, then the natural partitioning of the unit interval is uneven and this nonuniformity contributes to the $f(\alpha)$-spectrum, as will any nonuniformity in the distribution of iterates $\{P_i\}$.

There is, in the forward iteration of the map, the reverse of the contraction of the intervals that occurs in backward iteration, namely, the butterfly effect, whereby small differences in initial data are magnified exponentially fast. For example, if we apply the map to iterate a particular interval $l_i^{(n)}$ forward n times then this is more or less equivalent to the

Fig. 7.2   The n-times backward iteration of the unit interval by a map in our chosen class generates a complete binary tree of 2^n intervals $\{l_i^{(n)}\}$. The first two generations of intervals are shown here.

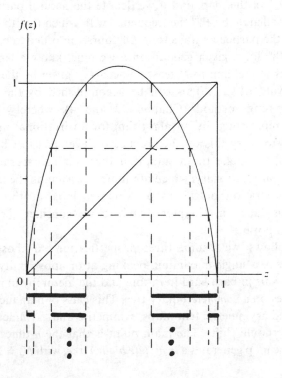

n-times forward iteration of an initial condition x_i that is covered by the interval and is specified to within a precision that is equal to the size of $l_i^{(n)}$. Here, the reverse of the construction of the interval $l_i^{(n)}$ by iterating the map n times backward, starting with the entire unit interval, is that we set the accumulated error δz_n equal to unity (equal to the maximum possible error) in the error propagation equation

$$\delta z_n \sim e^{n\lambda_i} \delta z_i, \qquad (7.3)$$

yielding, with the identification $\delta z_i = l_i^{(n)}$,

$$l_i^{(n)} \simeq e^{-n\lambda_i} \qquad (7.4)$$

as an asymptotic estimate for the size of the intervals, where

$$\lambda_i = \frac{1}{n} \ln|f^{(n)'}(z_i)| \qquad (7.4b)$$

is the Liapunov exponent for the orbit that evolves in time from the initial condition z_i.

We shall see that, when the map is asymmetric, there are infinitely many different Liapunov exponents and stationary distributions $\{P_i\}$ that occur for infinitely many different classes of initial conditions. Classes of initial

Fig. 7.3 The intervals are generated by iterating the unit interval backward by the map. If the map peaks at unity, then each generation of intervals precisely covers the unit interval, as is indicated for the two first-generation intervals ($l_1^{(1)} + l_2^{(1)} = 1$).

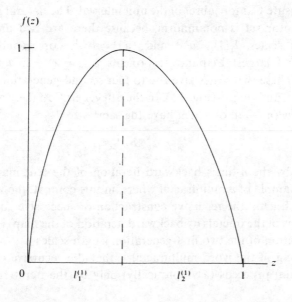

conditions will be precisely defined by using symbolic dynamics in the next two chapters: the point to be emphasized there is that a typical chaotic attractor or repeller *cannot* be characterized merely by specifying *one* Liapunov exponent and a single corresponding distribution of iterates. When a map gives rise to two Liapunov exponents, it typically has also a continuum of exponents, and this is illustrated by using the asymmetric tent maps as an example in the next sections. We shall also use the asymmetric tent map to illustrate the different meanings of the word *multifractal* that appear in the literature.

The Cantor sets that are covered naturally by 2^n equal scales $l_n = a^{-n}$ with $a > 2$ are examples of fractals that are geometrically selfsimilar: by an appropriate magnification of any infinite subset of this hierarchy of intervals, one can obtain a carbon copy of the original set of intervals. Geometrically selfsimilar Cantor sets provide mathematical examples of uniform Cantor sets: they are recursively constructed by the backward iteration of a single length scale $l_1 = a^{-1}$, yielding $l_n = a^{-n} = e^{-n\lambda}$ where $\lambda = \ln a$ is the symmetric tent map's only Liapunov exponent. There are also nonuniform fractals. Consider as the simplest example the invariant set of the tent map defined by

$$f(z) = \begin{cases} az, & z < b/(a+b), \\ b(1-z), & z > b/(a+b), \end{cases} \tag{7.5}$$

where we take $a > b$ and also $f_{max} = ab/(a+b) > 1$ (Fig. 7.4). When $ab > a + b$, then there is again an escape set, so that the invariant set is a zero measure Cantor subset of the unit interval. The natural partitioning of this Cantor set is nonuniform, because there are two unequal first-generation scales, $l_1(1) = a^{-1}$ and $l_2(1) = b^{-1}$, corresponding to the smallest and largest Liapunov exponents $\lambda_{min} = \lambda^1 = \ln a$ and $\lambda_{max} = \lambda^2 = \ln b$. These two scales give rise to four second-generation scales a^{-2}, $(ab)^{-1}$, b^{-2}, and $(ab)^{-1}$ (Fig. 7.5). In the nth generation there are 2^n scales, exactly $n!/m!(n-m)!$ of which have the same size

$$l_m = a^{-m}b^{-n+m}, \tag{7.6}$$

as is seen by the n-times backward iteration of the tent map. Here, we have an example of a multifractal where, in this context, the word means multiscale fractal: the recursive construction of successive coarsegrained descriptions of the fractal (by backward iteration of the map) is equivalent to the iteration of the two first-generation length scales a^{-1} and b^{-1}. So, one meaning of the word multifractal is that the recursive construction of the fractal proceeds (asymptotically) only by the iteration of two or more different length scales. The resulting Cantor sets are nonuniform: they are statistically but are not exactly geometrically selfsimilar, in the sense that magnified infinite subsets look, not exactly, but only more or less like the entire set. But this is not the only use of the word multifractal.

For multifractal sets, an entire *spectrum* of fractal dimensions is required for the description of the set. As an example, the fractal dimension of the set of intervals in the two-scale Cantor set with fixed ratio m/n ($n = 1, 2, 3, \ldots$) is given by $N(m) = l_m^{-D(m/n)}$, where $N(m) = n!/m!(n - m)!$ is the number of intervals with the same size l_m for a given integer n. By varying the ratio m/n from 0 to 1 as n is increased, we generate a sequence of single-hump $D(m/n)$ curves, whose peak (for large n) is the box-counting dimension of the entire two-scale Cantor set. This $D(m/n)$ spectrum is completely independent of any probability distribution that may be constructed by iterating the map forward. The generalization of this idea to maps with variable slope proceeds via entropies and Liapunov exponents for specific classes of initial conditions (Chapter 8).

There is also the idea of a 'multifractal measure' or, better stated, *multifractal probability distribution*. In a given generation, the coarsegrained distribution of iterates of the map $\{P_i\}$ provides the probability distribution, while the coarsegrained description of the invariant set, the intervals $\{l_i^{(n)}\}$, provide the coarsegrained 'support' of the distribution.

Fig. 7.4 Asymmetric tent map with slope magnitudes a and b, and where $ab > a + b$

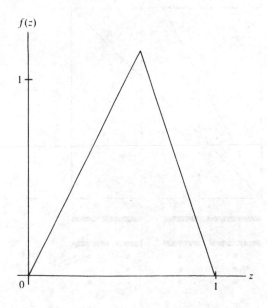

The main thing is that one can have a multifractal distribution whether or not the 'support' is fractal – all that is needed is a nonuniform, infinitely fragmented probability distribution. Consider as an example the symmetric tent map (set $a = b > 2$ in (7.5)), so that $l_n = a^{-n}$ and $D_0 < 1$. *Assume* next that we can find an initial condition of the tent map that generates a statistically independent set of iterates, where

$$P_m^{(n)} = P_1^m P_2^{n-m} \tag{7.7}$$

is the frequency with which the map visits the nth-generation bins of width l_m given by (7.6), and P_1 and P_2 are the first-generation frequencies, namely, the frequencies of visitation of the two first-generation intervals (exactly how such special initial conditions can be constructed for the tent map, and for other maps in our class, is described in Chapter 9). Then the nth-generation's generating function

$$\chi(q) = \sum_{i=1}^{N_n} P_i^q \tag{7.8}$$

Fig. 7.5   Two generations of natural intervals generated by the asymmetric tent map. The intervals, generation by generation, provide the optimal finite-precision description of the map's strange repeller.

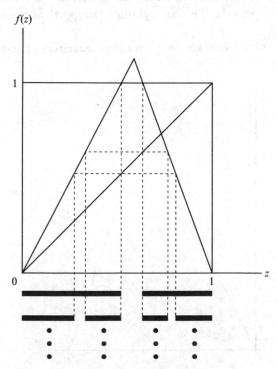

for the generalized dimensions D_q becomes a binomial distribution

$$(P_1^q + P_2^q)^n = a^{-n(q-1)D_q}, \qquad (7.9)$$

yielding

$$D_q = -\ln(P_1^q + P_2^q)/(q-1)\ln a. \qquad (7.9b)$$

Here, $D_\infty = -\ln P_1/\ln a$, $D_{-\infty} = -\ln P_2/\ln a$, and $D_0 = \ln 2/\ln a$ is the box-counting dimension, whereas the information dimension is given by $D_1 = (-P_1 \ln P_1 - P_2 \ln P_2)/\ln a$. The variation in D_q with q arises strictly from the probability distribution and not from any lack of uniformity of the support. It doesn't matter whether the support is fractal ($a > 2$) or nonfractal ($a = 2$). In either case, we have an example of a so-called *multifractal measure*. If, however, we set $P_1 = P_2 = \frac{1}{2}$, then we get $D_q = \ln 2/\ln a$ for all values of q. When $a = 2$, then we have a coarse-grained description of the (nonfractal) invariant density $p(x) = 1$ of the binary tent map, whereas if $a > 2$ then the distribution $P_i = 2^{-n}$ provides a coarsegrained description of the Cantor function (Chapter 4), which is a uniform measure on the uniform Cantor set. The Cantor function is everywhere nondecreasing, but is piecewise constant with zero slope, but it is not a multifractal probability distribution: the generalized dimensions D_q are all equal to D_0, for every value of the index q. Starting from any one of 'almost all' initial conditions defined on the fractal, the Cantor function is the probability distribution for a trajectory on the fractal. This means that the multifractal distributions, distributions where P_1 and P_2 are unequal in (7.7) and (7.8), must be generated by trajectories that start from any one of a set of initial conditions that occur with *zero measure* in the tent map's invariant set. We turn now to the $f(\alpha)$-spectrum as a convenient way to characterize multifractal sets and multifractal distributions.

7.2 The $f(\alpha)$-spectrum

We can think of a multiscale fractal as a hierarchy formed by a sequence of coarsegrainings that is described in the nth-generation by $N_n = \mu^n$ intervals $l_1^{(n)}, l_2^{(n)}, \ldots, l_{N_n}^{(n)}$ all with characteristic length $l_n < l_{n-1}$, and statistical selfsimilarity is reflected by the scaling law $N_n \simeq l_n^{-D_0}$, where D_0 is the fractal dimension (we remind the reader that the fractal dimension taken alone is inadequate to characterize an invariant set with a given distribution of iterates: an entire *spectrum* of scaling exponents D_q is required). The order of the tree onto which the intervals can be organized is the nearest integer greater than or equal to μ. Our choice of maps generates a complete binary tree, that is, $\mu = 2$.

Let us assume that scale $l_i^{(n)}$ has associated with it a probability P_i. In deterministic dynamics, P_i is the fraction of time that a dynamical system spends in the interval $l_i^{(n)}$ in the coarsegrained phase space. If the probability distribution is nonuniform, then we can begin with the generating function

$$\chi(q) = \sum_{i=1}^{N_n} P_i^q, \tag{7.10}$$

where

$$\chi(q) \approx l_n^{(q-1)D_q} \tag{7.10b}$$

defines the generalized dimensions D_q (cf. Chapter 4). In the special case where the dynamical system generates a uniform coarsegraining of phase space, so that all the intervals in one generation have the same size l_n, and, in addition, an initial condition is chosen such that the distribution of iterates is even over the $N_n = l_n^{-D_0}$ intervals, then we can write the visitation frequency in the special form $P_i = l_n^{D_0}$. We want to generalize this result in a way that avoids both the assumption of uniformity of the intervals and the assumption of uniformity of the distribution of iterates of the map. Therefore, we write the interval visitation frequency in the form

$$P_i = l_i^{\alpha_i}, \tag{7.11}$$

where the scaling index α_i may take on the same value α in

$$N(\alpha) = l_n(\alpha)^{-f(\alpha)} \tag{7.12}$$

of the N_n intervals, and l_n in (7.12) varies with α. Hence, (7.10) can be rewritten as

$$\chi(q) = \sum_\alpha l_n^{q\alpha - f(\alpha)} \tag{7.12b}$$

where $P_{\min} = l_n^{\alpha_{\max}}$ and $P_{\max} = l_n^{\alpha_{\min}}$, so that $D_\infty = \alpha_{\min}$ and $D_{-\infty} = \alpha_{\max}$.

The quantity $f(\alpha)$ is the fractal dimension of the subset of intervals that are visited with probability $P_i = l_i^\alpha$ by the map, for all generations n. $f(\alpha)$ can be related to the generalized dimensions as follows: Assume that (because $l_n \ll 1$) (7.12b) can be approximated by its $N(\alpha)$ largest terms,

$$\chi(q) \approx l_n^{q\alpha(q) - f(\alpha(q))}. \tag{7.12c}$$

Then $q\alpha(q) - f(\alpha(q))$ is the smallest of the exponents $q\alpha - f(\alpha)$, and $q = f'(\alpha(q))$ defines what we shall call $\alpha(q)$. It follows also that

$$(q - 1)D_q = q\alpha(q) - f(\alpha(q)), \qquad (7.13)$$

so that we can obtain the generalized dimensions D_q from the $f(\alpha(q))$ curve where q is the slope. Notice also that $f''(\alpha(q)) < 0$ for $|q| \ge 0$, so that $q = 0$ yields the maximum of $f(\alpha)$.

From the largest-term approximation (7.12c), it follows that $f(\alpha(q))$ is the fractal dimension of a certain *uniform* fractal that dominates the partition function that represents the nonuniform fractal for a given value of q. As q varies, different uniform approximating fractals are picked out, hence the variation in $f(\alpha)$ (the scale l_n in (7.12b) is fixed by a single value of q).

With $q = 0$, we have $f'(\alpha) = 0$, so that the box-counting dimension

$$D_0 = f(\alpha(0)) \qquad (7.14)$$

lies at the maximum of the $f(\alpha)$-curve (cf. Fig. 7.6).

Next, if we set $q = 1 + \delta q$ and denote $\alpha(1) = \alpha$, $\alpha'(1) = \alpha'$ when $q = 1$,

Fig. 7.6   The peak of $f(\alpha)$ yields the box-counting dimension D_0.

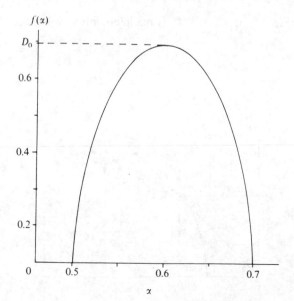

then we find that

$$D_1 \simeq \frac{1}{\delta q} [(1 + \delta q)(\alpha + \delta q \alpha') - f(\alpha) - \delta q \alpha' f'(\alpha)]$$

$$\simeq \alpha + \alpha'(1 - f'(\alpha) + \frac{1}{\delta q}(q\alpha - f(\alpha))$$ (7.15)

so that $\alpha(1) = f(\alpha(1))$, otherwise D_1 is infinite. Since $q = 1 = f'(\alpha(1))$, it follows that the information dimension

$$D_1 = \alpha(1)$$ (7.16)

lies on the $f(\alpha)$-curve at the point where $\alpha = f(\alpha)$ and $f'(\alpha) = 1$ (cf. Fig. 7.7).

For the correlation dimension D_2 (cf. Chapter 4), we have

$$D_2 = 2\alpha(2) - f(\alpha(2))$$ (7.17)

or $f(\alpha(2)) = 2\alpha(2) - D_2$, which corresponds to the geometric construction of Fig. 7.8. We remind the reader that the assumption of statistical independence of iterates of the map that is presumed in the definition of the correlation dimension cannot be taken for granted for arbitrary initial conditions of a dynamical system (see Chapter 9 for details).

Having introduced the relation of $f(\alpha)$ to D_q by using an approximation, we now show that the approximation is not completely necessary. We show in section 7.3, by an example, and also in Chapter 8, that α and

Fig. 7.7 $\alpha = f(\alpha) = D_1$ is the information dimension.

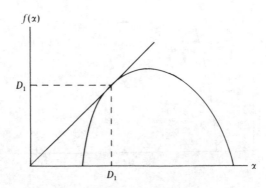

$f(\alpha)$ are, from the point of view of statistical mechanics of chaos, 'micro-canonical' quantities and provide a corresponding 'thermodynamic' interpretation of a chaotic dynamical system in statistical equilibrium. This point of view is illustrated conveniently in Chapter 8 for one-dimensional maps that generate fractal repellers. At this stage, we can simply define the canonical quantities $\langle \alpha \rangle$ and $\bar{f}(\langle \alpha \rangle)$ as follows: we define $\langle \alpha \rangle$ as the pseudo-average

$$\langle \alpha \rangle = \sum_{\alpha} \frac{\alpha l_n^{q\alpha - f(\alpha)}}{\chi(q)} = \tau'(q), \tag{7.18}$$

from which it follows that

$$\langle \alpha \rangle = \sum_{i=1}^{N_n} P_i^q \ln P_i \Bigg/ \sum_{i=1}^{N_n} P_i^q \ln l_n. \tag{7.18b}$$

We have not yet defined l_n in (7.10b), but it is clear that l_n must vary with q. In fact, there is no useful way to define l_n other than via the largest-term approximation to the generating function that we discussed above. With

$$\tau(q) = \left(\ln \sum_{i=1}^{N_n} P_i^q \right) \Bigg/ \ln l_n, \tag{7.19}$$

we can define the canonical fractal dimension

$$\bar{f}(\langle \alpha \rangle) = q\langle \alpha \rangle - \tau(q) \tag{7.20}$$

Fig. 7.8   Construction showing the geometric determination of the correlation dimension D_2.

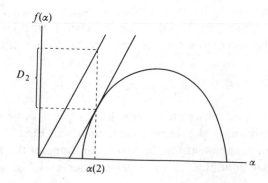

as a Legendre transform, so that the term-by-term microstructure of $P_i = l_n^{q\alpha - f(\alpha)}$ is reflected formally in the generating function

$$\chi(q) \approx l_n^{q\langle\alpha\rangle - f(\langle\alpha\rangle)}, \tag{7.21}$$

but this scaling behavior is guaranteed only if the largest-term approximation is valid in the first place.

The condition for using the $f(\alpha)$ formalism is that the probabilities P_i should be approximately constant, i.e., that a state of approximate statistical equilibrium holds. Because $\langle\alpha\rangle = d\tau(q)/dq$ in the Legendre transform (7.20), it follows that $q = df(\langle\alpha\rangle)/d\langle\alpha\rangle$ is the slope of the $f(\langle\alpha\rangle)$ curve, just as in the microcanonical case. The only catch in this otherwise nice formalism is that we must eventually resort to the largest-term approximation to the generating function in order to define the length scale l_n as a function of q. We illustrate this by an example in the next section.

7.3 The asymmetric tent map and the two-scale Cantor set ($f(\alpha)$ and entropy)

As an example of a chaotic dynamical system that generates a multiscale-fractal by the action of a continuum of different Liapunov exponents, consider the asymmetric tent map (Fig. 7.4)

$$f(z) = \begin{cases} az, & z < b/(a+b), \\ b(1-z), & z > b/(a+b), \end{cases} \tag{7.22}$$

where $\lambda_{min} = \ln a$ and $\lambda_{max} = \ln b$ if $a < b$, and where

$$f_{max} = ab/(a+b) \geq 1.$$

With $a = b > 2$, we generate recursively, by backward iteration of the unit interval, the uniform fractal with $D_0 = \ln 2/\ln a < 1$, but when $2 < a < b$, we obtain two first generation scales, $l_1^{(1)} = a^{-1}$ and $l_2^{(1)} = b^{-1}$. Then, when we iterate the unit interval twice backward, we obtain four intervals a^{-2}, $(ab)^{-1}$, b^{-2}, and $(ab)^{-1}$, and n-times backward iteration of the phase space covers the invariant set of the tent map by $N_n = 2^n$ intervals of the form

$$l_m = a^{-m}b^{-n+m}, \tag{7.23}$$

$N(m) = n!/m!(n-m)!$ of which are equal for a given value of m ($m = 0$, $1, 2, \ldots, n$). Two things should now be emphasized. First, the finite-time average Liapunov exponent for the first n iterations of the map starting with any initial condition z_i underneath interval l_m is, according to equation (7.46),

$$\lambda_m = \frac{m}{n} \ln a + \frac{(n - m)}{n} \ln b, \tag{7.24}$$

where $\lambda_{\min} = \ln a$ and $\lambda_{\max} = \ln b$, so that we can rewrite (7.23) as

$$l_m = e^{-m\lambda_{\min} - (n - m)\lambda_{\max}} = e^{-n\lambda_m}. \tag{7.23b}$$

Here we have an example where (7.4) holds for every generation n. We emphasize that the nonuniform fragmentation in phase space is caused in this case by the variation of the positive Liapunov exponent. The different possible Liapunov exponents are a consequence of different possible choices of initial conditions of the map.

The second point is that we can directly construct the microcanonical version of α and $f(\alpha)$. For that, we use a specific example that corresponds to a particular multifractal measure on the multiscale fractal invariant set.

If we can choose an initial condition[1] such that the map's iterates yield *deterministically* a statistically independent set of iterates then we can write

$$P_m = \bar{P}_1^m \bar{P}_2^{n-m}$$

$$\tag{7.25}$$

as the frequency with which the map visits any one of the $N(m)$ intervals l_m.

Writing

$$P_m = l_m^{\alpha_m} \tag{7.26}$$

permits us to define the microcanonical scaling index

$$\alpha_m = \frac{m \ln \bar{P}_1 + (n - m) \ln \bar{P}_2}{m \ln \bar{l}_1 + (n - m) \ln \bar{l}_2}, \tag{7.27}$$

with $l_1 = a^{-1}$ and $l_2 = b^{-1}$, which can be rewritten as

$$\alpha_m = \frac{m \ln \bar{P}_1 + (n - m) \ln \bar{P}_2}{-\lambda_m}, \tag{7.28}$$

showing that $\alpha_m \sim \lambda_m^{-1}$. When the Liapunov exponent is constant (as in the symmetric tent map when $\lambda = \ln a$), the result is that a *single* length scale $l_n = e^{-n\lambda}$ determines all of the higher-generation scales and the resulting fractal is geometrically selfsimilar. Whether or not $f(\alpha)$ is constant (equal to $\ln 2/\ln a$) for a geometrically selfsimilar fractal depends

---

[1] We discuss an example of this later. For arbitrary initial conditions, statistical independence will not hold.

upon whether initial conditions are chosen that generate an even distribution of iterates over the intervals. In other words, the variation of $f(\alpha)$
with α is not enough to tell you whether or not a given fractal is
nonuniform: variation in $f(\alpha)$ is produced by a nonuniform distribution
$\{P_i\}$ on a geometrically selfsimilar fractal, or even nonfractal, support.

What is the microcanonical $f(\alpha)$ for the two-scale Cantor set? We can
write

$$N(m) = \frac{n!}{m!(n-m)!} \simeq e^{nS(m)} \tag{7.29}$$

as the number of intervals with the same finite-time Liapunov exponent
and where Stirling's approximation for large n and m yields

$$S(m) = -m(\ln m/n)/n - (n-m)(\ln(1-m/n))/n \tag{7.30}$$

for the Boltzmann entropy per iteration of the map and the condition

$$q = \frac{\ln(1-x)\ln l_1 - \ln x \ln l_2}{\ln p_2 \ln l_1 - \ln p_1 \ln l_2} \tag{7.29b}$$

fixes q as a function of $m/n = x$ for $m = 0, 1, 2, \ldots, n$.

The combination of (7.29) with the box-counting condition

$$N(m) \simeq l_m^{-f(\alpha_m)} \tag{7.29c}$$

and with (7.24) yields

$$f(\alpha_m) = \frac{m \ln m/n + (n-m)\ln(n-m)/n}{m \ln \bar{l}_1 + (n-m)\ln \bar{l}_2}. \tag{7.31}$$

We can also write (7.31) in the very suggestive form

$$f(\alpha_m) = S(m)/\lambda_m, \tag{7.32}$$

with the interpretation that this is the fractal dimension of the set of initial
conditions that yields exactly the Liapunov exponent

$$\lambda_m = \frac{m}{n}\ln a + \frac{(n-m)}{n}\ln b \tag{7.33}$$

for the asymmetric tent map (fixed m corresponds to the class of initial
conditions where the slope of the tent map equals a for a fraction m/n
iterations of the map.

We note that in equation (7.30) there are two extreme limits: $m = n$
corresponds to zero entropy because there is only one way for the interval
a^{-n} to occur in generation n. This is also the $q \to \infty$ limit, where

$\chi(q) \sim P_{max}^q \sim l_n^{qD_x} \sim l_n^{q\alpha_{min}}$, so that (if we take $\bar{P}_1 > \bar{P}_2$) $\bar{P}_1 = \bar{l}_1^{\alpha_{min}}$ and $f(\alpha_{min}) = 0$. The zero entropy limit where $m = n$ corresponds to $\bar{P}_2 = \bar{l}_2^{\alpha_{max}}$, and it follows that $f(\alpha_{max}) = 0$.

In the largest-term approximation, $q = 0$ should yield the box-counting dimension $D_0 = f_{max}(\alpha)$, which means in our microcanonical formulation that

$$\frac{S'(m)}{\lambda'(m)} = \frac{S(m)}{\lambda(m)} \tag{7.34}$$

must be solved for m. This value of m then determines f_{max} by substitution into equation (7.31). If we write $x = m/n$ and $1 - x = (n - m)/n$, then (written as a function of x) the condition for the maximum of $f(\alpha)$ is

$$\ln x \ln l_2 = \ln(1 - x) \ln l_1. \tag{7.35}$$

If we begin with the definitions

$$\langle \alpha \rangle = \sum_{i=1}^{N_n} P_i^q \ln P_i \Big/ \sum_{i=1}^{N_n} P_i^q \ln l_n, \tag{7.19b}$$

$$\tau(q) = \ln \sum_{i=1}^{N_n} P_i^q / \ln l_n \tag{7.19}$$

and the Legendre transform

$$\bar{f}(\langle \alpha \rangle) = q\langle \alpha \rangle - \tau(q), \tag{7.20}$$

where $\langle \alpha \rangle, \tau, \bar{f}\langle \alpha \rangle$, and l_n are parametrized by q, then we can also compute the multifractal spectrum. For example, for a choice of initial conditions that yield statistical independence of the distribution of iterates of the asymmetric tent map, we can rewrite the generating function as

$$\chi(q) = (\bar{P}_1^q + \bar{P}_2^q)^n \tag{7.10b}$$

to obtain

$$\langle \alpha \rangle = \frac{\bar{P}_1^q \ln \bar{P}_1 + \bar{P}_2^q \ln \bar{P}_2}{(\bar{P}_1^q + \bar{P}_2^q) \ln l_n/n}, \tag{7.19c}$$

and

$$\tau\langle q \rangle = \frac{\ln(\bar{P}_1^q + \bar{P}_2^q)}{\ln l_n/n}, \tag{7.20b}$$

where l_n is to be determined by the largest-term approximation discussed above. The binomial $f(\alpha)$-spectrum for $\bar{P}_1 = \frac{7}{10}$, $\bar{P}_2 = \frac{3}{10}$, and $a = b = 2$ is shown as Fig. 7.9.

We know that $l_n \sim \bar{l}_1^n$ as $q \to \infty$ and $l_n \sim \bar{l}_2^n$ as $q \to -\infty$ (with $\bar{P}_1 > \bar{P}_2$) and we know from (7.20) that $q = \bar{f}'(\langle \alpha \rangle)$ is the slope of the $\bar{f}(\langle \alpha \rangle)$ curve, reflecting the fact that l_n in equations (7.19b) and (7.20) depends upon q. For $q \to \infty$, we obtain

$$\langle \alpha \rangle_{min} = \ln \bar{P}_1 / \ln \bar{l}_1, \tag{7.36}$$

and $q \to -\infty$ yields

$$\langle \alpha \rangle_{max} = \ln \bar{P}_2 / \ln \bar{l}_2, \tag{7.37}$$

if $\bar{P}_1 > \bar{P}_2$. It follows easily that $\bar{f}(\langle \alpha \rangle_{min}) = \bar{f}(\langle \alpha \rangle_{max}) = 0$, just as in the microcanonical case. Initial conditions that produce the equal-frequency result $\bar{P}_1 = \bar{P}_2 = \frac{1}{2}$ yield the maximum-entropy result $D_1 = 2 \ln 2 / \ln ab$.

In general, in order to determine the length scale l_n in (7.19c) and (7.20b) as a function of q, one must solve the microcanonical condition $q = \mathrm{d}f(\alpha_m)/\mathrm{d}\alpha_m$ for m as a function of q, substitute the result into (7.24) and identify the resulting scale $l_{m(q)}$ as l_n.

We can also introduce the generating function

$$\Gamma_n(q) = \sum_{i=1}^{N_n} \frac{P_i^q}{l_i^{\tau(q)}}, \tag{7.38}$$

which is either decreasing or increasing as n increases, but is of the order

Fig. 7.9 $f(\alpha)$-spectrum for a statistically independent distribution of iterates of the binary tent map for the case $P_1 = \frac{3}{10}$ and $P_2 = \frac{7}{10}$; distributions without statistical independence may yield qualitatively similar plots.

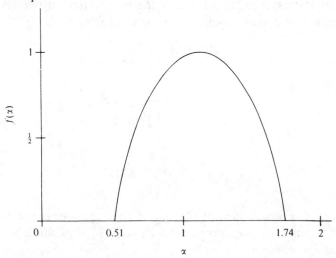

of unity if $\tau(q) = (q - 1)D_q$; this follows from the largest-term approximation to (7.38), using (7.11) and (7.12). Therefore, one can compute $f(\alpha(q)) = q\tau'(q) - \tau(q)$ and $\langle\alpha\rangle = \tau'(q)$ by solving first for $\tau(q)$. Our new generating function is a generalization of the condition

$$\sum_{i=1}^{N_n} l_i^{D_H} \approx 1 \tag{7.38b}$$

for the Hausdorff dimension D_H, which is itself a generalization to nonuniform intervals of the box counting condition $N(l_n)l_n^{D_0} \approx 1$. For the case of equal probabilities, setting (7.38) equal to unity yields

$$N_n^{q(\tau)} = \sum_{i=1}^{N_n} l_i^{-\tau} \tag{7.39}$$

and, as an example, for the two-scale Cantor set we obtain

$$q(\tau) = \ln(a\tau + b\tau)/\ln 2. \tag{7.40}$$

If we set $q = 0$, then the condition for the Hausdorff dimension (the peak of $f(\alpha)$) is simply $a^{-D_H} + b^{-D_H} = 1$, which can be solved for D_H if we specify a and b. Because we have assumed a statistically independent distribution where $P_1 = P_2 = \frac{1}{2}$, it follows from the limits where the magnitude of q goes to infinity (and using $\tau(q) = (q - 1)D_q$) that $\alpha_{min} = \ln 2/\ln b$ and $\alpha_{max} = \ln 2/\ln a$ if we take $a < b$. In general, one must solve (7.39) numerically for $\tau(q)$ in order to compute $\alpha(q)$ and $f(\alpha(q))$.

One can also get variation in $f(\alpha)$ from nonfractal distributions with nonfractal support: all that is needed is a singular probability density. Consider as an example the invariant density of the logistic map

$$p(z) = (\pi(z(1-z))^{1/2})^{-1}, \tag{7.41}$$

where we showed in Chapter 4 that

$$D_q = \begin{cases} 1, & q < 2, \\ \dfrac{q/2}{q - 1}, & q > 2. \end{cases} \tag{7.42}$$

It follows that $\alpha = 1$ when $q < 2$, which describes the region $(0, 1)$, where it follows from $f(\alpha) = (1 - q)D_q - q\alpha$ that $f(\alpha) = 1$ and the density is nonsingular. On the other hand, $\alpha = \frac{1}{2}$ if $q > 2$, and this index corresponds to the two singular points $z = 0$ and $z = 1$ with fractal dimension $f(\alpha) = 0$ where the probability density diverges as $z^{-1/2}$. The use of the $f(\alpha)$-spectrum is therefore not restricted to fractals or to multifractal probability distributions. One can generalize this viewpoint: sometimes, $f(\alpha)$

is described as a spectrum of singularities. This follows from defining a *coarsegrained* version of the density $\rho_i = P_i/l_i = l_i^{\alpha_i - 1}$, which diverges if $\alpha < 1$. For a nonsingular density, $\alpha = 1$ for every index i, and this must be true for every nonsingular invariant density (in general, we can write $\rho_i = c_i l_i^{\alpha_i - 1}$). Observe also that a uniform invariant density obeys $P_i = l_i$, so that the coarsegrained density $\rho_i = P_i/l_i = 1$ is independent of l_i.

For higher-dimensional maps, methods of analysis are also available. Two-dimensional maps have been studied by Cvitanović and co-workers (Cvitanović, Gunaratne, and Procaccia, 1988); there, the study of the unstable periodic orbits yields a hierarchy of intervals that are organized onto a tree, and the 'pruning' of orbits that are forbidden by the dynamics leads to incomplete trees. The incomplete trees that occur for higher-dimensional maps that include the Henon map are generated by one-dimensional logistic and related one-hump maps that peak below unity. One motivation for study the unstable periodic orbits is that they are topological invariants. However, they are not the only topological invariants: the nonperiodic symbol sequences of a class of maps are also topological invariants. The study of the hierarchy of unstable periodic orbits is an extremely interesting mathematical problem, which we do not cover in this text, where preference is given to the study of the unstable nonperiodic orbits. As we show in Chapter 9, the study of symbol sequences of nonperiodic orbits leads directly to universal statistics. Finally, it was Cvitanović who first emphasized the importance of organizing chaotic dynamics onto trees.

Finally, one way to generalize (7.10b) without adding any complication is just to increase the order of the tree onto which the fractal set is organized. If we ask for a two-scale fractal on a complete tree of any order $t = 2, 3, 4, \ldots$, and also assume statistical independence for the probability distribution, then the generating function is

$$\chi_n(q) = (bP_1^q + (t - b)P_2^q)^n \qquad (7.10c)$$

where b takes on any of the integral values $1, 2, \ldots, t - 1$ (the generation of higher-order trees by maps is illustrated at the end of Chapter 9, and the generalization of (7.10c) to include fractals with more than two scales is clear). It is interesting that a tree of higher order allows for qualitatively different behavior than in the binary case: with $P_1 > P_2$, then $f(\alpha_{\min}) = -\ln b/\ln l_1$ and $f(\alpha_{\max}) = -\ln(t - b)/\ln l_2$, so that $f(\alpha)$ is nonzero at both end points except for the special case $b = 1$ (or $t - b = 1$), where $f(\alpha)$ vanishes at one end point but not at the other so long as $t > 2$. An example is shown as Fig. 7.10 for the case $l_1 = l_2 = \frac{1}{3}$ with $t = 8$, $b = 6$, $P_1 = \frac{1}{12}$, and $P_2 = \frac{1}{4}$. In this case, $\alpha_{\min} = 1.26$, $f(\alpha_{\min}) = 0.63$, $\alpha_{\max} = 2.26$, and

$f(\alpha_{max}) = 1.63$. Incomplete binary trees can also yield qualitatively different behavior for $f(\alpha)$ than follows from the complete binary case (see the discussion of the circle map in Halsey *et al.*, 1986).

7.4 Multifractals at the borderlines of chaos

As one approaches the transition to chaos from the regime where orbits are stable, either for the case of period doubling or for that of instability of quasiperiodic orbits for circle maps, the method of backward interation cannot be used because the average Liapunov exponent is either negative or zero, and, at any rate, vanishes at criticality. That is, the invariant sets of interest are not attractors for backward iteration of the respective maps. However, length scales (intervals) can be assigned by a different means, and one can still organize the hierarchy of intervals onto a tree (which is binary if $N_n = \mu^n$ with $1 < \mu \le 2$). The complete theory has been worked

Fig. 7.10 $f(\alpha)$-spectrum for a two-scale Cantor set on an octal tree. The chosen distribution of probabilities is statistically independent, corresponding to the generating function (7.10c), with $P_1 = \frac{1}{12}$ and $P_2 = \frac{1}{4}$. The computation was made for $-30 < q < 30$; the required infinite slope magnitude of $f(\alpha)$ at α_{min} and α_{max} is beyond numerical resolution.

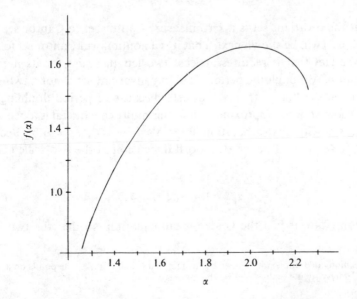

out by Feigenbaum (1980) and is a lengthy but readable exercise in binary arithmetic; we describe only the results in what follows.

At the period-doubling transition, the fixed-point equation for quadratic maps of the interval is

$$g(z) = -\alpha_{PD}g(g(z/\alpha_{PD})), \tag{7.43}$$

where $\alpha_{PD} = 2.5029\cdots$. In this case, it is clear from the bifurcation diagram (see Chapter 6) that the invariant set that we want to describe lies on a complete binary tree. It is possible to interpret α_{PD}^{-n} and α_{PD}^{-2n} as scales (see Halsey *et al.*, 1986):

$$l_i^{q\alpha_i} \to (\alpha_{PD}^{-n})^{q\alpha_{min}} \tag{7.44}$$

as $q \to -\infty$ and

$$l_i^{q\alpha_i} \to (\alpha_{PD}^{-2n})^{q\alpha_{max}} \tag{7.44b}$$

as $q \to \infty$, so that α_{PD}^{-n} and α_{PD}^{-2n} play the role of two separate length scales in the nth stage of coarsegraining. If we assume an equal frequency distribution $P_i = 2^{-n}$,[2] it follows that

$$\alpha_{min} = \ln 2/\ln \alpha_{PD} \simeq 0.38 \tag{7.45}$$

and

$$\alpha_{max} = \ln 2/2 \ln \alpha_{PD} \simeq 0.75. \tag{7.45b}$$

That the invariant set is a zero measure Cantor set (a Cantor set with $0 < D_0 < 1$) will be shown next. That it is a nonuniform Cantor set follows from the fact that it requires at least two length scales for its iterative description. Although the period-doubling invariant set is not a two-scale Cantor set, the binary tree is complete (because of period doubling), so that the *lowest* order approximation to this multiscale fractal is a two-scale Cantor set with first-generation intervals $l_1 = \alpha_{PD}^{-1}$, $l_2 = \alpha_{PD}^{-2}$. Because $l_2 = l_1^2$, the equation for the Hausdorff dimension of the two-scale Cantor set is

$$\alpha_{PD}^{-D_H}(1 + \alpha_{PD}^{-D_H}) = 1. \tag{7.46}$$

Equation (7.46) is just the golden-mean equation, so that the two-scale

---

[2] This is motivated but is not justified by the fact that, when $D \leq D_\infty$, the points on a stable 2^n cycle are visited equally often by the map as $n \to \infty$.

approximation yields

$$D_H \approx \frac{\ln(2/(\sqrt{5}-1))}{\ln \alpha_{PD}} \simeq 0.5245, \tag{7.47}$$

compared with the estimate $D_H \approx 0.537$ obtained by Halsey *et al.* by a numerical method.

At the critical parameter value where a circle map loses invertibility, the irrational winding numbers w that yield quasiperiodic orbits are confined, as a function of the bare winding number Ω, to a nonuniform Cantor set with fractal dimension $D_0 \approx 0.87$. The hierarchy of length scales can be organized onto an incomplete binary tree. In order to compare the theory with experimental time-series, it is of interest to know $f(\alpha)$ for the quasiperiodic orbit with golden-mean winding number at criticality. As $q \rightarrow -\infty$, the intervals are given by

$$l_i^{q\alpha_i} \sim (\alpha_{GM}^{-n})^{q\alpha_{min}}, \tag{7.48}$$

while

$$l_i^{q\alpha_i} \sim (\alpha_{GM}^{-3n})^{q\alpha_{max}} \tag{7.48b}$$

as $q \rightarrow \infty$, so that, again, we have a problem involving more than two length scales. In (7.43), α_{GM} is the scaling parameter in the circle map fixed-point equation

$$f(z) = \alpha_{GM}^2 f(\alpha_{GM}^{-1} f(\alpha_{GM}^{-1} z)). \tag{7.49}$$

According to Arnol'd, stable quasiperiodic orbits with golden-mean winding number yield a uniform invariant density on the circle as the number of iterations $n \rightarrow \infty$. If we also assume a uniform distribution on the nonuniform intervals at criticality, $P_i = 1/F_n$ for finite n, then $P_i \rightarrow w^n$ as $n \rightarrow \infty$, yielding

$$\alpha_{min} = -\ln w/3 \ln \alpha_{GM} \approx 0.6326 \cdots \tag{7.50}$$

and

$$\alpha_{max} = -\ln w/\ln \alpha_{GM} \approx 1.8980 \cdots. \tag{7.50b}$$

In this case, $D_0 = f_{max} = 1$ because we consider the distribution of iterates of a circle map, but in the determination of the $f(\alpha)$ spectrum, the required coarsegrained intervals are nonuniform, even for large n.

8
Statistical mechanics on symbol sequences

8.1 Introduction to statistical mechanics

In the last chapter, we found that the fractal dimension can be understood as the ratio of a Boltzmann entropy to a corresponding Liapunov exponent. We also met and used a certain Legendre transform that reminds us of thermodynamics. Are these results accidental, or are they the sign of a deeper analogy that we have yet to explore?

Let us begin with the statistical equilibrium of a chaotic dynamical system for the special case where the distribution of iterates is even over the natural coarsegraining: $P_i = N_n^{-1}$ is the frequency with which the interval $l_i^{(n)} = l_i$ is visited. Then the condition $\Gamma_n(q) \sim 1$ yields the result

$$N_n^{q(\tau)} = \sum_{i=1}^{N_n} l_i^{-\tau}, \tag{8.1}$$

where $\tau(q) = q\alpha - f(\alpha)$, and we can attempt formally to treat (8.1) as a classical statistical mechanical partition function

$$Z_n(\beta) = \sum_i l_i^{\beta} \tag{8.2}$$

if we make the identification of the inverse temperature

$$\beta = -\tau(q) \tag{8.3}$$

and Gibbs potential per iterate,

$$g(\beta) = -q \ln N_n/n, \tag{8.3b}$$

on the assumption that the 'thermodynamic limit' $Z_n(\beta) \sim e^{-ng(\beta)}$ exists. In Chapter 7 we argued that, for large n, we expect that

$$l_i \sim e^{-n\lambda_i}, \tag{8.4}$$

where $\lambda_i = \ln|f^{(n)'}(z_i)|/n$ is the Liapunov exponent of the map $z_{n+1} = f(z_n)$ that (through backward iteration) generates the natural coarsegraining $\{l_i\}$. If we substitute (8.4) into (8.2), then we are indeed reminded of a one-particle canonical partition function with single-particle energy levels given by $n\lambda_i$:

$$Z_n(\beta) = \sum_{i=1}^{N_n} e^{-n\beta\lambda_i}. \tag{8.1c}$$

In this case, we can use our analogy to predict the entropy per iteration in the canonical ensemble from the Legendre transform

$$S(\lambda) = \beta\lambda - g(\beta), \tag{8.5}$$

so that

$$\lambda = dg(\beta)/d\beta \tag{8.6}$$

follows if we require

$$\beta = dS(\lambda)/d\lambda. \tag{8.6b}$$

Before going further, we should check to see whether we can make sense of these equations for the simplest case, namely, the dynamics of the symmetric tent map. For the tent map with slope $a > 2$, we have $l_n = a^{-n}$ and $N_n = 2^{-n}$, so that

$$Z_n(\beta) = 2^n e^{-n\beta\lambda}, \tag{8.1d}$$

which leads to

$$g(\beta) = \beta\lambda - \ln 2, \tag{8.5b}$$

where $\lambda = \ln a$. The resulting entropy per iteration $S(\lambda) = \ln 2$ corresponds to the particular set of initial data where half of the iterates of the map fall into the left interval $[0, 1/a]$ and the other half fall into the right interval $[1 - 1/a, 1]$, in agreement with the original assumption that $P_i = 2^{-n}$. In this case $\beta = dS/d\lambda = 0$, which is the correct statistical mechanical result (infinite temperature) whenever all possibilities occur with equal likelihood. Also: ln 2 is the largest possible entropy, and the maximum entropy should only occur at infinite temperature. So far, so good, but does the analogy go further?

In our simple example, $D_q = D_0 = \ln 2/\ln a = \alpha = f(\alpha)$ because of the use of a *uniform* distribution on a *uniform* partitioning, in agreement with $\beta = -\tau = 0$. In the event that the partitioning $\{l_i\}$ is uneven (for any asymmetric tent map, e.g.) but the probability distribution is still even, then (with $N_n = \mu^n$) we can make the dictionary, called Feigenbaum's dictionary,

$$\alpha = \ln \mu / g'(\beta),$$

$$D_q = \beta(1 + g(\beta)/\ln \mu), \qquad (8.7)$$

$$f(\alpha) = \beta - g(\beta)/g'(\beta),$$

in addition to the identification of the inverse temperature as $-\tau$ in (8.3) and the connection of q with g in (8.3b).

Therefore, in accordance with classical thermodynamics, quantities of interest can be computed from the Gibbs potential and its derivatives. It follows that $f(\alpha)$ has the formal character of a thermodynamic quantity, but our formal thermodynamics is of an unexpected sort: it describes the statistical equilibrium of a dissipative dynamical system far from *ordinary* thermal equilibrium. That is because one-dimensional maps can only represent driven systems, systems that experience a net energy flow. The condition for statistical equilibrium is that the distribution $\{P_i\}$ is stationary and that the scales $\{l_i\}$ are also stationary. It is remarkable that the classical thermodynamic formalism is so general that it works here; in sections 8.3 and 8.4, we shall find that the formalism of one-dimensional statistical mechanics has the same broad applicability, far beyond the conservative dynamics that it was originally invented to describe.

In order to go further, we must first try to generalize the preceding results. Originally, we started with the Γ-generating function of Chapter 7 and then derived the canonical partition function $Z_n(\beta)$ from the assumption of an even distribution of iterates, $P = N_n^{-1}$. We now *abandon for the time being all underlying assumptions about probability distributions* and directly study the partition function

$$Z_n(\beta) = \sum_{i=1}^{N_n} l_i^\beta \qquad (8.2)$$

for what it is worth, namely, a quantity that is characterized by the hierarchy of intervals $\{l_i\}$, that is to say, by the underlying map that generates the intervals. With no connection to a particular distribution $\{P_i\}$, (8.3), (8.3b), and (8.7) do not hold, and we have at this initial stage *no* interpretation for the arbitrary parameter β – it is simply a free parameter in the generating function, at this stage of discussion. *Therefore, $f(\alpha)$ cannot be derived or even defined under these primitive circumstances.* However, we know that the interesting mathematical behavior of $f(\alpha)$ follows from the fact that it is proportional to a microcanonical Boltzmann entropy, and the same microcanonical entropy $S(\lambda)$ appears in our partition function: therefore *we can just as well study $S(\lambda)$ as a function of λ instead of $f(\alpha)$.*

In order to see this, consider the partition function in the form

$$Z_n(\beta) = \sum e^{-n\beta\lambda_i}, \tag{8.1c}$$

but rewritten as a sum over Liapunov exponents rather than as a sum over intervals:

$$Z_n(\beta) = \sum N(\lambda) e^{-n\beta\lambda}. \tag{8.8}$$

Then $S(\lambda) = \ln N(\lambda)/n$ is the entropy associated with the disorder of $N(\lambda)$ intervals with the *same* exponent λ, just as in the case of $f(\alpha)$. Given the fact that there is a minimum and a maximum Liapunov exponent in the sum (8.8), negative temperatures $\beta = dS(\lambda)/d\lambda$ will appear because $S(\lambda)$ will have a finite maximum at a finite value of λ, just as in a two-level system in classical statistical mechanics or in any classical system with a bounded phase space. The negative temperatures are discussed and explained in section 8.4, but the fact that $S(\lambda)$ increases to a single maximum and then decreases, quite typically, means that its qualitative behavior is similar to that of $f(\alpha)$. Also, the condition $\beta = S'(\lambda)$ is formally analogous to $q = f'(\alpha)$, although β and q are unrelated arbitrary parameters in two entirely different generating functions.

8.2 Introduction to symbolic dynamics

In order to develop the statistical mechanical analogy further, we need the idea of a lattice, and that is conveniently provided by the symbol sequences of chaotic dynamics. Nonperiodic symbol sequences are models of disordered one-dimensional lattices, and are in one-to-one correspondence with binary expansions of the irrational numbers in the unit interval. These are the major themes to be developed in what follows.

We expect that our hierarchy of intervals can be organized onto a tree of some order. The symbol sequences can then be used as address labels for the different branches of the tree, that is to say for the different intervals. The resulting digital labeling of intervals paves the way to a clear and deeper exploitation of the statistical mechanical analogy, because it allows us to think directly of our partition function as one for disordered lattices made up of several different sorts of atoms, or of a classical spin system, whichever analogy one prefers.

For the sake of clarity, we continue to illustrate the development of the theory by the use of the same class of models, chaotic maps of the unit interval that peak at or above unity and that generate a binary tree with $N_n = \mu^n$ intervals $\{l_i^{(n)}\}$ in the n-times backward iteration of the unit interval. If the tree is complete, then $\mu = 2$; otherwise $1 < \mu < 2$. There are two first-generation intervals $l_1^{(1)}$ and $l_2^{(1)}$. Consider now a definite orbit of the map that begins from a definite initial condition x_0. If we write a binary string with the alphabet (L, R), where we write an L every time that the orbit visits $l_1^{(1)}$ and an R whenever $l_2^{(1)}$ is visited (Fig. 8.1), then the orbit of n iterates generates a particular n-bit string, or word, e.g. RRLRLLLRR $\cdots$ L. Now for the assignment of binary addresses to the intervals.

If the binary tree is complete, then it is clear that the binary labels on all branches of the tree are in one-to-one correspondence with all possible binary strings. But how do we know which branch (i.e., which interval) gets which binary address? For that information, we must ask the map. Each possible initial condition of the map is a point in the invariant set, and vice versa, and the entire invariant set is covered, generation by generation, by the natural coarsegraining. To each initial condition in the invariant set corresponds exactly one infinite-length symbol sequence, an

Fig. 8.1   Assignment of symbols in the two-letter alphabet of maps that peak exactly once.

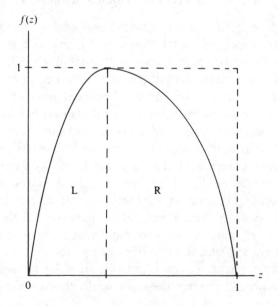

infinite binary string. All of the initial conditions that belong to the invariant set and are covered by the same interval $l_i^{(n)}$ begin with exactly the same n letters – they will differ in the remaining 2^∞ yet unspecified bits. In fact, the n-times forward iteration of an interval $l_i^{(n)}$ by the map is equivalent to the n-times forward iteration of any initial condition x_i that lies beneath that interval and is known or specified to within an accuracy $\delta x_i = l_i^{(n)}$. When this interval is iterated forward n times, the result is a single, definite n-bit symbol sequence. For the maps of our class, where the binary tree is complete, the resulting symbol sequences are shown for the first three generations in Fig. (8.2), that is to say, the correct address labeling is illustrated for the first three generations.

We can now introduce the following very useful notation: We write $l_i^{(n)} = l(\varepsilon_1 \varepsilon_2 \cdots \varepsilon_n)$, where (as is shown in Fig. 8.2) ε_i is either L or R and the string $\varepsilon_1 \varepsilon_2 \cdots \varepsilon_n$ is the symbol sequence for the first n iterations of the map, starting from any initial condition that is covered by this particular interval. The reader is reminded that this correspondence is possible only for the *natural partitioning*. Hence, if we iterate the interval $l_i^{(n)}$ n times forward with the map, then the result is

$$f^{(n)}: l(\varepsilon_1 \varepsilon_2 \cdots \varepsilon_n) = f^{(n-1)}: l(\varepsilon_2 \varepsilon_3 \cdots \varepsilon_n) = f: l(\varepsilon_n) = 1,$$

after n iterations. That is to say, the *same* symbol sequence is followed when we iterate the map n times, starting from *any* initial condition that is covered by the interval $l_i^{(n)}$. Now we are prepared to go further.

Fig. 8.2 For maps that peak at or above unity, the symbol sequences decorate a complete binary tree.

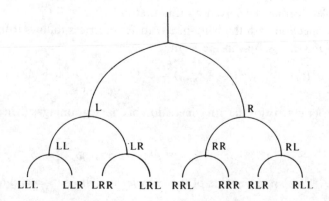

8.3 The transfer matrix method

We begin with the generating function

$$Z_n(\beta) = \sum_{i=1}^{N_n} l_i^\beta, \tag{8.2}$$

without further assumptions, except to assume that $Z_n(\beta) \sim e^{-ng(\beta)}$ for large n, yielding the condition

$$\sum_{i=1}^{N_{n+1}} l_i^{(n+1)\beta} \cong e^{-g(\beta)} \sum_{i=1}^{N_n} l_i^{(n)\beta} \tag{8.9}$$

for successive generations n and $n + 1$. At this stage, we need not restrict the theory to one-dimensional maps – we do not yet have to say *how* the natural partitioning is to be generated. In particular, we do not have to restrict to maps that peak at or above unity, but for definiteness we use the language of dynamical systems that generate either a complete or an incomplete binary tree where the symbolic dynamics can be described by a binary alphabet. The indices i and n in $l_i^{(n)}$ should be replaced systematically by an address on the binary tree. For maps of the interval that peak at or above unity, the correct address labeling is described in the last section – each class of maps generates its own postal address systems. Also, we can use either the (L, R) alphabet or, with L replaced by 0 and R by 1, the (0, 1) alphabet.

Next, we introduce the daughter-scale-to-mother-scale ratio

$$\sigma(\varepsilon_1 \cdots \varepsilon_{n+1}) = l(\varepsilon_1 \cdots \varepsilon_{n+1})/l(\varepsilon_2 \cdots \varepsilon_{n+1}), \tag{8.10}$$

where we have written $l_i^{(n)} = l(\varepsilon_1 \cdots \varepsilon_n)$. The procedure will make sense if, for large enough but finite n, no new length scales are introduced into the calculation. If the introduction of new and independent length scales continues without limit as n increases, then the formalism is useless and the problem cannot be solved by this method.

The connection with the Ising-like transfer matrices follows from using (8.10) to rewrite (8.9) as in the form

$$\sum_{\varepsilon_1 \cdots \varepsilon_{n+1}} \sigma^\beta(\varepsilon_1 \cdots \varepsilon_{n+1}) l^\beta(\varepsilon_2 \cdots \varepsilon_{n+1}) = e^{-g(\beta)} \sum_{\varepsilon_2 \cdots \varepsilon_{n+1}} l^\beta(\varepsilon_2 \cdots \varepsilon_{n+1}) \tag{8.11}$$

The transfer matrix for the nth generation of coarsegraining is then defined by

$$\langle \varepsilon_2 \cdots \varepsilon_{n+1} | T^{(n)} | \varepsilon_1 \varepsilon_2' \cdots \varepsilon_n' \rangle = \sigma^\beta(\varepsilon_1 \cdots \varepsilon_{n+1}) \delta_{\varepsilon_2, \varepsilon_2'} \cdots \delta_{\varepsilon_n, \varepsilon_n'}, \tag{8.12}$$

so that $e^{-g(\beta)}$ is an eigenvalue of the transfer matrix. As is well known in statistical physics, the largest eigenvalue of the transfer matrix completely determines the thermodynamics in the large n limit.

As the simplest example, we can write down the daughter-to-mother ratios for the asymmetric tent map directly from the binary tree constructed by inverse iteration of the map. For $n = 1$, we obtain for $\sigma(\varepsilon_1\varepsilon_2)$ the results $\sigma(00) = a^{-1}$, $\sigma(01) = a^{-1}$, $\sigma(10) = b^{-1}$ and $\sigma(11) = b^{-1}$, yielding the transfer matrix $T^{(1)}$ whose largest eigenvalue is $\lambda(\beta) = a^{-\beta} + b^{-\beta}$ and agrees with the result for $e^{-g(\beta)}$ derived from the binomial partition function

$$Z_n(\beta) = (a^{-\beta} + b^{-\beta})^n \qquad (8.13)$$

for the asymmetric tent map. For this model, higher-order transfer matrices give no new information, because of the statistical independence guaranteed by the binomial factorization.

At the transition to chaos in period doubling, the binary tree is complete and there are two scales asymptotically for large magnitude of q (Chapter 7), so the transfer matrix (8.12) provides the lowest-order approximation for the description of that model. In that case, however, higher-order transfer matrices contain more information. At the transition to chaos for the quasiperiodic orbits of the circle maps, the binary tree is incomplete and the lowest-order transfer matrix has exactly three nonzero entries. This system cannot be modeled at all as a two-scale Cantor set, and again, the lowest-order transfer matrix does not contain complete information about the thermodynamics.

Finally, we note that the general condition for the Hausdorff dimension is simply that the Gibbs potential vanishes at an inverse temperature β_c, $g(\beta_c) = 0$, where it follows that $\beta_c = D_H$. So far, we have an interpretation for β only in this special case. In order to provide an interpretation for β it is necessary to make a connection with a particular sort of stationary probability distributions, the Gibbs distribution.

8.4 What is the temperature of chaotic motion on a fractal?

We attempt next to make a more serious analogy with the microcanonical and canonical ensembles in classical statistical mechanics. In general, there will be (for large n) many different paths on the tree that correspond to the same Liapunov exponent λ (one can get the same exponent from different initial conditions). From the standpoint of statistical physics, the right word for this is 'disorder', because you can get the same parameter value for more than one 'lattice configuration' (symbol sequence). Denote by $nS(\lambda)$ the entropy that describes the disorder: $nS(\lambda) = \ln N(\lambda)$ where

$N(\lambda)$ is the number of different symbol sequences with the same λ. In this 'microcanonical ensemble', $n\lambda$ is the constrained parameter that is analogous to the energy in ordinary equilibrium statistical mechanics, so we can introduce the analog of the inverse temperature in the usual way for the microcanonical ensemble,

$$\beta_{\mathrm{micro}} = \mathrm{d}S(\lambda)/\mathrm{d}\lambda, \tag{8.14}$$

where $S(\lambda)$ is the entropy per iteration.

Because λ_i is the time-averaged Liapunov exponent for n iterations of the map starting from a particular initial condition x_i, the ensemble average

$$\bar{\lambda} = \sum_{i=1}^{N_n} P_i \lambda_i \tag{8.15}$$

is an average of the Liapunov exponent over the coarsegrained invariant set where $P_i = l_i^{\beta}/Z_n(\beta)$ is the analog of the Gibbs distribution. If the limit $Z_n(\beta) \sim e^{-ng(\beta)}$ exists for large n (the thermodynamic limit), then we can also write

$$\bar{\lambda} = g'(\beta). \tag{8.16}$$

We can introduce the entropy per iteration in the canonical ensemble

$$\bar{S}(\bar{\lambda}) = -\frac{1}{n} \sum_{i=1}^{N_n} P_i \ln P_i \tag{8.17}$$

and from (8.17) and the definition of $g(\beta)$ it follows that

$$\bar{S}(\bar{\lambda}) = \bar{\lambda}\beta - g(\beta). \tag{8.18}$$

This is a Legendre transform: with $\bar{\lambda} = g'(\beta)$, we therefore obtain

$$\beta = \mathrm{d}\bar{S}/\mathrm{d}\bar{\lambda} \tag{8.19}$$

and this defines the inverse temperature in the canonical ensemble in a way analogous to (8.15) for the microcanonical ensemble. As in ordinary statistical mechanics, these two ways to define β are in agreement only in the event that $Z_n(\beta)$ can, in the form (8.20) below, be approximated by its largest terms, so that the canonical ensemble can be approximated by a certain microcanonical ensemble. When that approximation does not hold, then (8.19) is the correct definition of β in the canonical ensemble.

We can also rewrite the partition function as a sum over microcanonical 'energy shells',

$$Z_n(\beta) = \sum_{\lambda} N(\lambda) e^{-\beta n \lambda} \tag{8.20}$$

where $N(\lambda) \sim e^{nS(\lambda)}$ is the number of different symbol sequences with Liapunov exponent λ. If we combine this Boltzmann formula with the box-counting definition of the fractal dimension $N(\lambda) \sim l_n^{-D(\lambda)}$ and use $l_n \sim e^{-n\lambda}$, then we get

$$D(\lambda) = S(\lambda)/\lambda \qquad (8.21)$$

for the fractal dimension of the subset of the fractal that (regarded as initial data of the map) yields the Liapunov exponent λ. $D(\lambda)$ is therefore a fractal dimension in a certain microcanonical ensemble.

In the thermodynamics of this statistical mechanics, the idea of 'heat' is not useful: the entropy is strictly the result of lattice disorder. As a result, the temperature $\sim \beta^{-1}$ is in no way connected with the kinetic energy of the dynamical system. Instead, the value of β is determined by *initial conditions*: if we push the analogy further, namely, we attempt to interpret the Gibbs distribution

$$P_i = l_i^{\beta}/Z_n(\beta) \qquad (8.22)$$

as the distribution of iterates of the map in the long time limit, i.e., in a certain state of statistical equilibrium. When we do this, we are forced to the conclusion that β is fixed by an entire class of initial conditions of the map. We can illustrate the truth of the above assertion by looking at specific, recognizable cases: first, set $\beta = 1$. The corresponding Gibbs distribution is

$$P_i = l_i/Z_n(\beta) = l_i/L_n, \qquad (8.23)$$

where $L_n = \sum_{i=1}^{N_n} l_i$ is the total length of all intervals in the nth generation. Each point in the invariant set is a possible initial condition of the map: there are no initial conditions that lead to stationary distributions in the case of a repeller *except* for the points that lie in the invariant set; all other initial conditions belong to the escape set. Therefore, the condition (8.23) requires us to choose initial conditions for the map that yield iterates that (in the long run) are distributed evenly over equal lengths (equation (8.23) states that equal lengths are equally probable).

If, instead, we set $\beta = 0$, then we are studying the equal frequency distribution

$$P_i = N_n^{-1}. \qquad (8.23b)$$

The initial conditions that generate this distribution are distinct from the ones that generate (8.23), except for the symmetric tent map where (8.23) and (8.23b) collapse into each other because there is only *one* length scale $l = 1/|f'| = 1/a$ in the first generation. In this case, the initial conditions that generate the equal frequency distribution are analogs for the fractal of the normal numbers. Normal numbers are defined in Chapter 9. In general, the initial conditions that generate the different distributions (8.23) and (8.23b) are nonoverlapping.

We can illustrate more explicitly some of the above results by analyzing a relatively transparent model, the asymmetric tent map

$$z_{n+1} = \begin{cases} az_n, & z_n < b/(a+b), \\ b(1-z_n), & z_n > b/(a+b) \end{cases} \tag{8.24}$$

where the condition $f_{\max} = ab/(a+b) > 1$ is required.

Let $n = m + (n-m)$ where m is the number of times that $f' = a$ in n iterations. Then the time-average Liapunov exponent for some initial condition x_{i0} is

$$\lambda_m = (m \ln a + (n-m) \ln b)/n \tag{8.25}$$

and with $a < b$, $\lambda_1 = \ln a$ is the smallest Liapunov exponent, while $\lambda_2 = \ln b$ is the largest. For n iterations, because all possible symbol sequences occur, there are exactly

$$N(m) = n!/(m!(n-m)!) \tag{8.26}$$

different intervals

$$l_m = a^{-m}b^{-(n-m)} \tag{8.27}$$

that yield exactly the *same* exponent λ_m. Note that

$$l_m = e^{-n\lambda_m}, \tag{8.27b}$$

and also that backward iteration of the map yields $l_1^{(1)} = a^{-1}$, $l_1^{(2)} = b^{-1}$; $l_1^{(2)} = a^{-2}$, $l_2^{(2)} = (ab)^{-1}$, $l_3^{(2)} = b^{-2}$, $l_4^{(2)} = (ab)^{-1}$, and so on, so that (8.27) is the general form of an nth-generation interval.

If we use Stirling's approximation ($\ln n! \sim n \ln n - n$) for large n and m, we obtain an entropy per iteration

$$S(\lambda_m) \cong -(m/n)\ln(m/n) - ((n-m)/n)\ln((n-m)/n) \tag{8.28}$$

that describes the 'disorder' of all symbol sequences that have exactly m L's and $n - m$ R's, and, hence, have the same Liapunov exponent (8.25).

The partition function is therefore binomial,

$$Z_n(\beta) = (a^{-\beta} + b^{-\beta})^n, \tag{8.29}$$

and the first-generation frequencies for visiting the intervals $l_1^{(1)} = a^{-1}$, $l_2^{(1)} = b^{-1}$ are given by

$$\bar{P}_1 = a^{-\beta}/(a^{-\beta} + b^{-\beta}) = 1 - \bar{P}_2, \tag{8.30}$$

which fixes β in terms of P_1, a, and b. P_1 is determined by initial conditions of the map.

The condition for the Hausdorff dimension $g(D_H) = 0$ is

$$a^{-D_H} + b^{-D_H} = 1, \tag{8.31}$$

but we can now show that there is no Gibbs distribution that describes the distribution of iterates for generations $n > 1$ for this particular inverse temperature. It is enough to analyze the easily solvable case where $b = a^2$, so that (with $w = a^{-D_H}$) we get

$$w(w + 1) = 1, \tag{8.32}$$

the Golden-mean equation: $w = (\sqrt{5} - 1)/2$. Hence

$$D_H = \frac{\ln(2/\sqrt{5} - 1))}{\ln a}. \tag{8.32b}$$

Consider next the Fibonacci sequence which we can construct as follows: start with the two-letter alphabet (L, R) and construct words of n letters by using the seed L and the substitutions L $\rightarrow$ LR, R $\rightarrow$ LRR. The result is the sequence of words L $\rightarrow$ LR $\rightarrow$

LRLRR

LRLRRLRLRRLRR

LRLRRLRLRRLRRLRLRRLRLRRLRRLRLRRLRR

$$\vdots \tag{8.33}$$

where, with $F_n = F_{n-1} + F_{n-2}$ and $F_0 = F_1 = 1$, each word has (for odd n) F_n letters, F_{n-2} of which are L's, F_{n-1} of which are R's. Hence, if we take these as F_n-letter symbol sequences of the tent map, the letter L occurs a fraction $\bar{P}_1 = F_{n-2}/F_n$ of the time, R a fraction $\bar{P}_2 = F_{n-1}/F_n$ of the time with no fluctuations. The absence of deterministic noise is a signal that there is no statistical mechanical interpretation in this case. $\bar{P}_1 \rightarrow w = a^{-\beta_c}$ and $\bar{P}_2 \rightarrow w^2 = b^{-\beta_c}$ as $n \rightarrow \infty$, in agreement with (8.22) and (8.30) when $n = 1$, but for $n \geq 2$ the interpretation of (8.22) as the distribution of iterates does not hold. The reason *why* it does not hold is that the combination LL cannot occur in (8.33), hence, the interval $l_1^{(2)} = a^{-2}$ (and all subintervals that grow from it) are not visited by the map. In other words, $\bar{P}_1$ and $\bar{P}_2$ are the frequencies with which the map visits the intervals $l_1^{(1)} = a^{-1}$ and $l_2^{(1)} = b^{-1}$, but *statistical independence* does not hold for the higher generation probabilities. The combination of (8.27) with (8.22) means that the Gibbs distribution for the tent maps must have the factorization

$$P_i = P_1^m P_2^{n-m}, \tag{8.22b}$$

which is a requirement for statistical independence that is simply not possible to satisfy whenever $\beta = D_H$. The classes of initial data for which (8.27b) and therefore (8.22) are satisfied are described in Chapter 9.

In any statistical mechanical system with finitely many states there must be negative temperatures $\beta^{-1} < 0$ that range from $\beta^{-1} = -\infty$, just above $\beta^{-1} = +\infty$, all the way to $\beta^{-1} = 0^-$, the highest negative temperature. The reason for this is the form taken by the microcanonical entropy combined with $\beta = dS/d\lambda$. While $\beta = 0$ yields the equal frequency distribution and corresponds to symbol sequences where (8.27b) holds with $P_1 = P_2 = \frac{1}{2}$, the zero temperature limit $\beta = \infty$ corresponds to $\lambda = \ln a$ and the zero-entropy (therefore perfectly ordered) lattice configuration LL$\cdots$L$\cdots$. The inverse temperature $\beta = 0^-$ corresponds to the zero-entropy state RR$\cdots$R$\cdots$ where $\lambda = \ln b$. Also, when $\beta = 0$, $\bar{P}_1 = \bar{P}_2 = \frac{1}{2}$ and

$$\bar{\lambda} = 2\ln 2/\ln ab. \tag{8.34}$$

The entropy of the corresponding lattice configurations is the maximum entropy

$$\bar{S}(\bar{\lambda}) = \ln 2. \tag{8.35}$$

8.5 $\bar{f}(\langle\alpha\rangle)$ as a thermodynamic prediction in the canonical ensemble

Having understood $nS(\lambda) \cong \ln N(\lambda)$ as the microcanonical entropy of all lattice configurations with the same time-average Liapunov exponent, we now show how to derive an $f(\alpha)$-spectrum from the canonical ensemble. The results are valid only for inverse temperatures such that the Gibbs distribution describes the distribution of iterates of the map.

In order to make contact with $f(\alpha)$, we start with the Gibbs distribution

$$P_i = l_i^\beta/Z_n(\beta) \tag{8.22}$$

as the distribution of the map's iterates and rewrite it in the form

$$P_i \simeq l_i^{\alpha_i}, \tag{8.36}$$

which yields the microcanonical result

$$\alpha(\lambda) = \frac{\bar{S}(\bar{\lambda}) + \bar{S}'(\bar{\lambda})(\lambda - \bar{\lambda})}{\bar{\lambda}}, \tag{8.37}$$

where $\beta = \bar{S}'(\bar{\lambda})$ is the inverse temperature in the canonical ensemble. The fractal dimension of the corresponding subset of the fractal is then

$$f(\alpha) = D(\lambda) = S(\lambda)/\lambda, \tag{8.38}$$

so that α and $f(\alpha)$ are defined by the microcanonical ensemble with 'energy' $n\lambda$ in the canonical partition function. In contrast, $\bar{f}(\langle \alpha \rangle)$ and $\langle \alpha \rangle$ are computed as canonical ensemble averages in what follows.

In order to understand that the generating function $\chi(q)$ is also an object in the canonical ensemble, we rewrite it as follows: for the case where P_i is the Gibbs distribution at inverse temperature β, we can write

$$\chi(q) = \sum_{\lambda} N(\lambda)\, e^{-q\beta n\lambda + qng(\beta)} = \frac{Z_n(\beta q)}{Z_n(\beta)} \tag{8.39}$$

as the ratio of two different canonical partition functions at temperatures β^q and q. Assuming (as above) that the large n limit $Z_n(\beta) \simeq e^{-ng(\beta)}$ makes sense, it follows directly that we obtain $\chi(q)$ in the form

$$\chi(q) \cong l_n^{q\langle \alpha \rangle - \bar{f}(\langle \alpha \rangle)}, \tag{8.40}$$

where

$$\bar{f}(\langle \alpha \rangle) = \frac{\bar{S}(\bar{\lambda}(\beta q))}{\bar{\lambda}}, \tag{8.41}$$

$$\langle \alpha \rangle = \frac{\beta \bar{\lambda}(\beta q) - g(\beta)}{\bar{\lambda}}, \tag{8.42}$$

$\bar{\lambda}(\beta q) = dg(\beta q)/d\beta q$ is the canonical average Liapunov exponent at temperature βq, $\bar{\lambda}(\beta) = \bar{\lambda} = g'(\beta)$ is the average exponent when $q = 1$, and the rough estimate $l_n = e^{-n\bar{\lambda}}$ was made in order to arrive at (8.40)–(8.42); to obtain more accurate results one must determine l_n as a function of q from a largest term approximation to the generating function χ. Notice that (8.42) can be rewritten to read

$$\langle \alpha \rangle = \frac{\bar{S}(\bar{\lambda}) + \beta(\bar{\lambda}(\beta q) - \bar{\lambda})}{\bar{\lambda}} \tag{8.42b}$$

where the numerator is the first-order approximation to the average entropy $\bar{S}(\bar{\lambda}(\beta q))$ at inverse temperature βq.

As a check, one can see that (8.41) and (8.42) yield correct results for the asymmetric tent map, namely that

$$\bar{f}(\langle\alpha\rangle) = q\langle\alpha\rangle - \tau(q), \tag{8.41b}$$

and

$$\langle\alpha\rangle = \frac{\bar{P}_1^q \ln \bar{P}_1 + \bar{P}_2^q \ln \bar{P}_2}{(\bar{P}_1^q + \bar{P}_2^q)(\bar{P}_1\lambda_1 + \bar{P}_2\lambda_2)} \tag{8.42c}$$

and where $\tau(q) = -\ln(\bar{P}_1^q + \bar{P}_2^q)/(\bar{P}_1\lambda_1 + \bar{P}_2\lambda_2)$. From these results, the entire $D_q(\beta)$ spectrum can be computed for all values of q, and for each inverse temperature β for which (8.22) describes the map's iterates. For example, $\beta = 1$ corresponds to choosing initial conditions where the map visits equal-length intervals equally often, and for these initial conditions the canonical average Liapunov exponent for the tent map is

$$\bar{\lambda}(1) = \frac{b \ln a + a \ln b}{a + b}. \tag{8.43}$$

The information dimension is

$$D_1(1) = \frac{(a + b)[(a + b) \ln(a + b) - a \ln a - b \ln b]}{b \ln a + a \ln b} \tag{8.44}$$

and the box-counting dimension is

$$D_0(1) = \frac{(a + b) \ln 2}{b \ln a + a \ln b}. \tag{8.45}$$

For the special case of an equal probability partitioning, the previous approach fails and one must calculate the $f(\alpha)$-spectrum from the Γ_n-generating function of Chapter 7.

In thermodynamics, phase transitions are identified as singularities in a thermodynamic potential, e.g. in the Gibbs potential. We now turn to an example of a phase transition. The example to be discussed is simple and can be formulated by the statistical mechanics method, but we have already computed D_q for the same example by a different method in Chapter 4. It is enough to note that a singularity in $g(\beta)$ implies a singularity in D_q. This is what is meant by a phase transition in the context of the D_q-spectrum.

8.6 Phase transitions

We return now to the logistic map at fully developed chaos, $f(z) = 4z(1 - z)$, where the invariant density is given by $p(z) = (\pi(z(1 - z))^{1/2})^{-1}$.

This density corresponds, as we show in the next chapter, to the class of initial conditions where $P_i = 2^{-n}$ over the logistic map's natural partitioning.

We computed the generating function $\chi(q)$ for this invariant density in Chapter 4, where we obtained

$$D_q = \begin{cases} 1, & q < 2 \\ \dfrac{q/2}{q-1}, & q > 2. \end{cases} \tag{8.46}$$

According to our thermodynamic interpretation of D_q in the last section, the discontinuity in slope of D_q signals a phase transition at $q = 2$. What is the meaning of this 'critical point'? From $\tau = (q-1)D_q = q\alpha - f(\alpha)$, with $\alpha = d\tau/dq$, we obtain

$$f(\alpha) = \begin{cases} 1, & q < 2, \\ 0, & q > 2, \end{cases} \tag{8.47}$$

and

$$\alpha = \begin{cases} 1, & q < 2, \\ \frac{1}{2}, & q > 2. \end{cases} \tag{8.48}$$

For $q < 2$, the result $\alpha = f(\alpha) = 1$ corresponds to the domination of $\chi(q)$ by the length scale $l = e^{-\lambda}$, with $\lambda = \ln 2$, the average Liapunov exponent for almost all initial conditions in $[0, 1]$. On the other hand, when $q > 2$, the scale $l^2 = e^{-2\ln 2}$ dominates $\chi(q)$, corresponding to the end points $x = 0$ and $x = 1$ where $\ln|f'(x)| = 2\ln 2$. Hence, $\alpha = \frac{1}{2}$ corresponds to the singularity $p(z) \sim z^{-1/2}$ when $z \sim 0$ or 1 and this singularity occurs on a point set with fractal dimension $f(\alpha) = 0$. In the former case, the entropy per iteration is $\ln 2$, corresponding to the class of maximum entropy symbol sequences, while $f(\alpha) = 0$ corresponds to the perfectly ordered (therefore zero-entropy) lattices $LL \cdots L \cdots$ and $RR \cdots R \cdots$ at $x = 0$ and 1 respectively.

We can also analyze the phase transformation in the statistical mechanics formalism, beginning with

$$Z_n(\beta) \approx 2l_0^{n\beta} + (N_n - 2)l^{n\beta} \tag{8.49}$$

or

$$Z_n(\beta) \approx 2l_0^{n\beta} + l^{n(\beta-1)}, \tag{8.49b}$$

where $l_0 = l^2 = \frac{1}{4}$. l_0 is the scale determined by the local Liapunov exponent $2\ln 2$, and l is determined by the average Liapunov exponent

$\lambda = \ln 2$. If $\beta < -1$ then

$$Z_n(\beta) \approx 2l_0^{n\beta} \tag{8.50}$$

and $g(\beta) = 2\beta \ln 2$ follows, with $S = \lambda\beta - g = 0$, describing the perfectly ordered lattices $LLLLL \cdots L \cdots$ and $RRRR \cdots R \cdots$. When $\beta > -1$ then

$$Z_n(\beta) \approx l^{n(\beta - 1)} \tag{8.50b}$$

and it follows that $g(\beta) = (\beta - 1)\ln 2$ and $S = \ln 2$, which is the entropy of the class of lattices where R and L occur with equal frequency – most of these configurations correspond to what the physicist calls totally disordered (see Chapter 9).

In contrast, the conjugate binary tent map has a nonsingular $f(\alpha)$-spectrum for its corresponding orbits. The reason for this difference in thermodynamics is the singularity in the coordinate transformation (see Chapter 2) that connects the two maps. For example, the unstable nonperiodic orbits of the binary tent map that generate its uniform density correspond exactly to those that generate the singular invariant density of the logistic map. Because of the singular nature of the coordinate transformation from one map to the other, a qualifying trajectory of the tent map yields a constant $f(\alpha)$-spectrum, $f(\alpha) = 1$, whereas the $f(\alpha)$-spectrum for the corresponding trajectory of the logistic map has the phase transition discussed above. Therefore, phase transitions do not provide a topologically invariant way to characterize statistical distributions in deterministic chaos. In the next chapter we use the nonperiodic symbol sequences to provide a universal characterization of the different statistical distributions that can be generated by a particular class of chaotic maps.

9
Universal chaotic dynamics

9.1 Invariant probability distributions from chaos

There are in nature both states and processes that are described in completely different contexts as disorderly, noisy, random, chaotic, or turbulent. An interesting example of disorder is provided by the energy-eddy cascade in fully developed turbulence which appears, at least superficially, to be a very different problem from that of the solid state disorder of a glassy state. However, there *is* a common thread that connects all disordered phenomena in nature: they appear as noisy, complicated patterns that require a statistical description, in stark contrast with the periodic patterns of a crystalline lattice or a laminar flow. Is the assumption of randomness fundamentally necessary for the description of certain kinds of disorder in nature? Let us begin by comparing the idea of randomness with deterministic chaos.

True randomness presumes indeterminism: given the state of a random system at one time, there is in principle and in practice no formula or algorithm from which one can predict which possible state will occur at the next instant. Or, given a pattern as the end result of a random process, there is in principle no way to predict what will be the pattern in an identical repetition of the process starting from the same initial state. The resulting uncertainty gives rise to statistics that are describable empirically by probabilities. In fact, probability distributions would provide the only way to organize the statistics of hypothetical truly random systems. Our interest is in whether it is possible to generate the different kinds of observed disorder deterministically. We shall show in this chapter that distributions that have the property of statistical independence, once regarded as the hallmark of randomness, can be produced algorithmically, hence deterministically.

Deterministic chaos is fundamentally pseudorandom: given initial conditions and control parameters in the form of algorithms for digit strings of numbers, the state of the system at the next instant can be computed to within a precision that is limited in the long run only by computation time. The limitation of computer time follows from a discretization of the condition for sensitivity with respect to small changes in initial conditions, the butterfly effect. The butterfly effect can account in principle for the *apparent* randomness of the different patterns observed for nonlinear systems in nature, where identical preparation of the initial state of a system is possible only to within finite accuracy: one can imagine a ball of initial conditions in the phase space of a deterministic chaotic system. Each attempt to repeat an experiment (to generate an experimental time series) consists in the attempt to prepare the experimental system in the same initial state, which means that each experiment starts from some initial condition within the ball, and the radius of the ball is fixed by the resolution of the experiment (we consider a gedanken experiment). The time evolution of the dynamical system from each different initial condition within the ball produces a different time series because of the butterfly effect, whereby a change in the least-significant digits in an initial condition is enough to produce a different long-time trajectory. If, then, one coarsegrains the phase space into $N = 1/\delta z_0^d$ different bins, where δz_0 is the resolution set by the experiment and d is the dimension of the phase space, then from M repetitions of the experiment (M different choices of initial conditions) one gets a set of frequencies $\{P_i\}$, where P_i is the frequency with which the system visits the ith bin. The frequencies $\{P_i\}$ are by necessity the combined statistics of M different chaotic trajectories of finite length where, to be systematic, we assume that each thought experiment runs for the same finite time t. One gets statistics from this collection of short-time orbits precisely because each chaotic orbit generates its own set of statistics (the invariant density reflects the statistics of a precise orbit for initial conditions belonging to a special class, as we have illustrated for the binary tent map in Chapter 2). Whether there is any *simple* way to describe the *combined* statistics of M different trajectories starting from completely arbitrary initial conditions, all taken together, is a very hard question that has not been addressed in this book or elsewhere. Typically, a writer will assume that this is an unimportant question; he will restrict his attention to the measure-one initial conditions, but we do not so restrict our considerations. What are the different kinds of statistics that a dynamical system can generate? Distributions other than those corresponding to invariant densities are possible, and we shall discover how to generate some of them.

As motivation for discovering how to generate statistical distributions other than the invariant density deterministically, let us first bring to light and discuss two misconceptions that commonly occur in the literature on deterministic chaos. The first is the notion that a dynamical system can be characterized in a useful way by a *single* Liapunov exponent. For example, standard textbooks discuss the Lorenz model and discrete maps as if, for a fixed set of control parameter values, each system could be described by a *single* Liapunov exponent *independently* of initial conditions. The second misconception is that one should be able, in experiment or in computation, to discuss 'attractor statistics' as if every nonperiodic orbit (or long-period unstable periodic orbit) of importance would generate the *same* statistics for a given set of control parameter values. Both of these misconceptions have their origin in the assumption that there is only one statistical distribution that we need consider, the one that occurs for 'almost all' initial conditions of the map in question. In other words, all of the distributions that follow from measure-zero initial conditions should be ignored, according to the traditional belief. This would, of course, eliminate the possibility of using deterministic chaos to describe how multifractal distributions are generated, which is surely wrong, so it is important to reconstruct the reasoning in favor of the 'measure-one' argument in order to see where the mistake lies. We shall discover that the prejudice in favor of the measure-one initial conditions follows from placing too great an emphasis upon the mathematical continuum, and from assuming that *truly* random choices are possible.

There is a *hidden assumption* that, if accepted, leads one to believe that one particular invariant distribution that solves the master equation should predict the outcome of either experiment or computation, but, in order that this should be true, one must assume that it makes sense to assume that nature (in experiment) or the human kind (in computation) can draw initial conditions randomly. A random draw of initial conditions from the uncountable point set that constitutes the middle-thirds Cantor set would, with probability (measure) one, yield a distribution of iterates of the ternary tent map that would be described by the Cantor function. The corresponding distribution of iterates for (symmetric and asymmetric) tent maps that peak at unity would be described by the uniform invariant density. It is widely assumed that this random-choice assumption makes sense, both in computations and in any possible application of chaos theory to nature. The random-draw assumption can be called into question by virtue of the fact that the continuum is not available in computation (cf. the discussion of Turing's ideas in Chapter 11), and it was argued by Borel, who contributed important ideas to measure theory and probability theory, that the assumption of the continuum should be avoided in *applications* of probability theory in order to avoid paradoxes. As we shall see in our discussion of the results of the turbulence experiments in the next chapter, nature seems not to be bounded by the requirement that she should choose from measure-one initial conditions either. In other words, what has been called the 'natural measure' in the literature is not necessarily natural for either computation or in nature. Once one admits that nonperiodic orbits that arise from initial conditions that occur with measure zero may be of interest, then there is a countable infinity of other interesting statistical distributions and (except for certain simple maps, e.g., the usual symmetric logistic map) a corresponding infinity of positive Liapunov exponents that must be considered for each map (for an example, see the discussion of the asymmetric tent map in Chapter 7). The measure-zero initial conditions include those that generate unstable periodic orbits, but we shall concentrate completely upon the nonperiodic orbits in all that follows, because they can be organized and understood by a universality principle. One can also study the problem from the standpoint of organizing the unstable periodic orbits hierarchically (see Cvitanović, Gunaratne and Procaccia, 1988).

Let us now reconstruct the argument that leads to the conclusion that the measure-one solutions of master equations should necessarily describe the statistics of chaotic systems that occur in nature. Let us return to our gedanken experiment, where experimental resolution δz_0 fixes the size of the ball in phase space from which nature chooses initial conditions in any attempt by the experimenter to prepare the same initial state and thereby repeat the same experiment M times. Assume for simplicity that the dynamical system is a map of the interval $z \to f(z)$ and that the map has an invariant density $p(z)$. In this case, the 'ball' is the interval $[z_0, z_0 + \delta z_0]$, where δz_0 is the resolution of the experiment. If we coarsegrain the phase space into $N = 1/\delta z_0$ bins, each of width δz_0, then under exactly which circumstances can we expect the N probabilities given by integrating the invariant density $p(z)$ over each bin to agree with the observed statistics $\{P_i\}$ that follow from the M different finite-length time series produced by the dynamical system, starting from M different initial conditions $z_{0,i}$ in the interval $[z_0, z_0 + \delta z_0]$? First, one must assume that the 'time' n is so large that the infinite-time limit applies, but that is not enough. In addition, it is necessary to assume that nature, in her choice of the M initial conditions, draws M times randomly from the continuum, or at least from a certain subset of the continuum that occurs with measure one. Under these circumstances, the gedanken experiment's statistics $\{P_i\}$ are exactly those of the coarsegrained invariant density precisely because each *separate* trajectory generates the invariant density. This is strictly an assumption, and an unjustified one. Therefore, we shall continue along our chosen path, whereby we set up the theory from a general standpoint that is not prejudiced toward any particular set of initial conditions. In section 9.2 we shall not consider balls of initial conditions in phase space, where *mixed* statistics can occur, but shall restrict ourselves to classes of initial data where each orbit in the class generates exactly the *same* statistics for each level of coarsegraining – the way that such initial conditions are found, for a given map, is discussed below. In that case, as in the case of the invariant density, the correct statistics for the entire class are generated by one *single* infinite-time trajectory that follows any initial condition in the class. Our classes of initial data include the measure-one class, but also all of the others as well. In particular, we show in what follows how to generate coarsegrained versions of the so-called multifractal measures completely deterministically, from measure-zero sets of initial data.

The usual argument in favor of the measure-one distributions is that, when the map is subjected to external noise, that distribution typically describes the effect of the map plus the noise. Because no real physical system is ever completely isolated, the argument sounds strong. However, it must be stated clearly that the noise source is implicitly presumed to have certain symmetry properties, and, when different symmetries are imposed, different distributions are the result of map plus external noise. In fact, we know of no example from nature where results of measurement on low-dimensional chaotic systems have required the use of an invariant density for their explanation. Therefore, we regard as weak the argument in favor of ignoring the measure zero distributions, and shall not do so.

Originally, the motivation for studying invariant densities was that they were supposed to represent properties of the map that are independent of initial data. Because the invariant density is generated by 'almost all' initial conditions, the underlying assumption is then that the measure-zero initial conditions are unimportant. However, the measure-one initial conditions constitute only a single possible class, and we shall show that there is a much broader notion of universality that follows from considering classes of initial data, namely, that the statistics for a particular class of initial conditions is even independent of the map. This is the symbol sequence universality that we introduce in the next section.

In order to reproduce theoretical predictions numerically, it is necessary to take into account our fundamental limitation to finite precision in computation. This can be accomplished if we work with a hierarchy of finite coarsegrainings of the theoretical invariant set of a dynamical system. In the multifractal formalism developed in Chapter 7 we coarse-grained the invariant set into N_n different intervals $\{l_i\}$. We then denoted $P_i = l_i^{\alpha_i}$ as the frequency with which the dynamical system visits the interval l_i. One must then allow for the possibility that $N(\alpha) \cong l_n^{-f(\alpha)}$ different intervals (all with the same characteristic size $\approx l_n$) may have the same index $\alpha_{i_1} = \alpha_{i_2} = \cdots = \alpha_{i_N} = \alpha$, for all generations n in the hierarchy, so that $f(\alpha)$ is the fractal dimension of that particular subset of intervals. Nonuniform coarsegrainings are generated by asymmetric maps, but the lack of uniformity of the intervals does not *necessarily* imply a corresponding lack of symmetry in the distribution of iterates that follows from a particular initial condition.

In both computed and experimental $f(\alpha)$-spectra there are two important independent underlying ideas: a hierarchy of finite partitionings $\{l_i^{(n)}\}$ of the phase space and a coarsegrained invariant distribution $\{P_i\}$ on each partitioning. For reasons that have to do with fundamentals, we want to restrict to what we have called natural partitioning generated by a map. A chaotic dynamical system, mathematically, has an invariant set that has the cardinality of the continuum and can consist of finite subsets of the phase space or a Cantor subset. Whether the invariant set is an attractor or a repeller is irrelevant, although the ideas below are developed for the case of repellers. By the *natural partitioning*, we mean a hierarchy of intervals that is generated by *the action of the dynamical system and is therefore an optimal coarsegraining of the invariant set*. We know how to use backward iteration to construct the natural partitioning for maps of the interval that peak at or above unity. Natural partitionings can also be constructed at the borderline of chaos (Feigenbaum, 1988). The main point is that when $\{l_i^{(n)}\}$ is a natural coarsegraining and the $\{P_i\}$ are the frequencies with which the system visits the different intervals, then universal properties follow that otherwise would remain hidden. An observable example of a natural coarsegraining in nature may be provided by the eddy cascade in turbulence (see Chapter 10), where the eddies in each generation of the cascade have a set of characteristic sizes $\{l_i^{(n)}\}$. This may not be a mathematically precise example, but it is the essence of the sort of fragmentation that we have in mind when we go beyond one-dimensional maps.

One may well ask: if we can write $P_i = l_i^{\alpha_{i'}}$ then why do we have to bother with initial conditions? The answer is that to write $P_i = l_i^{\alpha_i}$ does not change anything: the natural intervals are determined by the map and reflect the details of the map, but the set of indices $\{\alpha_i\}$ depends strongly upon initial conditions, because while the natural intervals do not, the probabilities P_i do. We are going to show that an $f(\alpha)$-spectrum generally tells us more about the class of initial conditions than it does about the details of the underlying map. We should think of a chaotic dynamical system as the map plus the set of initial conditions. The map is like a processor and the set of initial conditions is the set of possible programs. It is the program as well as the processor that determines the output. The main point is that the physics can be encoded not only in the map, but also in the initial conditions.

We restrict our considerations to the class of one-dimensional maps of the unit interval that peak at or above unity (Fig. 9.1), and, by n-times backward iteration of the unit interval, generate 2^n intervals $\{l_i^{(n)}\}$, as in Fig. (9.2). These intervals define the natural partitioning, the optimal coarsegraining of the map's invariant set. If the map peaks at unity, then for the box-counting dimension (the peak of $f(\alpha)$) we get $D_0 = 1$, whereas maps that peak above unity have invariant sets with $D_0 < 1$ (Cantor sets). The main point is that the natural partitioning lives on a complete binary tree: exactly 2^n intervals provide the nth-generation description of the invariant set (Fig. 9.2). Symmetric and asymmetric logistic-like and tent maps as well as Bernoulli shifts are included within this class. We expect that such maps can follow from higher-dimensional dynamical systems

Fig. 9.1 The universality class under consideration consists of maps of the unit interval that peak at or above unity and contract intervals in backward iteration. Shown here is the asymmetric logistic map $f(x) = Dx(1 - x) - \gamma x(1 - x)^2$, with $D = 5$ and $\gamma = 0.1$.

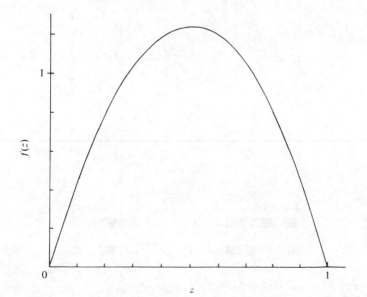

by ignoring information, although one does not know how to do this systematically. The main point is that we have a system that systematically generates both trees and statistical mechanics. The reason why statistical mechanics on trees is important will be brought out when we discuss the phenomenology of turbulence in the next chapter.

Let us now review, from our coarsegrained perspective, the statistics $\{P_i\}$ that correspond to several well-known invariant densities $p(z)$. The invariant densities are generated by the maps in our class that peak at unity, and

$$P_i = \int_{z_{i-1}}^{z_i} p(z)\, \mathrm{d}z \tag{9.1}$$

is then the probability for the map to visit the interval l_i. In (9.1),

$$z_i = f^{-m}(1), \tag{9.2}$$

Fig. 9.2   The *n*-times backward iteration of the unit interval by any map in the class generates exactly 2^n intervals, and the hierarchy of intervals so generated provides the *optimal finite-precision* descriptions of the map's invariant set.

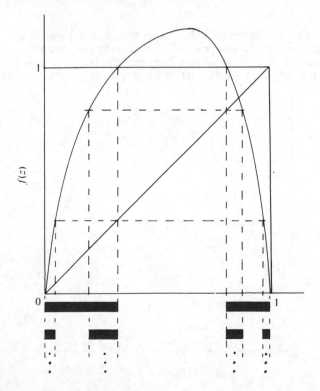

for $m = 1, 2, \ldots, n$ is any of the $2^n - 1$ preimages of the map's peak $f_{max} = 1$, and the $N = 2^n$ intervals $\{l_i\}$ are given by $l_1 = z_1, l_N = 1 - z_{N-1}$, otherwise $l_i = z_i - z_{i-1}, z_0 = 0$, and $z_N = 1$.

Consider first the binary tent map

$$g(x) = \begin{cases} 2x, & x < \tfrac{1}{2}, \\ 2(1 - x), & x > \tfrac{1}{2}. \end{cases} \tag{9.3}$$

The invariant density is uniform, so that $P_i = l_i$, which means that $\alpha_i = 1$ for all indices i, or that the coarsegrained density P_i/l_i is independent of l_i (for a uniform invariant density with a uniform natural coarsegraining, there is no multifractal spectrum). For the binary tent map, $l_i = 2^{-n}$, so that the even distribution $P_i = 2^{-n}$ follows. In other words, in order to generate the invariant density, initial conditions must be chosen so that the coarsegrained probabilities have statistical independence in the form $P_i = \overline{P_1^m} \, \overline{P_2^{n-m}}$ with $\overline{P_1} = \overline{P_2} = \tfrac{1}{2}$. This means that every interval in a given generation is visited equally frequently by the map, in the ideal limit of infinitely many iterations.

Consider next a map $z_{n+1} = f(z_n)$ that is conjugate to the binary tent map $x_{n+1} = g(x_n)$ by an *invertible* transformation $z = h(x)$. The new map is given by the conjugacy operation

$$f(z) = h(g(h^{-1}(z))). \tag{9.1b}$$

If the transformation h is differentiable, then the map g has the invariant density

$$p(z) = dx/dz = 1/h'(h^{-1}(z)), \tag{9.1c}$$

where the right-hand side has been evaluated at the point $x = h^{-1}(z)$. For the symmetric logistic map $f(z) = 4z(1 - z)$, we get the well-known invariant density $p(z) = (\pi(z(1 - z))^{1/2})^{-1}$, as was shown in Chapter 2. Here, the corresponding coarsegrained distributions show statistical independence $P_i = \overline{P_1^m} \overline{P_2^{n-m}} = 2^{-n}$, with $\overline{P_1} = \overline{P_2} = \tfrac{1}{2}$, just as does the binary tent map's invariant density, even though the natural partitioning is made up of *uneven* intervals. This result follows from combining (9.1), (9.1b), and the tent map result $P_i = 2^{-n}$, while using the fact that the preimages of the maxima of the two maps are also conjugate to each other. It follows, by the same reasoning, that the invariant density of *every* map that is smoothly conjugate to the binary tent map is realized for initial conditions that yield iterates that visit each interval in a given generation equally often. *For all such maps, the even distribution $P_i = 2^{-n}$ is just the coarsegrained way to describe those maps' invariant densities.*

The logistic map's invariant density has a (trivial two-point) multifractal spectrum (see Chapter 4), even though the tent map's density is not multifractal, because the transformation h to the logistic map is singular at the end points of the unit interval.

Consider next the asymmetric tent maps that peak at unity. They can be written in the form

$$f(z) = \begin{cases} az, & z < b/(a + b), \\ b(1 - z), & z > b/(a + b), \end{cases} \tag{7.5}$$

with $ab = a + b$, but where a and b are unequal. We showed in Chapter 2 that the corresponding invariant density is uniform. By combining this result with (9.1) and the natural intervals $l_i = a^{-m}b^{-n+m}$, we discover that the uniform invariant density demands initial conditions that generate statistical independence in the form $P_i = l_i = a^{-m}b^{-n+m}$, for each generation n, in the limit of infinitely many iterations of the map. In this case, the map's natural intervals are *not* visited with equal frequency for initial conditions that generate the map's invariant density. Therefore, we conclude that the condition (9.1c) and smooth conjugacy to the binary tent map cannot hold. The reason for the breakdown of the expected smooth transformation of probability densities is that the transformation $z = h(x)$ from the binary tent map to an *asymmetric* tent map, where $ab = a + b$, is a kind of Cantor function, hence has vanishing derivative almost everywhere (see Grossmann and Thomae, 1977), so that (9.1c) cannot hold. However, because $P_i = l_i$, it again follows that $\alpha_i = 1$ for all indices i: the uniform invariant density produces no multifractal spectrum. There are also symmetric maps that are not smoothly conjugate to the binary tent map; invariant densities for some of those maps have been classified and derived by Györgi and Szepfalusy (1984).

The reader should observe that single-hump maps that peak at unity, but not above, generate invariant densities *because* they partition the unit interval in backward iteration in such a way that the sum of the intervals in each generation adds up to exactly one (phase space is hierarchically partitioned but is not fragmented into a fractal). *For example, the condition* $1/a + 1/b = 1$ *that we derived in Chapter 2 as the condition for existence of a uniform invariant density of the asymmetric tent map* (7.5) *is precisely the condition that the two first-generation intervals* $1/a$ *and* $1/b$ *of the tent map add up to unity.* Whenever $ab > a + b$, the tent map peaks above unity, and the *pointwise* existence of an invariant density is impossible because the unit interval is partitioned into a fractal set (multifractal set, if a and b are unequal), with first-generation intervals satisfying

$1/a + 1/b < 1$. At any rate, the solutions of the master equation that yield invariant densities seem to be intimately connected with a class of initial conditions that generates *statistical independence* of one sort or another. We shall discover that other forms of statistical independence are also possible for the same maps, but for other choices of initial conditions. For an *arbitrary* choice initial condition, a given map will not produce statistical independence of *any* sort. By an arbitrary choice we mean a real choice where no forethought is given *systematically*. An arbitrary choice is by no means the same as a random choice. It is doubtful whether truly random choices are possible in reality, and the assumption that the brain can make even 'effectively random' choice, much less a *truly* random choice, is unproven and is highly questionable: as Knuth (1981) has shown, the set of instructions that must be followed deterministically in order to simulate randomness effectively is both highly nontrivial and numerically involved, and there is so far no evidence that the human brain can effectively simulate any such set of instructions.

Next, we generalize the master equation to show that one should also expect the occurrence of invariant distributions that are nondifferentiable, hence cannot be described by an invariant density.

Let $P(z)$ denote the probability that the map's iterates all fall to the left of the point z, where $z_{n+1} = f(z_n)$ is the map. That probability $P^{(n)}(z) = P(z)$ is given by the hierarchy of sums

$$P(z) = \sum_i P_i, \qquad (9.1d)$$

where x is restricted to the interval end points x_m ($x = x_n$ in (9.1d)), and the intervals are given by $l_m = x_m - x_{m-1}$, and the sum extends over the interval that covers z as well as all intervals to the left of it. By increasing n, where $n = 1, 2, 3, \ldots$, one generates finer- and finer-grained descriptions of the cumulative probability $P(z)$ for more and more points z in the phase space. We can generalize the Frobenius–Peron equation and write down the condition that $P(z)$ should be an *invariant* distribution:

$$|P(z) - P(y)| = |P(z_1) - P(y_1)| + |P(z_2) - P(y_2)|, \qquad (9.1e)$$

where the two points z_1 and z_2 iterate to z, and the points y_1 and y_2 iterate to y. When $P(x)$ is differentiable, then setting $z = y + dy$ yields $P'(y) = p(y)$, with the P_i given by (9.1) above, and (9.1e) reduces to the condition for the solution of the Frobenius–Peron equation for the invariant density $p(y)$ that we derived in Chapter 2. However, (9.1e) holds whether or not $P(y)$ can be differentiated, and we can interpret (9.1e) as follows: if we choose $|z - y| = l_i$ to be an nth-generation mother interval, then the magnitudes of $z_1 - y_1$ and $z_2 - y_2$ are the lengths of the two daughter intervals that are born from l_i in the passage to the $(n+1)$th generation. (9.1e) simply states that the probability for the map to hit the mother interval is the sum of the two probabilities to hit the daughter intervals, which is a true statement about the interval probabilities. Notice that the even distribution $P_i = 2^{-n}$ *always* satisfies the condition (9.1e), because $2^{-(n+1)} + 2^{-(n+1)} = 2^{-n}$. Hence, the even distribution is always an invariant distribution, whether the map peaks at or above unity, is symmetric or asymmetric. This result holds irrespective of whether the distribution $P(z)$ that is constructed from the even distribution can be differentiated to yield an invariant density. For symmetric maps that peak at unity and are *smoothly* conjugate to the binary tent map (the logistic map $f(z) = 4z(1-z)$ provides an example), the even distribution $P_i = 2^{-n}$ *is* the coarsegrained description of that map's invariant density. For the asymmetric tent map (7.5) with $ab = a + b$, the even distribution is *still* a solution, but it is *not* the one that corresponds to that map's uniform invariant density: as we have seen above, the uniform invariant density for the asymmetric tent map corresponds to an *uneven* coarsegrained distribution given by $P_i = a^{-m} b^{-n+m}$ in generation n. The main point here is that these two different invariant distributions must be generated by iterating the *same* asymmetric tent map from two entirely *different*, nonoverlapping sets of initial data, both of which are subsets of the unit interval. Furthermore, there are invariant distributions *other* than the even one that we have *yet* to discover, not only for the asymmetric tent map, but for every map in our class! The other side of the coin, as we shall see, is that perfectly symmetric maps like the binary tent map can also generate very uneven distributions: it is all a question of nonoverlapping classes of initial condition: each class of initial data can be characterized by its own hierarchy of coarsegrained statistics.

It is interesting to ask when it makes sense to look for a distribution of the form $P_i = l_i/L$, which represents a uniform coarsegrained density $\rho_i = 1/L$. This particular distribution typically does not satisfy the master equation whenever $f(z) = 4z(1-z)$. For the tent maps that peak at or above unity, however, this distribution always exists, even if the tent map generates a multifractal invariant set. We can understand this by looking at the first- and second-generation interval probabilities. When $ab = a + b$, the sum of the daughter intervals whose mother interval is $1/b$ is $1/ab + 1/b^2$, which is equal to $1/b$ because $1/a + 1/b = 1$. In this case, $\rho_i = 1$ even holds pointwise, in the limit where l_i goes to zero, but the $f(\alpha)$-spectrum collapses onto a single point, D_0. When $ab > a + b$, then $L < 1$ and the invariant set is fractal. In this case $P_i = l_i/L$ satisfies the master equation. For example, it is easy to check that the first-generation mother interval probability $b^{-1}(b^{-1} + a^{-1})$ is equal to $(b^{-2} + a^1 b^{-1})/(a^{-2} + 2a^{-1}b^{-1} + b^{-2})$, which is the sum of the corresponding two daughter probabilities. Here, the coarsegrained uniform density $\rho_i = 1$ has no pointwise limit on the mathematical fractal, and so it generates a nontrivial $f(\alpha)$-spectrum, where (with $a > b$) $\alpha_1 = (\ln a + \ln L)/\ln a$, $\alpha_2 = (\ln b + \ln L)/\ln b$, and $L = a^{-1} + b^{-1}$ is the length of the first-generation intervals. However, the solution $P_i = 2^{-n}$ also exists for these tent maps, as do many other distributions that we have not discussed.

Finally, if we consider the ternary tent map (set $a = b = 3$ in (7.5)), then we observe that the Cantor function for the middle-thirds Cantor set satisfies the condition (9.1e) for the invariant distribution (cf. Chapter 4 for the Cantor function). That is because the coarsegrained description of the Cantor function, with intervals given by $l_n = 3^{-n}$, is given by the even distribution $P_i = 2^{-n}$. Here, there is no invariant density because the invariant set is a zero-measure Cantor set. The Cantor set has zero measure *in the unit interval*, but because that set is uncountable, a measure, the Cantor function, can be defined on the set.

The statistics of all of the invariant distributions of our class of maps that we have discussed above can be summarized by observing that the generating function

$$\chi_n(q) = \sum_{i=1}^{N_n} P_i^q \tag{9.4}$$

collapses into the binomial form

$$\chi_n(q) = (\bar{P}_1^q + \bar{P}_2^q)^n, \tag{9.4b}$$

where the P_i correspond either to an invariant density or to a Cantor function. In the next section, we are going to show that, by choosing initial conditions carefully, every map in our class can generate a very wide spectrum of nontrivial multifractal distributions (9.1d). Naturally, the required initial conditions are not the ones that generate the invariant densities. In the limit of infinitely many iterations and infinite precision, the limiting statistical distributions are called multifractal measures.

The main point is that the metric properties of deterministic chaos are not universal. To discover universal properties, it is necessary to study symbolic dynamics. Then one discovers topological properties that are universal because they are independent of the particular map that one starts with.

9.2 Symbol sequence universality

In order to see the same problem from a different perspective, we note that we could also have begun our discussion with the question: To what extent can an experimental $f(\alpha)$-spectrum be used to infer the underlying dynamical system? This is an important question, because we are going to prove the following result: for a special class of initial conditions in the invariant set of *each* map in our class, all of the maps produce exactly the *same* generating function,

$$\chi_n(q) = \sum_{i=1}^{N_n} P_i^q, \qquad (9.4)$$

where q is a free index, P_i denotes the frequency with which the interval $l_i^{(n)}$ is visited by a map. N_n is the number of nth-generation intervals, and we have written $P_i^{(n)} = P_i$ for convenience. As we noted at the beginning of Chapter 7, the *intervals* provide a *finite* description of the underlying invariant set of a particular map. As we have seen, each map generates its own hierarchy of intervals. When the observable generations of intervals $\{l_i^{(n)}\}$ that are generated by the different maps are to within experimental error the same, then an $f(\alpha)$-spectrum derived from (9.4) cannot be used to distinguish one map from another. The main point is that it is *not* an invariant density (or its generalization) that is the signature of a dynamical system; instead, it is the *intervals* that characterize a particular map.

P_i is the frequency with which the dynamical system visits the interval $l_i^{(n)}$, during its time evolution, starting from a definite initial condition. By a *class* of initial conditions, we mean all initial conditions of the system that generate the *same* hierarchy of coarsegrained distributions. When a Gibbs distribution can be constructed, then the inverse temperature is the same for all initial conditions that generate that particular Gibbs distribution, but, as we showed in Chapter 8, there are classes of initial conditions for which the Gibbs distribution cannot be constructed. In what follows, we shall show how to construct distributions with statistical independence; these distributions can all be rewritten as finite-temperature Gibbs distributions, although we shall not discuss them using that language.

In contrast, the generating function (with $l_i^{(n)} = l_i$) of Chapter 8,

$$Z_n(\beta) = \sum_{i=1}^{N_n} l_i^\beta \cdot \tag{9.5}$$

is based strictly upon the hierarchy of intervals $\{l_i\}$, but without further assumptions (9.5) has absolutely nothing to do with a distribution of iterates $\{P_i\}$. (9.5) follows from the Γ generating function by assuming an equal frequency distribution, but dynamics is not restricted to initial conditions that generate even distributions.

Our considerations can be motivated further by asking the question: *when* can (9.4) be rewritten in multiplicative form, for example, as

$$\chi_n(q) = (\bar{P}_1^q + \bar{P}_2^q)^n, \tag{9.4b}$$

where $\bar{P}_1$ and $\bar{P}_2$ denote the frequencies with which two first-generation intervals $l_1^{(1)}$ and $l_2^{(2)}$ are visited? For the asymmetric tent map $z_{n+1} = f(z_n; a, b)$,

$$f(z; a, b) = \begin{cases} az, & z < b/(a + b) \\ b(1 - z), & z > b/(a + b) \end{cases}, \tag{9.6}$$

the generating function (9.5) can *always* be written in multiplicative form

$$Z_n(\beta) = (a^{-\beta} + b^{-\beta})^n, \tag{9.5b}$$

but (9.4b) does not hold for *arbitrary* initial conditions of the tent map. We shall show how to construct examples of symbol sequences for which (9.4b) holds not merely for the tent maps, but universally even for asymmetric maps that produce hierarchies of nonuniform intervals. Clearly, (9.4b) follows from (9.4) whenever $P_i = \bar{P}_1^m \bar{P}_2^{n-m}$ occurs for $n!/m!(n - m)!$ different intervals, and this factorization is usually regarded as the signature of *randomness*. It is not: it is only a condition for *statistical independence*. We shall generate both the statistical independence and the probabilities $\bar{P}_1$ and $\bar{P}_2$ *completely deterministically*, by using algorithms. An unnecessary appeal to randomness would only mask the underlying order from which stems the apparent disorder described by (9.4b).

The main thing about our class of maps, irrespective of all other details, except that they must contract intervals in backward iteration, is that the natural partitioning lives on a *complete binary* tree: exactly 2^n intervals provide the nth-generation description of the invariant set. Symmetric and asymmetric logistic-like and tent maps as well as Bernoulli shifts are included within this class. By starting from any initial condition z_i in the invariant set that is covered by the interval $l_i^{(n)}$, the n-times forward iteration of the map produces an n-bit symbol sequence constructed from the two-letter alphabet (L, R) (see section 8.2). We write an L whenever the map visits the first-generation interval $l_1^{(1)}$, and R whenever $l_2^{(1)}$ is visited. The symbol sequences are addresses on the complete binary tree (cf. Fig. 9.3). The completeness of the tree means that *every* binary string is a symbol sequence of the map for some initial condition in the invariant set. Given a symbol sequence, the corresponding initial condition can be constructed by backward iteration of the map. For a very good reason, we concentrate upon the nonperiodic symbol sequences, and therefore upon the nonperiodic, or so-called 'true chaotic orbits'. Because our binary tree is complete, our symbol sequences are in one-to-one correspondence with all possible numbers in the unit interval written as binary strings, and this fact is what makes chaos theory relatively easy for this class of map.

The important detail of the maps is that the backward iteration must be a contraction mapping. In order to develop the proof that yields our universality classes, we shall assume that to each infinite-length symbol sequence there corresponds exactly one infinite-precision initial condition z_i. In order that this is true, backward iteration of the map must contract the intervals so that $l_i^{(n)} \to 0$ as $n \to \infty$ for each of the 2^n intervals. We should think of $l_i^{(n)}$ as the precision within which any initial condition z_i

Fig. 9.3 The construction of the complete binary tree and the assignment of symbol sequences are a *universal*, as they are the same for *every* map in the class.

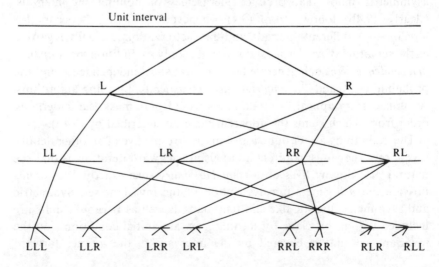

that lies beneath the interval $l_i^{(n)}$ and belongs to the invariant set is specified. This contraction will hold, for example, if, as we argued in Chapter 8, $l_i^{(n)} \sim e^{-n\lambda_i}$, where λ_i is the Liapunov exponent for the trajectory that starts from z_i, if λ_i is positive for every initial condition in the invariant set.

The condition that $P_i = 2^{-n}$ for the 2^n intervals $\{l_i^{(n)}\}$ for all generations n of the binary tent map, and all maps that are conjugate to it by a smooth transformation of coordinates, is equivalent to the condition that the invariant density describes the distribution of the map's iterates. For maps that peak above unity and thereby have repelling Cantor sets as invariant sets, the continuous probability distribution that corresponds to the equal frequency condition $P_i = 2^{-n}$ is (for a given Cantor set) a certain Cantor function (see Chapter 4 for an example). That Cantor function provides the uniform measure that forms the basis for the decision whether a particular set of initial conditions exists in the Cantor set with zero or with finite measure. For the ternary tent map and its corresponding middle-thirds Cantor set, the Cantor function corresponds to the even coarsegrained distributions $P_i = 2^{-n}$, but we expect that asymmetric tent maps that generate repelling fractal invariants sets have a Cantor function that corresponds to the coarsegrained probabilities $P_i = a^{-m}b^{n-m}$, where $ab > a + b$, in analogy with the nonfractal case where $ab = a + b$, which is discussed above.

The basis for our universality principle is that *all maps in the class have exactly the same set of symbol sequences*. The consequence of having exactly one initial condition for each map for each symbol sequence is that when two different maps start from different initial conditions in their own invariant sets and produce the same symbol sequence, then each map generates exactly the same distribution of iterates over its own natural coarsegraining. We shall prove this in the next paragraph. We shall also prove that complete information about each orbit over the map's own natural partitioning for every generation is encoded in the symbol sequence, which we can think of as a one-bit orbit constructed from the binary (L, R) alphabet. The details given later in this section for the two-bit and three-bit orbits of the binary tent map in binary arithmetic provide a paradigm for what happens with more complicated maps like asymmetric tent maps and logistic-like maps where there is no 'natural' base of arithmetic. For the binary tent map in binary arithmetic there is a one-to-one correspondence between the occurrence of 'words' made of different combinations of blocks of m 0's and $N - m$ 1's in the symbol sequence (one-bit orbit) and specific N-bit words in the first N bits of the map's iterates. The frequency with which different N-bit words occur as in the symbol sequence determines the frequency with which intervals that are labeled by those words are visited during forward iteration of the map. The right generalization of the binary tent map's two-bit orbit, for more complicated maps where there is no natural base of arithmetic, is to construct the two-bit symbol sequences using the alphabet (LL, LR, RR, RL): we write LL whenever the orbit visits the interval $l_1^{(2)}$, LR whenever $l_2^{(2)}$ is visited, and so on. The three-bit orbit constructed from the eight-letter alphabet (LLL, LLR, ..., RRR) describes the visitation of the eight third-generation intervals $\{l_i^{(3)}\}$, and a corresponding N-bit word orbit can be used to describe the Nth-generation coarsegrained dynamics. As we shall show below, we are saved from that labor by the fact that, just as with the binary tent map, complete information about the N-bit symbol sequence orbits is already encoded in the one-bit symbol sequence. In other words, complete information about any complicated trajectory over any generation of coarsegraining of an asymmetric logistic map is encoded in the exact initial condition (typically an infinite-length non-periodic binary string) that generates the orbit, but so long as two entirely different maps have the same symbol sequence, they must also yield (for that sequence) the same frequencies $\{P_i\}$ for every generation of coarse-graining. Had we chosen any partitioning other than the natural one, this universality would have remained hidden. Chaotic maps do not generate a single measure, a 'natural measure', that includes all of the important statistical properties that can be generated by the map, but chaotic maps generate an optimal partitioning of phase space, which we have called the 'natural partitioning'.

We explain next why complete dynamical information about the orbit is contained in the one-bit symbol sequence. The idea is to show that two different maps in the class with the same symbol sequence must generate exactly the same coarsegrained distribution $\{P_i\}$, each over its own 2^n intervals. This can be understood as follows: the intervals are constructed by iterating the unit interval backward and the construction of each interval corresponds to a certain trajectory on the binary tree. The forward iteration of the map applied to any interval $l_i^{(n)}$ is essentially the same as applying the map to any initial condition that is *covered* by that interval (and is therefore specified to a precision that is given by the width of the interval) and yields a certain symbol sequence (see Fig. 9.3). In $n - 2$ forward iterations, the frequencies with which the map visits its four second-generation intervals (labeled by the addresses LL, LR, RR, RL) are exactly the same as for any other map that is iterated exactly $n - 2$ times from its own corresponding nth-generation interval $l_i^{(n)}$. The easiest way to understand this is as follows: assume for convenience that the nth-generation intervals $\{l_i^{(n)}\}$ and frequencies $\{P_i\}$ are labeled so that l_{i-1}

and l_{i+1} lie to the left and right of l_i, respectively, where $i = 1, 2, \ldots, 2^n$. Now, we introduce a binary labeling of these same intervals (see also Chapter 8): we write $l_i^{(n)} = l(\varepsilon_1 \varepsilon_2 \cdots \varepsilon_n)$ where ε_i is either L or R and the n-bit word $\varepsilon_i \cdots \varepsilon_n$ is the symbol sequence that the interval $l_i^{(n)}$ gives rise to in forward iteration of a map (see Fig. 9.3). In other words, $l_4^{(3)} = l(\text{LRL})$, because every initial condition that lies beneath $l_4^{(3)}$ has the symbol sequence LRL in the three forward iterations of the map that expand this particular interval to unit length. When the map is applied to an initial condition covered by a particular interval $l_i^{(n)} = l(\varepsilon_1 \varepsilon_2 \cdots \varepsilon_n)$, then in the next $n - 2$ iterations the map visits the intervals $l(\varepsilon_2 \cdots \varepsilon_n) \to l(\varepsilon_3 \cdots \varepsilon_n) \to l(\varepsilon_{n-1} \varepsilon_n)$, each of which is covered by one of the four second-generation intervals $l_i^{(2)}$. Therefore, each map generates exactly the same second-generation frequencies $\{P_i\}$ as every other map when each map is applied to its *own* corresponding nth-generation interval $l(\varepsilon_1 \cdots \varepsilon_n)$ that is labeled by the *same* symbol sequence $\varepsilon_1 \varepsilon_2 \cdots \varepsilon_n$. For the mth-generation of coarsegraining, if we use a particular map to iterate one of its nth-generation intervals forward $n - m$ times, then, for the same reason, the resulting mth-generation coarsegrained distribution will be the same as that produced by any other map in the class that is applied $n - m$ times in forward iteration to its own corresponding nth-generation interval. The general result then follows from taking $n \gg m$, but we do not attempt to pass to the limit where n becomes infinite: it is unnecessary and irrelevant for both computation and experiment. The important thing is that both the time sequence and the statistics of the orbit encoded completely in the symbol sequence. This means that, seen as a function of q, the generating function (9.4) is the same for all maps that generate the same symbol sequence. When, in addition, two different maps generate scalings that are the same to within experimental error, then $f(\alpha)$ spectra obtained from experiment cannot be used to infer exactly which is the correct underlying map. This is not a disaster: one can at least pinpoint a qualifying class of maps, and, *because of the universality, the precise details of the map do not matter. All that matters is that a symbol sequence from the right class is chosen.* The hierarchy of intervals $\{l_i^{(n)}\}$ is the signature of a particular chaotic map, but the frequencies $\{P_i\}$ are the signature of a particular class of symbol sequences (equivalently, of a set of binary irrational numbers) that exist *independently* of the details of any particular map and which occur for a special set of initial conditions in the invariant set of each map in our class. Therefore, the study of the statistics of all of these maps is reduced to the study of the different distributions of bits that are generated by algorithms for numbers, that is, to arithmetic.

It is not enough to know that complete information about the visitation frequencies for all generations is encoded in a symbol sequence: we must be able to *decode* that information in order to make use of it. For this, it is enough to decode the binary tent map in binary arithmetic, because (i) binary arithmetic produces directly the distribution of that map's iterates over its natural partitioning and (ii) the resulting frequencies will be the same for every other map in our class (universality) for the same symbol sequence. In what follows (see also section 9.3) we shall give digital proofs of certain properties of the tent map, some of which were covered earlier in Chapter 2.

The binary tent map follows from (9.6) when we set $a = b = 2$ and can be thoroughly understood by studying the iterates as binary strings,

$$z_n = \sum_{i=1}^{\infty} \frac{\varepsilon_i(n)}{2^i} = 0.\varepsilon_1(n)\varepsilon_2(n)\cdots, \tag{9.7}$$

where $\varepsilon_i(n) = 0$ or 1. For nontruncating strings that are either periodic or nonperiodic, the tent map can be replaced by the automaton

$$\varepsilon_i(n+1) = \begin{cases} \varepsilon_{i+1}(n), & \varepsilon_1(n) = 0 \\ 1 - \varepsilon_{i+1}(n), & \varepsilon_1(n) = 1 \end{cases}. \tag{9.8}$$

Given a binary string

$$\varepsilon_1(0)\varepsilon_1(1)\varepsilon_1(2)\cdots\varepsilon_1(n)\cdots \tag{9.9}$$

made up of the first bit of each iterate $z_n = 0.\varepsilon_1(n)\cdots$, the symbol sequence is given by replacing each 0 in (9.9) by an L and each 1 by an R. However, it is more convenient to use the (0, 1) 'alphabet' rather than the (L, R) alphabet in what follows.

Given a particular string (9.9), the automaton (9.8) can be iterated backward to produce the iterates and the corresponding initial condition z_0 bit by bit. The accuracy to within which z_0 can be computed is limited by the length n of the known symbol sequence. Backward iteration with the automaton is the inverse of the parallel processing method. By restricting to nonperiodic symbol sequences, we shall guarantee that the corresponding orbits are truly chaotic (unstable-nonperiodic). By restricting to nonperiodic symbol sequences that are generated by algorithms for computable irrational numbers, we thereby guarantee that our true chaotic orbits are computable (see also Chapter 11) – the procedure works because we can iterate the automaton (9.8) backward. By this method shall we generate computable properties of deterministic chaos that hold universally. We turn now to the decoding of the symbol sequence.

Table 9.1. *The frequency with which the binary tent map visits each of its four second-generation intervals is encoded in the symbol sequence*

Frequency of occurrence of $\varepsilon_1(n)\varepsilon_1(n+1)$ in the symbol sequence	Frequency of occurrence of $\varepsilon_1(n)\varepsilon_2(n)$ in z_n	Second-generation frequencies
00	00	P_1
01	01	P_2
11	10	P_3
10	11	P_4

Table 9.2. *The frequency with which the binary tent map visits each of its eight third-generation intervals is encoded in the symbol sequence*

Frequency of occurrence of $\varepsilon_1(n)\varepsilon_1(n+1)\varepsilon_1(n+2)$ in the symbol sequence	Frequency of occurrence of $\varepsilon_1(n)\varepsilon_2(n)\varepsilon_3(n)$ in z_n	Corresponding third-generation frequencies
000	000	P_1
001	001	P_2
011	010	P_3
010	011	P_4
110	100	P_5
111	101	P_6
101	110	P_7
100	111	P_8

Consider the pair of bits $\varepsilon_1(n)\varepsilon_1(n+1)$ in the symbol sequence (9.9). According to (9.8), $z_n = 0.\varepsilon_1(n)\varepsilon_1(n+1)\cdots$ if $\varepsilon_1(n) = 0$, while it follows that $z_n = 0.\varepsilon_1(n)(1 - \varepsilon_1(n+1))\cdots$ if $\varepsilon_1(n) = 1$. The correspondence in Table 9.1 shows how the frequency with which the binary tent map visits its four second-generation intervals, each of width $l_n = 2^{-n}$, is encoded in the symbol sequence. Clearly, we can obtain the four second-generation frequencies $\{P_i\}$ merely by reading the frequency with which different blocks of bit pairs appear in the symbol sequence. For higher-order generations, there are corresponding rules, and the rules for the third generation are shown in Table 9.2.

In other words, the frequency with which the tent map or any other map in the class visits any one of its 2^n intervals can be computed directly from the symbol sequence by sliding a window of n bit width to the right along the sequence (9.9) and recording the frequency with which a particular combination of m 0's and $n - m$ 1's appears as the window is moved to the right, one bit at a time. In practice in our approach, the sequence (9.9) is not generated by the map, but by an algorithm for an irrational number: because our maps generate a complete binary tree, every such algorithm yields an admissible symbol sequence for our maps; an example is shown in Fig. 9.4. The motivation for starting with a symbol sequence rather than with an initial condition is that the former *may* be extracted from certain experimental time series while the latter cannot be obtained directly from experiment. The difficulty – and it is a nontrivial one – is that one must have a way of extracting the natural partitioning unambiguously from the data in order to carry out the procedure.

The encoding of all information about the frequencies onto a symbol sequence is interesting, because it illustrates that there is no chaos that can be produced, e.g., by as many functional compositions of an asymmetric logistic map as you can perform, that cannot be reduced to the nonperiodicity of the binary string of an irrational number! The consequence of this is that initial conditions should be carefully chosen and carefully treated: for a particular map, the information required to generate a particular symbol sequence by forward iteration is encoded in the initial condition, which is typically of infinite length and therefore cannot be handled *except* by an algorithm. In other words, the *information* that tells the map where to go as the iterations proceed is completely stored, *in digital form*, in the initial condition.

In order to emphasize the nontriviality of using the distribution of iterates of the map to generate correctly finer and finer coarsegrained approximations to the invariant distribution of a map that is smoothly conjugate to the binary tent map, let us demonstrate what is required. In this case, we must use an initial condition for each map from a very special class, such that the resulting iterates are distributed uniformly, with frequency $P_i = 2^{-n}$, for every interval in the nth generation of the map's natural partitioning. Therefore, by the correspondence illustrated for the second and third generations in Tables 9.1 and 9.2, the required symbol sequences are equivalent to normal numbers: L occurs as often as R; LL, RL, LR, and RR occur equally often, and so on. In spite of the fact that normal numbers exist with measure one in the mathematically defined continuum, we are aware of only one *independent* algorithm by means of which one can recursively construct a string that is normal to base-2, namely, write the natural numbers in sequence (see Niven, 1956):

$$1101110010111011110001001 \cdots . \qquad (9.9b)$$

Fig. 9.4   Illustration of the universal rule for reading the statistics of
a map from its symbol sequence: for the 37-bit symbol sequence given
by RRLRRRLLRLRRRLRRRRLLLRLLRRLRLRLRRRRLL, e.g.,
one obtains: histogram (*a*) with relative frequencies $\frac{15}{37}$ and $\frac{22}{37}$ by
reading the symbol sequence with a 1-bit window, (*b*) $\frac{5}{36}, \frac{1}{4}, \frac{1}{3}$, and $\frac{5}{18}$
follow and yield the four-bin histogram by reading a 2-bit window,
and (*c*) the three-bit sliding-window yields the eight-bin histogram
with relative frequencies $\frac{1}{35}, \frac{3}{35}, \frac{1}{7}, \frac{4}{35}, \frac{6}{35}, \frac{1}{35}, \frac{6}{35}$, and $\frac{4}{35}$.

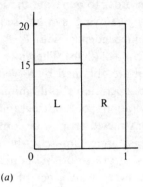

(*a*)

RRLRRRLLRLR $\boxed{R}$ RLRRRRLLLRLLRRLRLRLRRRRLL

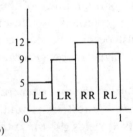

(*b*)

RRLRRRLLRL $\boxed{RR}$ RLRRRRLLLRLLRRLRLRLRRRRLL

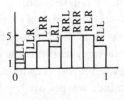

(*c*)

RRLRRRLLRLRRRLRRRRLL $\boxed{LRL}$ LRRLRLRLRRRRLL

The initial condition for the binary tent map that generates (9.9b) cannot likely be constructed by pure guesswork (by either an 'arbitrary' or a 'random' choice of initial conditions), but can be constructed bit by bit by the use of (9.8) in reverse. The result is

$$z_0 = 0.1001011100101101011110001\cdots. \tag{9.10}$$

The main point is that in order to generate the invariant distribution by iterating the map, every qualifying symbol sequence must satisfy the condition for statistical independence, with $\bar{P}_1/\bar{P}_2 = 1$, that is guaranteed by the method of construction of (9.9b). Therefore, in order that computer-generated histograms that are obtained by forward iteration of a map in the floating-point mode of operation should mimic a theoretical invariant density for several generations of coarsegraining, the corresponding symbol sequence (as constructed from the computer-imposed pseudo-orbit) must satisfy, at least approximately, the condition for statistical independence: for maps that are smoothly conjugate to the binary tent map, the condition is that every n bit-block of m L's and $n - m$ R's must appear with approximately the same frequency $P_i = 2^{-n}$ in the symbol sequence for some finite number of generations. On the other hand, when computer-generated orbits produce histograms that resemble those that would follow from a coarsegraining of the map's invariant density, then those *same* computer-generated pseudo-orbits *cannot* produce the sort of statistics that must occur in order to mimic the multiplicative *multifractral* distributions that can be generated by iterating the same map.

We shall not consider further the most general universality based upon arbitrary symbol sequences. We restrict in what follows to the classes of symbol sequences where (9.4) yields (9.4b), so that each universality class is labeled by the ratio $\bar{P}_1/\bar{P}_2$.

We now have two powerful results: first, merely by reading the distribution of blocks of bits in a symbol sequence we can discover the distribution of iterates of a particular map (for that particular symbol sequence) over its own natural partitioning. For each map in the class, the support (the invariant set) varies from map to map, but the frequencies do not. Second, when we generate frequency distributions by forward iteration of the binary tent map in binary arithmetic, we are universally generating the frequencies for every map in our class for the *same* symbol sequence. With this in mind, we go on to the main result. Because all possible binary strings are admissible symbol sequences, we can generate the symbol sequences from algorithms for irrational numbers – *only* the intervals $\{l_i^{(n)}\}$ must be provided by the map!

9.3 Universal deterministic statistical independence

In order to produce multiplicative multifractal generating functions of the form

$$\chi_n(q) = (\bar{P}_1^q + \bar{P}_2^q)^n \qquad (9.4b)$$

from (9.4), the condition for statistical independence

$$P_i = \bar{P}_1^m \bar{P}_2^{n-m} \qquad (9.11)$$

must be satisfied, where $\bar{P}_1/\bar{P}_2$ is the relative fraction of 0's to 1's in the symbol sequence. When this ratio is a rational number P/Q that is less than one, we can accomplish our aim by a trick that is equivalent to the redundant organization of binary strings onto a tree of order $P + Q$. If we start with a number that is normal to base $P + Q$ and then replace any P of the $P + Q$ digits $0, 1, 2, \ldots, P + Q - 1$, by the bit 0 and the remaining Q digits by the bit 1, then the result is a nonnormal binary string where $\bar{P}_1 = P/(P + Q) = 1 - \bar{P}_2$ and the condition for statistical independence (9.11) is guaranteed to hold. For example, to generate statistical independence with $\bar{P}_1/\bar{P}_2 = \frac{1}{2}$, we can start with a number that is normal to base 3, namely, we write the base three numbers in their natural order:

$$12101112202122100\cdots. \qquad (9.9c)$$

If next we replace the ternary digits 1 and 2 by the bit 1, we have an example of a nonnormal binary symbol sequence

$$11101111101111100\cdots \qquad (9.9d)$$

that is guaranteed to yield the generating function

$$\chi_n(q) = ((\tfrac{1}{3})^q + (\tfrac{2}{3})^q)^n \qquad (9.4c)$$

from which $f(\alpha)$ can be computed, given the intervals generated by a particular map!

Given a symbol sequence with the property of statistical independence, it is an interesting question what are the statistical properties of the bit distribution of the initial condition that produces it for a given map. We can show that, even for the binary tent map, there is no easy correspondence. The (incomplete) results are shown as Tables 9.3 and 9.4, where $P(\varepsilon_1\varepsilon_2\cdots\varepsilon_N)$ denotes the frequency with which the N-bit block $\varepsilon_1\varepsilon_2\cdots\varepsilon_N$ occurs as one reads the binary string of an initial condition of the tent map from left to right.

Table 9.3. *Pairs of bits in the initial condition determine the first generation frequencies in the symbol sequence*

$\varepsilon_1(n)$	Frequency of occurrence of $\varepsilon_{n-1}(0)\varepsilon_n(0)$ in z_0	Frequency of occurrence of $\varepsilon_1(n)$ in the symbol sequence
0	00 or 11	$P(00) + P(11)$
1	01 or 10	$P(01) + P(10)$

Table 9.4. *Second generation coarsegrained frequencies are determined by the frequency of occurrence of triplets of bits in the initial condition*

$\varepsilon_1(n-1)\varepsilon_1(n)$	Frequency of occurrence of $\varepsilon_{n-2}(0)\varepsilon_{n-1}(0)\varepsilon_n(0)$ in z_0	Frequency of occurrence of $\varepsilon_1(n-1)\varepsilon_1(n)$ in the symbol sequence
00	000 or 111	$P(000) + P(111)$
11	010 or 101	$P(010) + P(101)$
10	100 or 011	$P(100) + P(011)$
01	001 or 110	$P(001) + P(110)$

One can use Tables 9.3 and 9.4 to check that statistical independence in the distribution of iterates of the symmetric tent map is inconsistent with statistical independence in the distribution of blocks of bits in the initial condition, except for the special case $\bar{P}_1/\bar{P}_2 = 1$, where both initial conditions and the corresponding symbol sequences have the property of normal numbers.

The symbol sequence (9.9d), constructed by an algorithm that is an example of perfect order, has the statistical properties that we associate with total disorder. It is a paradigm for the deterministic production of disorder. Whether you see the order or the disorder in (9.9d) depends upon what you choose to concentrate on. We are led by this example to ask whether what we mean by disorder is a deterministic pattern that we are not familiar with. The use of words like normal and disorder confers a ranking that may prove to be misleading upon closer examination. Normal numbers may be a prime example: algorithms that produce them seem to be rare in computation, and we have as yet no evidence that the corresponding symbol sequences occur frequently in applications of low-dimensional chaos theory to nature. In physics, the use of the word disorder stems from a historical development where crystalline solids were treated as the normal ones and patterns that did not fit into that rigid framework were called disordered. We now know that the 'disordered' structures are far more common in nature and are also the more interesting ones.

The initial condition that yields (9.9d) for the binary tent map can be constructed bit by bit by backward iteration

$$z_0 = 0.1011010100\cdots. \tag{9.12}$$

As we have shown above, the string (9.12) cannot have the property of statistical independence in its bit-block distribution. At present, we can only suggest one way to produce different symbol sequences with the same statistics as (9.9d), and that is to start with (9.9d) and discard the first N bits, where $N = 1, 2, 3, \ldots$. This procedure trivially corresponds in each case to starting the iterations of the map from an initial condition different from that of (9.12).

Any experimental time series corresponds to a finite-length symbol sequence. This corresponds to iterating the tent map n times, where n is relatively small. The first few terms in (9.9d) yield estimates for $\bar{P}_1/\bar{P}_2$ that are not very close to $\frac{1}{2}$. So far as the infinite-time limit is concerned, the first N terms in a symbol sequence can always be discarded, because this procedure merely starts the iterations from a different initial condition in the invariant set. Because all possible combinations of blocks of one 0 and two 1's occur with equal frequency somewhere in the sequence, every nonperiodic experimental sequence of exactly n terms is contained within the infinite-length version of the sequence (9.9d). The more terms that one knows in the symbol sequence, the more digits one can pin down in the corresponding initial condition. More simply, one can use a symbol sequence constructed from an observed time series to iterate any one of our maps backward to construct, to within experimental accuracy, the first N bits of the initial condition that generates the observed nonperiodic sequence. More important, the expected behavior (9.4c) arises from (9.4) by the application of a law of large numbers that holds in the limit of infinite-length symbol sequences. In computation, as in experiment, we are confined to finite-length symbol sequences, so that deviations from the ideal 'expected' behavior must be expected. Such deviations from expectations are what we discuss next.

9.4 Deterministic noise

You can construct the visitation frequencies $\{P_i\}$ simply by counting the frequency with which N-bit blocks of 0's and 1's occur in a given finite-length symbol. For this, no map is needed, but with or without a map you should not expect to get *exactly* (9.4b) with a given ratio $\bar{P}_1/\bar{P}_2$. The reason for this is simple: deterministic noise. The exact ratio $\bar{P}_1/\bar{P}_2 = \frac{1}{2}$ for (9.9d), e.g., holds only in the sense of the law of large numbers (see Kac, 1959) as the length of the string goes to infinity, and for finite string lengths (important for *observation*) there are deterministic fluctuations in the ratio of 0's to 1's. In other words, $f(\alpha)$-spectra, as computed from (9.4) for different values of n, will exhibit deterministic fluctuations, and (9.4b) follows from (9.4) only as a 'most probable generating function' in the sense of the law of large numbers as n goes to infinity. The analogy with the justification of the Boltzmann distribution for the statistical equilibrium of a gas is obvious, but we cannot arbitrarily apply the limit of large n to either experiment or computation in this case. It would be a mistake to imagine that (for finite n) these deterministic fluctuations should be eliminated by an artificial 'smoothing' technique: they reflect the fact that the symbol sequence has the right statistical properties.

The finitely constructed first-generation frequencies P_1 and P_2 (set $N_n = 2$ in (9.4)) fluctuate deterministically as the length n of the symbol sequence is increased. If you ask what is the most likely ratio of P_1/P_2 as n goes to infinity, the result is what we have heretofore called $\bar{P}_1/\bar{P}_2$. In other words, $\bar{P}_1$ and $\bar{P}_2$ follow from the law of large numbers and can be interpreted as theoretical probabilities (as opposed to the 'empirical' frequencies P_1 and P_2) for the occurrence of L's and R's in the deterministic sequence as n goes to infinity. So long as the symbol sequence is constructed by an algorithm, these probabilities have absolutely nothing to do with true randomness: because the deterministic sequence has certain statistical properties, the correct probability rules can be derived from the sequence. These probabilities can then be used to describe the 'expected' statistical properties, but they cannot be used to generate the sequence itself. Thus does determinism give rise to a stochastic description; the stochastic description of the dynamics is incomplete, but the deterministic description that gives rise to it is not. The probabilistic description is an idealization whose applicability is confined to the law-of-large-numbers limit. For small n, the best one can do is to compute the 'empirical' probabilities (the frequencies) from the dynamics, which requires that the symbol sequence must be known. Otherwise, nothing useful can be predicted with any degree of confidence. We interest ourselves in probabilities to the extent that they can be shown to follow from an algorithmic description of deterministic dynamics. It makes sense that the appearance in nature of well-defined empirical probabilities, corresponding to the regular and stable frequencies of occurrence of (nonquantum) events, should not be the signature of indeterminacy, but reflects an *underlying order*. It is the underlying order that is responsible for the stabilization of the frequencies with which different events occur.

The mathematician's ideal fractal and nonfractal measures are constructed from the theoretical probabilities $\bar{P}_1$ and $\bar{P}_2$, and therefore neglect the deterministic noise that occurs in both experiment and in computation. In both experiment and computation one gets frequencies, not measures, from time series. The idea of a measure is a mathematical abstraction that has no place in experimental physics, where noise cannot be avoided and should in fact be described: the noise should not be 'smoothed', but should be described deterministically because scientists typically work in a limit where the law of large numbers cannot be used arbitrarily to rule out the likelihood of long runs of 'improbable' blocks of symbols.

When $\bar{P}_1/\bar{P}_2 \neq 1$, both the finitely constructed coarsegrained distributions and their mathematical idealizations, the so-called fractal measures, are infinitely fragmented. When $\bar{P}_1/\bar{P}_2 = 1$, the limiting frequencies are even; in other words, when we assert that $P_i = 2^{-n}$ for almost all initial conditions of the logistic map $f(z) = 4z(1 - z)$, the result is guaranteed to be true *only* in the law-of-large-numbers limit. This is equivalent to the statement, e.g., that the binary tent map for almost all initial conditions has a uniform invariant density, or to the statement that the Cantor function is the probability distribution for iterates of the ternary tent map, for almost all initial conditions. The corresponding symbol sequences are binary normal numbers (replace L by 0, R by 1). However, the corresponding finite-length symbol sequences may sometimes produce apparently fragmented coarsegrained distributions because of deterministic noise in the finitely constructed first-generation frequencies P_1 and P_2. Regardless of whether $\bar{P}_1/\bar{P}_2$ is equal to unity or not, the question of how much noise is there is related to that of how effective a pseudorandom number generator the corresponding symbol sequence is. By our method of construction, this reduces to the question of how good a pseudo-random number generator the digit string of a normal number is. A 'good' pseudo-random number generator is one that does not contain long runs of 'improbable' blocks of symbols. We refer the interested reader to Knuth (1981) for a discussion of both points.

We can apply our method to the analysis of experimental time series. Given a time series *and approximate knowledge of the first-generation scalings* $l_1^{(1)}$ *and* $l_2^{(1)}$, a finite-length symbol sequence can be constructed. The idea is then to use the symbol sequence to look for evidence that statistical independence holds approximately. When one finds a fraction

P_1 of L's and a fraction P_2 of R's, then the idea is to ask if $P_1^m P_2^{n-m}$ is approximately the frequency with which any block of m L's and $n - m$ R's occurs as one reads the n letter blocks bit by bit from one end to the other in sequence. 'Attractor reconstruction' from experimental time series cannot be achieved without knowledge of at least the first generation of coarsegrained scalings or their multidimensional equivalent: the first-generation scalings are the crudest possible level of description of an invariant set. In the next chapter, we discuss an experiment where both the intervals and the probabilities were inferred from the fit of an experimental $f(\alpha)$-spectrum by a binomial generating function. Generally, the big challenge is to extract the intervals from experiment – we have argued that that is the closest you can come to extracting information about a particular chaotic map from experiment.

9.5 Trees of higher order and incomplete trees

We can generalize our results to include a class of fully chaotic maps that generates a complete tree of any order. The generalization is illustrated by example for maps that generate a complete ternary tree. Consider the map

$$f(z) = \begin{cases} 3z, & z < \frac{1}{3}, \\ 3(z - \frac{1}{3}), & \frac{1}{3} < z < \frac{2}{3}, \\ 3(1 - z), & \frac{2}{3} < z < 1, \end{cases} \tag{9.13}$$

which is shown in Fig. 9.5. The map generates a complete ternary tree in backward iteration, with intervals given by $l_n = 3^{-n}$. The symbol sequences are formed from the three-letter alphabet (L, M, R), as is indicated in Fig. (9.5), and all possible symbol sequences are in one-to-one correspondence with all numbers in the unit interval written as ternary expansions (base three is the natural base of arithmetic for deducing the statistical properties of the trajectories of (9.13)).

In order to generate symbol sequences that yield an $f(\alpha)$ spectrum based upon statistical independence,

$$(P_1^q + P_2^q + P_3^q)^n = 3^{-n(q\alpha - f(\alpha))}, \tag{9.14}$$

we need a base-three normal number. To get an even distribution we can use the Champernowne number in base-3,

$$1210111222021221001011102\cdots, \tag{9.4d}$$

and then replace the digits (0, 1, 2) by the symbols (L, M, R) or any permutation thereof. In general, if $P_1 = P/(P + Q + S)$, $P_2 = Q/(P + Q + S)$, and $P_3 = S/(P + Q + S)$, where P, Q, and S are integers, then we can generate a symbol sequence by writing down a base-$(P + Q + S)$ normal number and making replacements as follows: P of the digits are to be replaced by L, Q by M, and the remaining S are replaced by the symbol R. Now, it is possible to construct $f(\alpha)$ spectra where either $f(\alpha_{min}) \neq 0$ or $f(\alpha_{max}) \neq 0$: whenever $P_1 = P_3 > P_2$, then $f(\alpha_{min}) = \ln 2/\ln 3$ while $f(\alpha_{max}) = 0$. The roles are reversed if $P_2 > P_1 = P_3$ (see equation (10.17) for a generalization to trees of order $t = 4, 5, 6, \ldots$).

Incomplete trees are generated by maps of the unit interval that peak beneath unity. The logistic map $f(z) = 4z(1 - z)$, with $D_c < D < 4$, is an example, as is the circle map just beyond the transition to chaos of quasi-periodic orbits. These cases are much more difficult to handle mathematically than are maps that generate complete trees (see Cvitanović, Gunaratne, and Procaccia, 1988).

Fig. 9.5 Example of a map of the interval that generates a complete ternary tree; this example is purposely constructed to generate 3^n intervals of exactly equal size in n backward iterations of the unit interval.

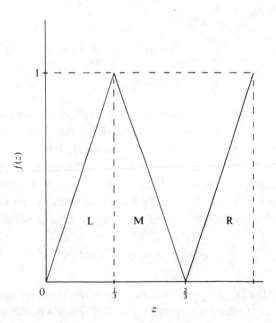

10
Intermittence in fluid turbulence

10.1 Fluid turbulence in open flows

We begin with the traditional continuum considerations and then return to our finite-precision framework in order to analyze a set of experiments in section 10.6 below.

Consider the flow of an incompressible fluid past an object whose characteristic size is L and where the fluid velocity $v(x, t)$ of the incoming flow is uniform, with magnitude U far from the obstacle. The Navier–Stokes equations with the incompressibility condition (the fluid density $\rho = $ constant) are

$$\frac{\partial v}{\partial t} + v \cdot \nabla v = -\nabla P + \nu \nabla^2 v, \tag{10.1}$$

$$\nabla \cdot v = 0,$$

where ν is the kinematic viscosity and has units of a diffusion coefficient, $\text{cm}^2\,\text{s}^{-1}$, while P is the pressure divided by the fluid density. In the Navier–Stokes equations there is a complicated competition between two terms, the nonlinear term $v \cdot \nabla v$ and the dissipative term $\nu \nabla^2 v$. The relative magnitude of these two terms is called the Reynolds number, $R = (U^2/L)/\nu U/L^2 = UL/\nu$. The Reynolds number is the control parameter in the dimensionless form of the Navier–Stokes equations,

$$\frac{\partial v'}{\partial t'} + v \cdot \nabla' v' = -\nabla' P' + R^{-1} \nabla'^2 v', \tag{10.1b}$$

so that it is at the large Reynolds limit where nonlinearity dominates that turbulence occurs, whereas at small enough Reynolds numbers dissipation

260

dominates and all disturbances of the flow are damped. To obtain the dimensionless equation of motion from (10.1), we rescaled the space and time variables so that $r = Lr'$, $v = Uv'$, $t = t'L/U$, and ∇' is the dimensionless gradient.

For the flow past a cylinder, the transition to turbulence can be described as follows (we assume that the cylinder is long relative to its diameter $2L$): when $R < 5$, there is only a thin laminar wake behind the cylinder. The formation of the wake is due to the build up of vorticity $\boldsymbol{\omega} = \nabla \times \boldsymbol{v}$ in the boundary layer that surrounds the cylinder as a result of the no-slip boundary condition whereby the velocity must vanish at the surface of the cylinder. The vorticity is then convected downstream and forms the wake. Quantitatively, the boundary layer is the region near the object where the diffusion term must be taken into account in order to satisfy the no-slip boundary condition at the surface of the obstacle (outside the boundary layer, the free stream is then approximated as an ideal fluid flow). When $5 < R < 10$, a symmetric pair of vortices (eddies) forms behind the cylinder owing to boundary layer separation as a result of the too-rapid buildup of vorticity in the boundary layer (Fig. 10.1b). As R increases, these eddies grow in size, until R is about 40. In this range of Reynolds numbers, the flow is both stable and steady: time-dependent transient perturbations are damped effectively by viscosity. At $R > 40$, the wake becomes periodic and can be modeled by von Karman's vortex street; the motion is stable but oscillatory and reminds us of a stable limit cycle. At $R > 90$, the vortices are alternately shed from the cylinder, and by $R = 150$ one can certainly call both the wake and boundary layer turbulent: the vortices (eddies) that make up the wake are chaotic rather than laminar. The sequence is shown in Fig. 10.1a–e. For phenomenological descriptions of the transition, we refer the reader to Goldstein (1965), Prandtl (1952), and Schlichting (1979).

What is the mechanism for turbulence? If we rewrite the Navier–Stokes in the vorticity transport form

$$\frac{\partial \boldsymbol{\omega}}{\partial t} + \boldsymbol{v} \cdot \nabla \boldsymbol{\omega} = \boldsymbol{\omega} \cdot \nabla \boldsymbol{v} + \nu \nabla^2 \boldsymbol{\omega} \tag{10.1c}$$

then the vortex-stretching term $\boldsymbol{\omega} \cdot \nabla \boldsymbol{v}$ provides the mechanism for the transfer of energy to smaller and smaller scales, until finally so small a length scale is reached that viscosity becomes effective in dissipating the energy. Until that scale is reached, the vorticity that composes the turbulence is not damped by viscosity. Here we have the first indication that we need a hierarchy of coarsegrained descriptions that distinguishes small-scale vorticity from large vortex structures. We will argue more strongly in favor of this idea below, by the introduction of local Reynolds numbers based upon a hierarchy of different eddy sizes in the turbulent flow.

The starting point for the description of the energy cascade is the relation of the vorticity to the dissipation of kinetic energy in the fluid,

Fig. 10.1   The change in qualitative behavior of the wake behind a very long circular cylinder is indicated for several different *approximate* Reynolds number ranges: (*a*) When $R < 5$, there is only a thin, narrow wake behind the cylinder; the thickness of the wake grows asymptotically as the square root of the distance behind the cylinder. (*b*) For $R > 5$, there is boundary-layer separation; consequently, a pair of bound vortices forms symmetrically within the separation region. For $R < 40$, the vortices only grow in size as the flow remains steady. (*c*) When $R > 40$, the wake becomes approximately periodic and is modeled by von Karman's ideal-fluid vortex street. (*d*) At $R > 90$, vortices are alternately shed from the cylinder. (*e*) When $R > 150$, both the boundary layer and the wake have become turbulent, which (according to Prandtl) means that the vortex street becomes irregular.

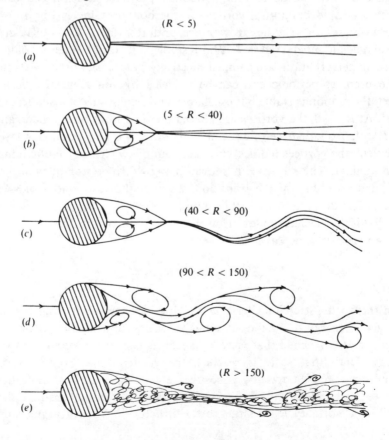

$$\frac{1}{2} \int \frac{\partial \rho \boldsymbol{v}^2}{\partial t} = -\rho v \int \omega^2 \, \mathrm{d}^3 x = \rho L^3 v \langle (\nabla \times \boldsymbol{v})^2 \rangle. \tag{10.2}$$

The cascade of energy to smaller and smaller length scales goes via the instability and rapid splitting of larger eddies into more and more smaller unstable and rapidly splitting eddies. The eddy cascade can be seen by letting a droplet of ink or dye fall into a still glass of water; the result is shown as Fig. 10.2. In this case, the initial large vortex ring forms owing to the shearing of two incompressible fluids past each other, the ink droplet and the water. The large eddy is unstable and gives birth spontaneously to four to six smaller eddies; these are themselves unstable, so that each cascades into four to six even smaller eddies, and so on. The flow is complicated: all of the eddies are interconnected by a very thin vortex-sheet-like structure. If we concentrate upon the eddies, we have a nice illustration of what is probably a multifractal, and it is nature that organizes the hierarchy of scales (different eddy sizes in different generations) for us onto a tree! In the ink-drop cascade, after only a few generations of development of the tree, each small eddy splits into only two smaller eddies, and the splitting then ceases: the last daughter eddies are stable against further splitting (they have not enough energy; the local Reynolds number is by that time roughly unity) and are only later eaten up

Fig. 10.2 Low–Reynolds-number eddy cascade as a result of letting a droplet of ink or food-coloring fall into a container of water (photo by A. Skjeltorp).

by viscosity. Here, the initial Reynolds number (L is the size of the initial large vortex ring) is about 15, so the flow is not turbulent. However, the flow is probably chaotic in both space and time. We shall see that there is evidence that for a truly turbulent flow the order of the tree is at least eight, not six. Any reader performing the ink droplet experiment should notice that the turbulence mixes the ink and water together and that repeated attempts to get exactly the same eddy-cascade pattern will fail. We expect that this is an example of the generation of a multifractal by deterministic chaos, although the speculation remains to be proven. However, without large positive and negative Liapunov exponents, it is difficult to understand the rapid vortex sheet formation in the cascade from one generation to another in the ink drop experiment. Fully developed turbulence is rapidly and very effectively mixing, and this is only a stronger argument that high-Reynolds-number vorticity dynamics is deterministic chaotic. We therefore offer the following definition: turbulence is chaotic vorticity in both space and time, with the occurrence of eddies of all possible sizes l_n, where $L \geq l_n \geq l_\kappa$, and l_κ is the scale that limits the smallest possible eddy size (to be estimated below).

10.2 Scale invariance and broken symmetry

Fluid turbulence follows from an extremely rapid buildup of vorticity $\boldsymbol{\omega} = \nabla \times \boldsymbol{v}$ in the boundary layer of the flow past a solid object, or in the shear layer of two fluids in relative motion at high enough Reynolds numbers. As a result, chaotic vorticity and vortex structures appear in the wake downstream from the object or in the shear layer. The velocity field $\boldsymbol{v}$ satisfies the Navier–Stokes equations,

$$\frac{\partial \boldsymbol{v}}{\partial t} + \boldsymbol{v} \cdot \nabla \boldsymbol{v} = -\nabla P + \nu \nabla^2 \boldsymbol{v}, \tag{10.1}$$

where P is the internal pressure divided by the fluid density and ν is the kinematic viscosity. That the fluid is incompressible means that cavitation is eliminated from consideration. Above, we have introduced Reynolds-number scaling, but that is not the most general (broken) scaling symmetry of the Navier–Stokes equations.

We now make a transformation that amounts to rescaling the variables $\boldsymbol{r}$, $\boldsymbol{v}$, and t:

$$\boldsymbol{r} = \boldsymbol{r}'/\lambda,$$

$$\boldsymbol{v} = \boldsymbol{v}'/\lambda^\alpha, \tag{10.3}$$

$$t = \lambda^{\alpha-1}t',$$

where α is a scaling index. The transformed Navier–Stokes equations

$$\frac{\partial \boldsymbol{v}'}{\partial t'} + \boldsymbol{v}' \cdot \nabla' \boldsymbol{v}' = -\nabla' P' + v' \nabla'^2 \boldsymbol{v}' \qquad (10.1d)$$

are formally scale invariant if we choose $P' = \lambda^{2\alpha - 2} P$ and $v' = \lambda^{\alpha + 1} v$. This choice of transformed viscosity leaves the Reynolds number $R = UL/v$ invariant, where U is a characteristic velocity of the fluid and L is the size of the solid object that obstructs the flow, hence causes the turbulence. Note that the exponent α is so far arbitrary, but there are also boundary conditions to be considered: energy is 'injected' into the cascade at a length scale L. Note also that v/x^α is formally invariant, so that we expect to obtain a scaling law where the velocity or velocity differences scale as a length raised to the power α.

The main idea of older studies of fully developed turbulence has been to ignore the boundary conditions representing impediments that generate the turbulence and to concentrate upon the resulting turbulence downstream, which was assumed to be homogeneous at small scales. Experimentally, a more or less homogeneous flow can be generated by a fine wire mesh. In what follows, the details of boundary conditions are replaced by the 'stirring length' L which can be regarded as the characteristic size of the largest eddies, hence of the object that created the largest eddies. Kolmogorov's reasoning was based upon the idea of homogeneous turbulence (Kolmogorov, 1941), but we shall see that the assumption of homogeneous turbulence may be replaceable by a less stringent requirement of approximate statistical equilibrium in a wake or boundary layer. As we described above, the underlying idea is that large unstable eddies form and decay spontaneously into smaller unstable eddies, so that the energy-eddy cascade continues until the eddies reach a size so small that the cascade is damped effectively by viscosity.

It is the phenomenology of the cascade of energy from a macroscopic scale (on the order of a 'stirring length') to a microscopic one (the Kolmogorov length scale) that we concentrate upon in what follows, ignoring the details of the Navier–Stokes equations, which we do not know how to solve. The singling out of these two length scales by solutions of the Navier–Stokes equations breaks the formal scaling symmetry (10.3) of the Navier–Stokes equations.

In two dimensions, there are special conservation laws (plus the impossibility of vortex stretching) that cause a cascade from smaller to larger vortices. The two-dimensional theory was first considered by Onsager (1949) and is important for understanding how larger vortices can form from smaller ones, e.g. in shear layers. It also provides an example of a negative temperature system in classical statistical mechanics, but we do not consider the two-dimensional case further in what follows.

10.3 Kolmogorov's 1941 theory (K41 theory)

The K41 theory can be derived in several different ways. Here, we follow a line of thought that is directly connected with the formulation of the previous section.

. The starting point is the breaking of formal scale invariance by solutions of the Navier–Stokes equations. The symmetry is broken, because in turbulence there is a large length scale L (the 'stirring length') at which energy is fed into the cascade; the energy then cascades through the eddy scales $L \geq l_n \geq l_\kappa$ and finally is dissipated at the Kolmogorov length scale l_κ by viscosity. If we write $l_\kappa = U^\alpha L^\beta v^\gamma$ and use dimensional analysis, then we find that $\alpha = -\gamma$, $\beta = 1 - \gamma$ and $\gamma = \frac{3}{4}$, yielding the prediction $l_\kappa = LR^{-3/4}$. The K41 theory (or model) follows from the assumption that the energy dissipation rate per unit mass

$$\varepsilon \sim v \langle (\nabla \times \boldsymbol{v})^2 \rangle \tag{10.4}$$

is maintained constant via local nonlinear interactions in wavenumber space, where $\langle \cdots \rangle$ here denotes a spatial average (see equation (10.2) above), although in the K41 theory it denotes an unspecified ensemble average. Even simpler, if we consider the energy dissipation ε under the transformation (10.3), we find that

$$\varepsilon = v' \lambda^{-3\alpha+1} \langle (\nabla \times \boldsymbol{v}')^2 \rangle = \lambda^{-3\alpha+1} \varepsilon', \tag{10.5}$$

so that the assumption that $\varepsilon = \varepsilon'$ is invariant under rescaling amounts to requiring that

$$\alpha = \tfrac{1}{3}, \tag{10.6}$$

and this, as we shall see below, is the K41 prediction.

From equation (10.6) we can infer that the differences in the velocity field $\boldsymbol{v}(r)$ over a distance l_n should scale as a $\frac{1}{3}$-power of l_n. This prediction can be tested. As we shall soon see, it leads to the prediction that the eddy cascade is not spatially intermittent, which runs into trouble with certain results from experiment.

Onsager showed in 1949 that the K41 prediction can be derived from dimensional analysis in wavenumber space. The fractal model in the next section introduces the possibility of spatial intermittency, but goes only

---

† This follows from assuming that l_κ is determined by ε and v alone and that ε is independent of v when R is large. This yields $l_\kappa = (v^3/\varepsilon)^{1/4}$, so that $\gamma = \frac{3}{4}$.

one step beyond dimensional analysis: there, an argument is made based upon the assumption of characteristic length and time scales for generations of eddies in the cascade.

10.4 The β-model of the inertial range cascade: a model of spatial intermittence

The K41 prediction follows from a certain selfsimilarity, invariance of the 'dissipation' under the rescaling transformation. Landau suggested that the deviations from the K41 theory are due to intermittence, or 'spottiness' in the turbulence, but there existed at that time no adequate theoretical framework for the description of intermittence. As we shall see, there are at least two different kinds of intermittence: intermittence in space, and intermittence in the dissipation ε.

Later, a suggestion was made by Mandelbrot (1983), based upon the Novikov–Stewart model: he suggested that the intermittence in turbulence is due to singularities of the infinite Reynolds number limit of the hydrodynamics equations, the Euler equations, that are concentrated upon a set with nonintegral fractal dimension. In Mandelbrot's approach, the energy transfer rate ε is assumed to fluctuate in both space and time. The β-model of fully developed turbulence that was introduced later by Frisch, Sulem, and Nelkin (1987) describes a certain spatial fluctuation, or spatial intermittence. The β-model is transparent and is based upon the assumption of a uniform distribution of energy over the intervals (eddies) in each generation of a geometrically selfsimilar fractal. We shall study the β-model, but we shall not concern ourselves with the much more difficult mathematics question of whether the Euler equations have a fractal distribution of singularities in velocity ($l_n = 0$ limit of an idealized, nonviscous cascade). Whether this is or is not true has little to do with understanding the results of *measurements* of turbulence, or with building a coarsegrained model that describes those measurements, because the measurements are always performed at *finite* length scales $l_n \geq l_\kappa$ (in experiments, R is *never* large enough to justify the extrapolation $l_\kappa = 0$). From our standpoint, what is needed is not a description of an idealized zero-length scale mathematical limit, but rather a hierarchy of coarsegrained descriptions of chaotic vorticity dynamics. At this stage, it is difficult to see how to extract the essential features of a correct coarsegrained set of descriptions systematically from the Navier–Stokes equations.

In the β-model, one assumes detailed balance of energy transfer per unit mass ('dissipation') within the inertial range, $L \gg l_n \gg l_\kappa$, because the idea is that essentially no energy can be dissipated before the small scale

l_κ is reached. Dissipation occurs essentially only at the Kolmogorov scale l_κ, owing to the action of viscosity (the diffusion time $t_{\text{diff}} \approx l_n^2/\nu$ is too large compared with laboratory time scales, unless $l_n \approx l_\kappa$), while energy enters the cascade only at the macroscopic length scale L.

The geometry of the model is simple and is qualitatively illustrated by a uniform Sierpinski carpet in two dimensions (cf. Fig. 10.3). There, we consider a sequence of length scales (sides of shaded boxes in Fig. (10.3)) $\{l_n\}$, where $l_n = 3^{-n}$, and if we denote by β the ratio of the area occupied by the $(n+1)$th-generation 'eddies' (shaded boxes) divided by that occupied by the shaded boxes in generation n, then $\beta = \frac{8}{9}$. Each shaded box corresponds symbolically to one eddy in the β-model in the nth-generation of the cascade model to be introduced next. Clearly, this is a model of spatial intermittence because the shaded boxes occupy less and less space as the generation number n is increased, although the octal tree of the model is complete. In the β-model we shall obtain spatial intermittence primarily through an incomplete tree.

In order to proceed, we make a model with a hierarchy of eddy sizes $l_n = La^{-n}$, where L is the size of the largest eddy. We assume also that there occurs an average number N_{n+1} of eddies at length scale l_{n+1} because of the unstable deterministic dynamics of the energy cascade at length scale l_n, as was observed in the ink drop experiment. The eddies (vortices) in the $(n+1)$th generation are supposed to occupy only a fraction $0 < \beta < 1$ of the volume occupied by the eddies in the nth generation, so that the volume occupied by the nth generation in the cascade is proportional to β^n. If $\beta < 1$, then the model has spatial intermittence.

Let V_n be the order of magnitude of the velocity *difference* across a typical nth-generation eddy. This idea could be made more precise by

Fig. 10.3   Third generation in the construction of a uniform Sierpinski carpet (see also Fig. 4.3).

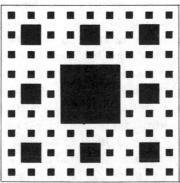

introducing definitions based upon velocity correlation functions, but we shall not take the trouble here: the refinement is not needed at this crude stage of development. Then the energy per unit mass at length scale l_n is on the order of

$$E_n \sim \beta^n V_n^2 \tag{10.7}$$

and with the characteristic time scale $t_n \sim l_n/V_n$ (there is no diffusion time scale because viscosity does not act in the inertial range), the energy transfer rate, the 'dissipation', $\varepsilon_n \sim E_n/t_n$, is given by

$$\varepsilon_n \sim \beta^n \frac{V_n^3}{l_n}. \tag{10.8}$$

If we denote by $N(l_n) \sim (l_n/L)^{-D_0}$ the number of eddies per unit volume in generation n, it follows that $\beta L^3 = N(l_n)l_n^3$ is the volume occupied by the above-mentioned nth-generation eddies, so that

$$\beta^n = (l_n/L)^{(3-D_0)} = a^{-n(3-D_0)}, \tag{10.9}$$

and so

$$V_n \sim \varepsilon^{1/3}l_n^{1/3}(l_n/l_0)^{-(3-D_0)/3} \approx l_n^{(D_0-2)/3}, \tag{10.10}$$

which shows that

$$\alpha = (D_0 - 2)/3. \tag{10.11}$$

Note that the volume contraction rate is given by

$$\beta \simeq 2^{D_0-3}. \tag{10.12}$$

The velocity structure functions of order p have the form

$$\langle |\delta V(l_n)|^p \rangle = \langle l_n^{\alpha p} \rangle \approx \beta^n \cdot l_n^{\alpha p} = l_n^{\zeta_p}, \tag{10.13}$$

where a spatial average is denoted. The factor β^n represents the assumption of an even distribution of energy over each of the N_n scales l_n for each generation n in the inertial range: there is no intermittence in the energy transfer rate ε_n in this oversimplified model. The spatial intermittence comes in two parts: the order of the tree is a^{D_0}, so that the tree is incomplete (certain branches are missing) if D_0 is nonintegral. Also, if $a > 2$, then this feature contributes to the spatial intermittence (with $a = 2$ and $D_0 = 3$, there is no intermittence: the nth-generation eddies occupy as much space as did the original large eddy).

The structure function of order p has the scaling exponent $\zeta_p = \alpha p + 3 - D_0$, and with $\alpha = (D_0 - 2)/3$, and we retrieve the K41 theory with $\alpha = \frac{1}{3}$ if we assume that $D_0 = 3$. The known scaling exponents (cf. Fig. 10.4) are linear in p for small p, but the slope is less than $\frac{1}{3}$, hence the K41 theory is wrong. With $D_0 \approx 2.91$, one obtains a fit to the linear region with the β-model, but nonlinear behavior occurs in ζ_p for $p \geq 8$, so the β-model is not the whole story. Unfortunately, the error bars in the data for $p > 8$ (the bars are not shown in Fig. 10.4) are so large that no firm conclusion can be drawn inside the nonlinear region. However, note that $N_n \sim a^{nD_0}$, so that with $a \geq 2$ and $D \approx 2.91$, an *incomplete* tree of at least *octal* order is indicated ($7 < 2^{D_0} < 8$).

At present, no theory exists which predicts the value of D_0 from first principles, but this is not the worst missing link, because the ability of a theory merely to predict one fractal dimension is no real test of that theory. More impressive will be the theory that tells us the order and degree of completeness of the tree. Next should come the eddy sizes and the statistics of the dissipation for each eddy. Then, we would have a complete description of both spatial intermittence and intermittence in the energy dissipation. A high order of business is to derive all of this from a hierarchy of coarsegrained vorticity dynamics – the hierarchy truncates at the Kolmogorov length scale.

10.5 Multifractal models of intermittence

We have seen how the cascade of energy from larger to smaller and smaller eddies leads to the idea of scaling exponents for the structure functions.

Fig. 10.4   Scaling exponents for velocity structure functions as a function of the moment index p. The data are presented naively without the error bars, which are very large whenever $p > 8$.

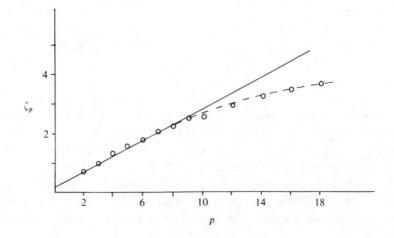

The idea was developed for the oversimplified case where the energy dissipation per eddy in uniform and the fractal, except perhaps for the incompleteness of the tree, is geometrically selfsimilar. It is easy to generalize the formulation of section 10.4 to include the idea that dissipation takes place on the nonuniform fractal (one that is not geometrically selfsimilar), and that allows the energy transfer rate to occur either uniformly or nonuniformly over N_n eddies with different characteristic sizes $l_1, l_2, \ldots, l_N$ in generation n. We begin by assuming that $|\delta V(l_n)^p| \approx l_n^{p\alpha}$ in the inertial range, and we assume also that the scaling index α occurs for a hierarchy of intervals with fractal dimension $f(\alpha)$. This is a direct generalization of the equation $\beta L^3 = N(l_n)l_n^3$ of the last section. The goal is to compute the exponent ζ_p in the velocity structure function $\langle |\delta V(l_n)|^p \rangle \sim l_n \zeta_p$ for the inertial range, where

$$\langle |\delta V(l_n)|^p \rangle \approx \sum_\alpha l_{i(\alpha)}^{p\alpha - f(\alpha) + 3}. \tag{10.14}$$

Application to the largest-term approximation to (10.14) for small but finite l_i yields

$$\zeta_p = 3 - f(\alpha(p)) + p\alpha(p), \tag{10.15}$$

where

$$p = f'(\alpha(p)) \tag{10.15b}$$

determines the scaling index $\alpha(p)$. It follows then that (10.15) is a Legendre transform, so that

$$\alpha(p) = d\zeta_p/dp \tag{10.15c}$$

is the slope in Fig. 10.4. If we assume that $f(\alpha) = D_0$ and $\alpha = (D_0 - 2)/3$, then we retrieve the uniform fractal of the β-model, but, more generally, we can write

$$f(\alpha(p)) = p\alpha(p) + 3 - \zeta_p, \tag{10.16}$$

where α and f vary with the index p.

If we ask for a model where a two-scale Cantor set is organized onto a complete tree of order t and the invariant distribution is statistically independent then the most general generating function is

$$\chi(p) = (b\bar{P}_1^p + (t - b)\bar{P}_2^p)^n, \tag{10.17}$$

where $b\bar{P}_1 + (t - b)\bar{P}_2 = 1$ and b is an integer, $1 \le b \le t - 1$. There are two scales, l_1 and l_2: $P_1 = l_1^{\alpha_1}$, $P_2 = l_2^{\alpha_2}$. When $t = 8$ and $0 < l_i \leq \frac{1}{3}$, (10.17) describes a planar Sierpinski carpet where $1 < D_0 \leq 2$. If $t = 8$ and $\frac{1}{3} < l_i \leq \frac{1}{2}$, then $2 \leq D_0 \leq 3$ and we have the fragmentation of a chunk of three-dimensional space into eight separate pieces. In other words, the fragmentation is organized onto an octal tree in both cases. The construction of models with an incomplete tree is also possible, but we shall soon discover that the experimental data are inadequate to distinguish between different models that predict the behavior of the scaling exponents in the nonlinear regime.

In general, then, we could write

$$\langle |\delta V(l_n)|^p \rangle \cong l_n^3 \chi(p) = l_n^{\zeta_p}, \tag{10.18}$$

where $\zeta_p = 3 - f(\alpha) + p\alpha$, but the problem that we face is that the data on velocity structure functions are inadequate to pin down $f(\alpha)$. Even with (10.17), there are *four* independent parameters, e.g. α_1, α_2, $f(\alpha_1)$, and l_1 (l_2 and $f(\alpha_2)$, e.g., can be eliminated in principle by use of the theoretical condition $bP_1 + (t - b)P_2 = 1$ and the knowledge of D_0 from experiment, leaving four of the original six unknowns), where $f(\alpha_1) = -\ln b/\ln l_1$ and $f(\alpha_2) = -\ln(t - b)/\ln l_2$. But the known data on ζ_p correspond only to a small portion of the $f(\alpha)$ just to the left of the peak $f_{max}(\alpha) = D_0$. That information is not enough to determine more than two of the unknowns accurately. In order to determine α_1 and $f(\alpha_1)$, for example, one must know $\alpha_1 = d\zeta_p/dp$ as $p \to \infty$. It would be necessary, therefore, to know whether ζ_∞ is infinite (finite α_1) or whether $\zeta_\infty = 3 - f(\alpha_1)$, in which case $\alpha_1 = 0$. This information, along with the portion of the spectrum $p \ll 0$ where α_2 could be determined, appears to be inaccessible. As matters now stand, infinitely many different generating functions $\chi(p)$ can be inserted into (10.18) and will reproduce the known structure function scaling exponents with total ease, so long as they merely reproduce a small neighborhood about the peak of the turbulence $f(\alpha)$ curve correctly. Furthermore, each such generating function is in principle consistent with infinitely many different dynamical systems (universality). Therefore, we conclude that it is best to measure objects other than the velocity structure functions in order to better understand turbulence. In particular, it would be very useful for experimenters to discover the order and degree of completeness of the eddy-cascade tree, along with the characteristic eddy sizes, as a function of Reynolds number: if t and b could be determined from experiment, from some sort of flow visualization, then measurements of structure functions at large values of p would become unnecessary as a means for pinning down $f(\alpha)$ and ζ_p, and we would have better guidance for future mathematical modelling of the eddy cascade as well.

10.6 Intermittence of energy dissipation at the small scales of turbulence

We now consider the evidence that the statistics and eddy sizes of energy dissipation at small scales as measured in one-dimensional cuts of turbulence are characteristic of fully developed chaos.

We have seen in Chapter 9 that generating functions of the form

$$\chi(q) = (\bar{P}_1^q + \bar{P}_2^q)^n \tag{10.19}$$

are the result of symbol sequences of one-dimensional fully chaotic maps that peak at or above unity, hence generate a complete binary tree in backward iteration of the map, if one chooses initial conditions that produce a statistically independent set of iterates. The class of maps includes symmetric and asymmetric logistic and tent maps and Bernoulli shifts.

According to the results of Meneveau and Sreenivasan (1987), which were confirmed by Saucier (1991), the energy dissipation (Fig. 10.5) that

Fig. 10.5   Energy dissipation as a function of position, inferred from a velocity time series on the basis of Taylor's 'frozen turbulence hypothesis' (from Saucier).

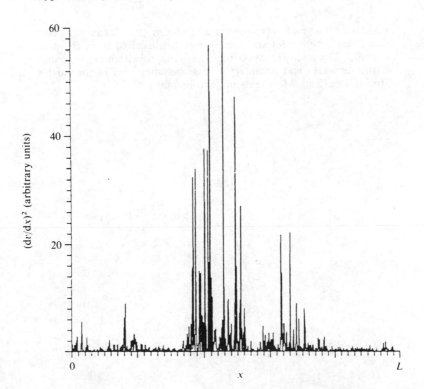

is measured for one-dimensional cuts of turbulence due to wakes, grids, and boundary layers can be described in every case by an $f(\alpha)$ spectrum given by

$$\chi(q) = (\bar{P}_1^q + \bar{P}_2^q)^n = l_n^{q\alpha - f(\alpha)}, \tag{10.19b}$$

with $l_n \approx 2^{-n}$ and $\bar{P}_1/\bar{P}_2 \approx \frac{3}{7}$ (Fig. 10.6). This generating function requires statistical independence on a complete binary tree. The scaling $l_n \approx 2^{-n}$ is characteristic of a Bernoulli shift or symmetric tent map that peaks at or near unity. It is possible to construct both the class of maps and the class of initial conditions that can reproduce the observed result. One weakness in our reliance upon the experiment as a guide to theory is that the fit to (10.19b) was made by invoking Taylor's frozen-turbulence hypothesis, whereby a time series for the energy dissipation in a small spatial region was reinterpreted as a spatial distribution of the dissipation at one time; one does not know if that assumption is justified. Still, we are free to ask whether there is a possible *deterministic* basis for the reported binomial statistics and coarsegrained intervals.

In Fig. 10.5, the portrayed dissipation $\varepsilon_r \approx (dv/dx)^2$ was computed from a measured velocity field by taking finite differences, and represents an

Fig. 10.6 Binomial $f(\alpha)$-spectrum deduced by Meneveau and Sreenivasan from data statistically indistinguishable from those of Fig. 10.5. The same $f(\alpha)$-spectrum was found, to within experimental error, for wakes and boundary layers, including that of the earth's atmosphere (from Meneveau and Sreenivasan).

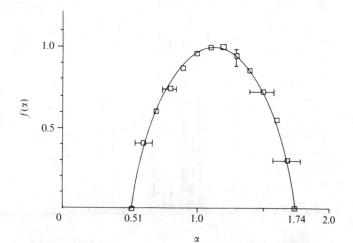

$L = 5$ m section of a 2400-m one-dimensional cut. According to Saucier, the $f(\alpha)$-curves obtained from the entire 2400-m sample, as well as those from smaller parts of it such as Fig. 10.5, match each other and Fig. 10.6, although better statistics are obtained from taking a section that is longer than 5 m. In order to compare the following theoretical predictions with the data, one needs a dictionary to connect our notation to that of Meneveau and Sreenivasan: E_r/E_L, r/L, and $\varepsilon_r/\varepsilon_L$ in the latter correspond to P_i, l_i, and $\rho_i = P_i/l_i$ in our notation, where E_r is the flux of kinetic energy from vortices of size r to smaller eddies, $\varepsilon_r = E_r/r/L$ is the corresponding density, E_L is the total flux in a cut of length L, and ε_L is the average dissipation in the cut.

Consider the binary tent map

$$f(x) = \begin{cases} 2x, & x < \tfrac{1}{2} \\ 2(1 - x), & x > \tfrac{1}{2} \end{cases} \qquad (10.20)$$

along with any symbol sequence with a ratio (in the long run) of $\bar{P}_1/\bar{P}_2 = \tfrac{3}{7}$ 0's to 1's, and where blocks of m 0's and $n - m$ 1's occur with average frequency $\bar{P}_1^m \bar{P}_2^{n-m}$. We can construct an example. Start with the base-10 normal number

$$123456789011121314\cdots \qquad (10.21)$$

and replace any three of the base-10 digits by L, the remaining seven digits by R. For example, let $0, 4, 8, \rightarrow$ L, while $1, 2, 3, 5, 6, 7, 9, \rightarrow$ R; the result is a binary symbol sequence of (10.21),

$$\text{RRRLRRRLRRLRRRRRRRL}\cdots, \qquad (10.21b)$$

that has the required statistical properties: it yields statistical independence with $P_1/P_2 = \tfrac{3}{7}$ in the sense of the law of large numbers ($P_1/P_2 = \tfrac{3}{7}$ is the fraction of L's to R's in an infinitely long sequence (10.21b).

Because of the finite uncertainty in the experimental result $l_n \approx 2^{-n}$, it is not possible to conclude that the binary tent map is the map that generated the observed $f(\alpha)$-spectrum – to within experimental error, infinitely many maps in the same universality class will pass the test. All that is needed is a map that, in backward iteration, produces the approximate scalings $l_n \approx 2^{-n}$ *to within experimental accuracy*. Statistical independence of iterates and the binomial generating then follow for a choice of symbol sequences from the right class. While we cannot say *exactly* which is the map that generated the experiment, it is clear that the $f(\alpha)$-spectrum that fits the turbulence experiments can be generated by fully developed deterministic chaos.

In practice, the symbol sequence (10.21b) yields the ratio $P_1/P_2 \approx \frac{3}{7}$ very slowly: 4000 bits are inadequate to reproduce that asymptotic behavior. The experimental data are reported to correspond to a Kolmogorov length scale $l_\kappa = l_n \approx 2^{-12}$, so that many more than 4000 bits in a symbol sequence will be required to reproduce the data. Although we do not know whether the irrational number π is normal to any base of arithmetic, it is known that π is *effectively* normal to base-10 for at least several million digits. In fact, the digits of π are approximately evenly and statistically independently distributed after as few as the first 500 digits. Therefore, in practice, the use of π in base-10 rather than (10.21) will very accurately reproduce the turbulence $f(\alpha)$-spectrum (Fig. 10.6) for a number of digits that realistically corresponds to the length of the experimental time series. We leave it as a nontrivial exercise (in the writing of computer programs that construct and read strings) to determine whether the first 8000 bits of the symbol sequence (10.21b) yields approximate statistical independence with $P_1 P_2 \approx \frac{3}{7}$.

Why should the tent map with a particular class of symbol sequences be able to simulate the observed dissipation? It is certain that the tent map has nothing whatsoever to do with large eddies, including the inertial range, but it is possible that chaotic vorticity dynamics may become relatively simple at the small scales where the one-dimensional cuts were measured. What follows next is only qualitative, a starting point for further analysis.

Recall that the β-model leads us to expect an eddy cascade on a tree that is incomplete and is of at least octal order. This is a high-Reynolds-number picture where the flow is fully turbulent, the inertial range. The ink drop experiment (Fig. 10.2) seems to require chaotic but nonturbulent vortex dynamics and begins at a low Reynolds number of about 15. In this case, one large eddy splits into 4 to 6 smaller eddies, but after a few generations (3 to 6, more or less), each eddy divides only once and the last eddies that are formed by this procedure are both laminar *and* nonchaotic: they do not split at all.[1] They are dynamically metastable and are only later eaten up by viscosity. The ink drop experiment is very likely representative of the small scales of the eddy cascade in fully developed turbulence. The reason for this assertion is: the local Reynolds number $R_n \approx l_n V_n/v$ for eddies in generation n is decreasing as the cascade proceeds, as do both the energy per eddy and the energy per density per eddy. The cascade likely stops *before* the local Reynolds number equals

---

[1] In the presence of larger chaotic eddies, or even several small-scale vortices, the small-scale eddies may remain chaotic but not turbulent, but this will not affect their inability to divide further.

unity ($R_n \approx 1$ will correspond to $l_n \approx l_\kappa$, if one can equate the eddy splitting time l_n/v_n to the diffusion time l_n^2/v when $l_n \approx l_\kappa$), because the next-to-last generation of unstable eddies splits only once (local binary tree). This is because those eddies have such low energy that they are *nearly* stable against further decay, and each of their two daughters *is* stable against further splitting. Therefore, the binary energy dissipation deduced from experiment by Meneveau and Sreenivasan has a possible physical explanation, and the chaotic vorticity dynamics may have become so simple that the statistics of the energy transfer and the eddy sizes can be reproduced by a map that generates a *complete binary* partitioning. More generally, the description of the entire cascade will require dynamics that generates a tree whose order decreases with decreasing local Reynolds number, a complexity that we are not yet familiar with from the study of relatively simple chaotic dynamical systems. The statistical description provided by the tent map can be correct *only* for the smallest scales of turbulence, and its validity does not extend to the nonturbulent but chaotic dynamics of vortices that give birth to five or six smaller vortices at the onset of the cascade in the ink droplet experiment (note that the inertial range lies well above this experiment in local Reynolds number).

What about statistical independence? Is it likely to be a feature of turbulence at larger scales? If we can take the symbol sequence given by the digits of π as a guide, one can say that it is reasonable to expect that several generations of the eddy cascade should occur before statistical independence in the energy dissipation should set in. Whether one should always observe statistical independence in the inertial range, or even at small scales, may well be a question of deterministic noise, large deterministic fluctuations in the interval probabilities P_i, correspondingly, in the energy flux per vortex.

Where might the tent map or a map that is near to the tent map (intervals $l_n \approx 2^{-n}$ and complete binary tree) come from? Perhaps from the differential equations or coupled chaotic maps that describe some sort of simplified or oversimplified vorticity dynamics in space and time at small scales, in a way analogous to that in which the Lorenz model leads to an approximately one-dimensional map when you ask the right question.

We close this chapter by observing that the most interesting questions remain unanswered: what are the order and degree of completeness of the eddy cascade tree, and how does one model the eddy cascade?

11

From flows to automata: chaotic systems as completely deterministic machines

11.1 The arithmetic of deterministic chaos

We began this book with the qualitative viewpoint: the formulation of nonlinear dynamics as flows in phase space and replaced the study of the flow by the study of an iterated map, namely a stroboscopic map, a Lorenz plot, or a Poincaré section. An invariant set of a map is the point set in the map's phase space that is left invariant under arbitrarily many iterations of the map – mathematically, it is the closure of the trajectory as the number of iterations goes to infinity. In other words, if you start iterating from an initial condition that belongs to the invariant set and imagine a gedanken computation where the arithmetic could be done exactly, then, as the time goes to infinity the map would never leave the invariant set. Infinite precision is impossible in arithmetic, as it is in experiment, so one main problem for us has been to understand systematically how to discover coarsegrained descriptions of invariant sets as well as different possible statistical distributions that can be generated by the dynamics on those coarsegrained sets by doing finite-precision arithmetic for finitely many iterations of a map. For that reason, we arrived at the algorithmic method that was used throughout the text. It is the aim of this chapter to explain why the algorithmic, or finite-precision, approach to chaos theory is not lacking in depth and is, from a certain standpoint, fundamental.

It would also be possible to introduce our considerations by asking the following basic question: what does it mean to be able to compute an orbit for a chaotic system where, because the system is deterministic, the future *is* fixed by the past, but the fact that the system is chaotic means that the future is in *some* sense unpredictable? When one starts from a continuum rather than from a digital formulation of chaos, this is a very

278

hard question to analyze, whereas from the fully digital standpoint it becomes transparent. Furthermore, the use of digital mathematics takes us much closer to the point where we can discuss chaos in the context of computations made on digital computers, and one thing that is essential is to contrast the continuum and computer-based viewpoints. We shall do exactly that later in this chapter. As always, for maximum clarity and for simplicity of discussion, we shall carry out our comparison by using one-dimensional maps, but there is no restriction of the method to such maps. In particular, if we should conclude that the continuum description poses fundamental difficulties for this simplest of all possible cases, that of one-dimensional maps, then it is clear that we will not have to ask the same question again for harder systems, like the Lorenz model, or the Navier–Stokes equations.

In general, a chaotic map has, mathematically, uncountably many different nonoverlapping invariant sets. Examples of invariant sets are fixed points, periodic solutions (limit cycles of dissipative maps and exact periodic orbits of conservative maps), and quasiperiodic orbits. When these orbits are all stable, then that is the same thing as saying that the motion is nonchaotic. We have seen that fractals (Cantor-like sets) can also be invariant sets, and these occur either at the transition to chaos or beyond. In the chaotic regime, the invariant sets that are made up of nonperiodic orbits can be either attracting or repelling, but the motion *on* the invariant set is always unstable in the sense of a positive Liapunov exponent. In this book, we have given examples of both cases: tent maps with $f_{max} > 1$ generate fractals that are chaotic repellers under forward iterations of the map but are attractors under backward iteration. At fully developed chaos, the entire phase space or else some fractal subset is left invariant by the map. In the chaotic regime, therefore, we can say that the map's elementary nonoverlapping invariant sets are of two types: unstable periodic orbits, orbits of order μ^n, with μ and n integers; and unstable nonperiodic orbits, orbits of cardinality μ^∞ with μ an integer. Whether the nonperiodic orbits occur on a fractal or have a closure that is the entire phase space (or only a finite part of it, as is the case if one studies chaotic regions with KAM boundaries in a conservative system, or if one studies the logistic map in the region between criticality and fully developed chaos) is not central to what we are about to consider. The central question in every case is: to what extent can one discover correct statistics of the different possible orbits on the so-called 'strange' invariant sets (i.e., sets with cardinality of order μ^∞) by computation? Also, how complete a description of the dynamics is the information that is obtainable from computation? Here, we allow ourselves to consider purely theoretical infinite-time gedanken computations that are restricted to finite but arbitrarily large precision as the theoretical limit of the case of real interest to us: *finite-precision* computations made during *finite times*. Normally, in theoretical physics, one takes infinite-time and infinite-precision limits for granted, because these approximations usually do not cause a problem and are convenient. In deterministic chaos, they are precisely the assumptions that cause trouble and must be analyzed: chaos theory is the subject where ideas of computability and noncomputability enter physics in a natural and nontrivial way, but that comes later. Let us first start on the path that leads to the idea of fundamental limits on computability in chaos theory.

When we want to understand a chaotic system, there is first the question of the geometry of the invariant set: is it fractal or nonfractal? Then, there is the question, for a given initial condition, of what is the distribution of iterates of the map on the invariant set. We know from our study in Chapter 9 that the question of invariant distributions of a chaotic system *cannot be separated from the choice of classes of initial conditions of a chaotic dynamical system*, except perhaps at a borderline of chaos.

In the preceding chapters, we have emphasized working with a hierarchy of natural coarsegrainings of phase space and with corresponding coarsegrained invariant distributions. The phase space was partitioned into N_n cells $l_1, l_2, \ldots, l_{N_n}$ and we denoted by P_i the fraction of iterates of the map falling into cell l_i as a map is iterated from a given initial condition. Early in the book, in sections 2.7, 3.7, and 4.7, we showed how to compute 'coarsegrained orbits'. These orbits are uniform approximations to exact chaotic orbits (unstable nonperiodic orbits) and can be computed to within any desired precision in finite-precision arithmetic by the parallel-processing method. One thereby predicts the frequencies $\{P_i\}$ directly for a specific initial condition for some large but finite number of iterations of the map. Even better, we showed in Chapter 9 how to dispense with the map, for the class of fully chaotic maps, and how to read the statistics of a nonperiodic orbit directly from a symbol sequence, which is just the binary expansion of an irrational number. In other words, and in direct contrast with the advice of most other texts on deterministic chaos, we have not advocated the forward iteration of a chaotic map on a digital computer (in the floating-point mode) as a method for discovering the statistical distributions of deterministic chaos. But *why* should one not be satisfied to carry out the numerics by using floating-point or fixed-precision real arithmetic on a digital computer, and then let measure theory be our guide as to what we should expect as computer output? For example, the uniform invariant density is the measure-one prediction of any asymmetric tent map that peaks at unity. That prediction follows from a random choice of initial conditions, for 'almost all' initial conditions. Does this mean that we should forget about all of the other possible distributions (including multifractal distributions, e.g.), and assume that an arbitrary choice of initial conditions, in computation, will *always* yield a distribution whose limiting form is described by the uniform invariant density? To phrase the question more generally, what is the overlap between computer output and coarsegrained versions of the measure-one predictions of chaos theory? We know from Chapter 9 that the measure-one orbits are not the whole story, and we learned from Chapter 10 that they may not even be the main story. Still, it is interesting and obligatory to analyse the question of whether certain known theorems from continuum mathematics, when combined with speculations about the effects of computer roundoff/truncation, permit one, as many authors have claimed, to conclude that invariant densities *should* be mimicked by using floating-point arithmetic on a computer. We warn the reader that the question of whether they *may* be mimicked in some cases is completely different from that of whether there are known theorems from continuum mathematics that provide us with any compelling reasons for *expecting* them to be mimicked in computation. There are two separate questions here: (i) Can theorems from *continuum* mathematics predict the output of a *digital* computer? (ii) How do digital computers behave, given certain programs? First, we need to compare the requirements for the validity of continuum mathematics with the limitations implicit in the construction of any digital computer.

The invariant sets of chaotic dynamical systems are made up of μ^∞ points that are 'almost everywhere' irrational and are therefore defined by infinite-length, nonperiodic decimal expansions.[1] This is the land where measure theory rules in mathematics, but these are precisely the numbers that a digital computer cannot handle, even approximately, except in the special case where algorithms and special programming methods are used and the precision of the computation is increased as the map is iterated. The parallel-processing method, which requires special programming, is an example of just such a method. There, irrational numbers are computed algorithmically to higher and higher precision, and the resulting chaotic orbits are nonperiodic, but that is not the way that most chaos researchers computer at present. Before going further, let us review the reason why the

---

[1] We remind the reader that even for the symmetric tent map with $a = 3$, rational points on the Cantor set belong to unstable cycles. Nonperiodic orbits are made up of irrational Cantor set elements.

parallel-processing method was introduced in the first place: information flow in digit strings of chaotic maps as the map is iterated.

In order to concentrate more clearly upon the formulation of chaotic systems as automata, we must now state what is meant by the word automaton in computation theory: an automaton is a machine or a mathemetic model of a machine whereby you get discrete information ('output') as a result of the transition of the machine through a *finite* number of internal states, starting from some definite initial state. The initial state is determined by the *finite* information that you gave the machine as 'input'. In particular, a deterministic machine is a deterministic dynamical system and the input, or program, is the initial condition. The information that constitutes the program is written in terms of a finite-length alphabet, and it is irrelevant whether one uses for symbols (labels) a letter alphabet (a, b, c, ...) or numbers $(0, 1, 2, \ldots, \mu^{-1})$ in some integral base-μ of arithmetic. The information that is programmed into the machine and the information that is produced by the machine are always letters or finite-length words in the chosen alphabet. It is easy to write down the substitution rule that generates, e.g., the Thue Morse sequence: consider a two-letter alphabet $\{a, b\}$ and the rule P whereby a $\rightarrow$ ab and b $\rightarrow$ ba. The initial condition a generates the Thue Morse sequence:

$$a \rightarrow ab \rightarrow abba \rightarrow abbabaab \rightarrow abbabaabbaababba \rightarrow \cdots.$$

Note also that the substitutions a $\rightarrow$ ab and b $\rightarrow$ ba are a fixed point of the rule. There are two other ways to generate this sequence. First, if we write the natural numbers in sequence in binary, 0 1 10 11 100 101 110 111 1000···, and then count the number of 1's, modulo 2, in each binary string in the sequence, then the result is (with a represented by 0 and b by 1) the Thue Morse sequence: 0 1 1 0 1 0 0 0 1 ···. Second, in order to illustrate the operation of a simple model of a machine (a special purpose computer), the Thue Morse sequence can be generated as the output of a sequence of programs of a certain finite automaton. In Fig. 11.1, the schematic of the required machine is displayed: it has two states: a and b. If you begin in state a and feed in a binary number (the program), the machine ends in state a or b, and that is the output for the program. Consider the sequence of programs 0, 1, 10, 11, 100, 101, 110, 111, 1000, 1001,···. We start the machine every time in the same state a, feed in the binary string that is the program, let the machine compute, and then record the output (state a or state b). Then, repeat the process for the next binary string in the sequence of programs. In this way, the automaton of Fig. 11.1 computes the Thue Morse sequence: a, ab, abba, abbabaab, abbabaabbaababba, ..., . Some sequences can be computed with finite automata, while others require 'infinite' automata for their computation. Addition of integers, for example, can be performed on a finite automaton, while the multiplication of integers of arbitrary size requires an arbitrarily large (infinite) automaton. By an 'infinite' automaton, one means only that the precision of the computation must be increased as you proceed, which means that the machine needs more and more internal states in order to do the job correctly. More generally, every digitization of a computable iterated map yields a kind of automaton, or a rule that can be carried out on an automaton, and this need for increasing precision in computation was illustrated by the automaton for the logistic map in Chapter 2, which is little more than the rule for multiplication.

In the parallel-processing method, the automaton can be written down in closed form for a simple piecewise linear map, the tent maps (cf. Chapters 2 and 4), and we obtain as output a string of digits (a word) of length N as the nth iterate of the map in base-μ arithmetic, starting from an initial condition that was written as a string of approximate length $N + N\lambda/\ln \mu$, and where λ is the average Liapunov exponent for the particular initial condition. This is merely a discrete description of the butterfly effect, whereby the Nth digit in x_0 is transformed at an average rate set by λ into the first digit of x_n after only a few, namely $n \sim N \ln \mu/\lambda$, iterations of the map. This *right-to-left irreversible information flow in digit strings* is the essence of deterministic chaos and is the reason for the necessity of the parallel-processing method, or something qualitatively similar to it: given the first N digits of any iterate x_m, you *can* compute at least the first digits of approximately the next $n \sim N \ln \mu/\lambda$ iterates correctly, which means that you know at least the first digit of the orbit for the next n iterations (one-bit orbit, exactly computed). This conclusion is not restricted to piecewise linear maps – it is true for all chaotic maps.

Fig. 11.1   This two-state machine, when properly programmed, produces the Morse–Thue sequence.

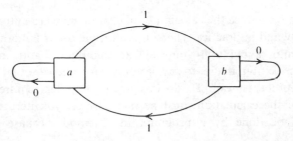

For a stable cycle, in contrast, the average Liapunov exponent,

$$\lambda = \frac{1}{n} \sum_{i=0}^{n-1} \ln|f'(x_i, D)|, \tag{11.1}$$

is negative (try it for the stable two cycle of the logistic map, e.g.) *and the irreversible information flow in digit strings is from left to right*: least-significant digits in an initial condition become less and less significant as $n \to \infty$, in contrast with chaos. For a conservative map with stable cycles or stable quasiperiodic motion, *there is no information flow*: the Liapunov exponents all vanish. A good example is given by the linear circle map $\theta_{n+1} = \theta_n + \Omega$ for irrational winding numbers Ω (yielding a quasiperiodic orbit that covers the circle densely as $n \to \infty$). Here, for any rational approximation to Ω you get a cycle and for better rational approximations you get longer cycles (cf. the continued fraction method in Chapter 5). For a given rational approximation to Ω, you neither approach the quasi-periodic orbit more closely, nor do you diverge from it as $n \to \infty$ because the Liapunov exponent is zero in this case. The point is this: the presence of a positive Liapunov exponent means that the attention that must be paid to error introduction and propagation is a fundamentally *different* problem from what it is in the case of stable motion, because the speed with which the errors accumulate rapidly brings you to the point where you will not understand the meaning of the numbers that the machine produces.

In order to avoid the introduction of unknown or difficult-to-track errors, our approach to chaotic dynamics has been to discretize both the control parameter D and the initial condition x_0 in some integral base-μ of arithmetic (Niven (1956) shows that mixed bases are also possible, but we have not considered this option). If you feed finite-length strings for D and x_0 into the map, the result will be a digit string for the nth iterate $x_n = f^{(n)}(x_0, D)$, where the number of digits that is given *correctly* in the string that represents x_n is determined by the Liapunov exponent for that particular initial condition,

$$\lambda_i \approx \ln|f^{(n)'}(x_i, D)|/n. \tag{11.2}$$

In this way, we reduce the calculation of chaotic maps to automata, both theoretically and in practice. The automaton picture is fundamental: you usually cannot avoid it if you want to compute chaotic orbits with controlled precision, that is to say, if you want to be sure that you know what the numbers mean. To do this on a computer requires that the precision of the calculation must be increased as you iterate the map. That must be done by a programming method, because no digital computer is constructed to do that *automatically* for you. Next, we analyze the outcome of a particular uncontrolled computation on a digital computer, where floating-point arithmetic is used.

A digital computer is itself a finite automaton, a finite-state deterministic machine. If you use 'real arithmetic' (no floating-point operations) in base-μ, then the computation is carried out in such a way that every number in the unit interval that you get from the machine has the finite form $0.\varepsilon_1\varepsilon_2\cdots\varepsilon_N$, where $\varepsilon_i = 0, 1, 2, \ldots, \mu - 1$, so that the machine operates on an alphabet of μ symbols and can produce μ^N different words (we restrict to numbers $x \in [0, 1]$ for convenience of discussion). Therefore, every initial condition is truncated to one of these μ^N numbers, as is every control parameter and every iterate x_n. Even if the map has no periodic orbit for the initial condition you have chosen, the computed orbit must have a period that is no longer than μ^N. Single precision in binary typically means $N \sim 25$ ($\mu = 2$ for typical digital computers: Apple, Commodore, CRAY, IBM, SUN, and VAX), whereas single precision on a typical pocket calculator means $N \sim 10$ because most pocket calculators do finite-precision arithmetic in base-10.

When floating-point operations are used, the computed orbits can have a much longer period, but the resulting accumulated errors are *extremely* difficult to track (see Knuth, 1981). As an example of what typically happens in a floating-point computation, suppose that you want to study the ternary shift map $x \to 3x \bmod 1$ on a digital computer by using single-precision floating-point operations. $x_0 = \frac{1}{10} = 0.001100110011\cdots$, an infinite periodic string in binary arithmetic. The computer holds about $N \sim 25$ bits correctly and then 'floats' the rest. The resulting pseudo-orbit (false orbit) on the computer is one with a very long period, whereas the exact orbit is merely a 4-cycle: $\frac{1}{10} \to \frac{3}{10} \to \frac{9}{10} \to \frac{7}{10} \to \frac{1}{10}$! For the first few iterations, the computed orbit stays near this repelling 4-cycle and then, because $\lambda = \ln 3 > 0$, diverges from it. The result is that, from the standpoint of an approximation to the correct 4-cycle, the computed orbit consists overwhelmingly of roundoff/truncation errors. In stark contrast, a typical pocket calculator will give you a 4-cycle *exactly*, because (i) it does arithmetic in base-10 and (ii) floating-point operations are not called into play by the numbers that make up that particular 4-cycle. Computation in chaotic dynamics is not at all the same as it is in nonchaotic dynamics, where one can identify least-significant digits and then truncate and round without changing the orbit calculation into something altogether different from what you set out to compute in the first place. This brings us to the next point.

There are two ways to describe computed orbits that are not exact orbits of the map. In the literature, they are called pseudo-orbits. A pseudo-orbit is defined to be any computed orbit that is not an exact orbit of the map. Taken literally, the prefix 'pseudo-' means 'false'. The above example produces a false orbit in the sense that is has, in the end, nothing whatever to do with the initial condition $x_0 = \frac{1}{10}$ for the tent map. In addition, one does not know, *without further analysis*, what class of statistics one has generated from a particular pseudo-orbit or whether it has anything to do with the statistics that *are* generated by some particular class of *exact* orbits, even for some small number of generations of coarsegraining. There are also trajectories that fit the definition of a pseudo-orbit that are not false orbits: for example, every uniform approximation to an exact chaotic orbit (every 'coarsegrained' orbit) fits the definition for a pseudo-orbit *because* it is an approximation to an exact infinite-precision orbit. Here, the prefix pseudo- applies where it should not (a coarsegrained orbit is in no qualitative sense a 'false' orbit), because the people who defined the term pseudo-orbit were not thinking about finite-precision arithmetic; they implicitly used the assumption of *infinite precision* in their definition of 'exact' orbit, as is common in mathematics.

Finally, there is another way to think of the pseudo-orbits like the one produced by a binary computer for $x \to 3 \bmod 1$ when $x_0 = \frac{1}{10}$: it is an exact orbit of an automaton that the computer *constructs* for you when you program it to compute $x \to 3x \bmod 1$ with $x_0 = \frac{1}{10}$! The computer cannot compute exactly what you want, because the fraction $\frac{1}{10}$ would require infinitely many binary digits for its representation in the computer, so it does something else, but what it does is perfectly systematic and reproducible: barring the failure of a switch, or some change in the program, the computer will give you *exactly* the same pseudo-orbit every time you program it in the same way to compute $x \to 3x \bmod 1$ with $x_0 = \frac{1}{10}$. It must: every digital computer is a completely deterministic finite-state machine, with absolutely no element of randomness in its construction. *Different* digital computers with *different* roundoff/truncation decisions will therefore produce different pseudo-orbits. Differently stated, each such computer invents a different automaton as a replacement for your map. When you do not understand the roundoff/truncation algorithm that was built into your computer, then you have no way to track the errors that accumulate and propagate as you iterate a chaotic map.

It was von Neumann (1966) who first observed that any discretization of a system of differential equations for computation is a replacement of the system by an automaton, and he advocated studying the statistics generated by 'artificial automata' in order to gain insight into the workings of real or model systems that are too complex for ordinary analysis (he had specifically in mind models that can reproduce themselves). He also noticed that the chaotic map $x \rightarrow 4x(1 - x)$ could be used as a pseudo-random number generator on a computer, were it not for the fact that after a few iterations you see only the effects of roundoff/truncation. Good pseudo-random number generators are, in practice, chosen to operate only on the integers. They are, from the start, formulated as automata.

When you discretize chaotic differential equations for computation, there is a further problem. Suppose that you keep derivatives up to some fixed order in your discretization method (fourth-order Runge–Kutta, e.g.): you have replaced the flow by a certain iterated map. Even if you should compute the orbits of this map exactly, the computed orbits will not be guaranteed to agree with those of the original system except at short times. The reason is that the higher-order terms that were ignored when you wrote down the map represent an error that is eventually magnified by the positive Liapunov exponent. So, you must take higher- and higher-order derivatives into account and also increase the precision of the computation as you iterate, if you want to avoid generating a pseudo-orbit that is dominated by mistakes. Typical power series solutions of chaotic systems are not likely to have an infinite radius of convergence, so analytic continuation may be necessary if one insists upon using methods that are based upon Taylor series. One of the challenges of chaos theory is to construct maps or classes of maps that correctly represent a given dynamical system (see Chapter 10 for an example). This particular problem has been almost entirely neglected in the literature on deterministic chaos.

One main idea of this monograph is that, by careful computation, it is possible to compute the coarsegrained orbit of a map for a definite initial condition. The ability to predict the future from the past is limited by computation time. In nature, dynamical systems are never ideally closed and are always subject to perturbations from their surroundings. One can find the following sort of statement in the literature: roundoff/truncation error on a computer is like a random perturbation of the map. The resulting pseudo-orbits are like those observed in real, open physical systems. The asserted basis for this claim is the so-called shadowing lemma (the Bowen–Anosov lemma). We shall analyze the shadowing lemma for its computational implications, if any, in order to see whether it forms the basis for such a broad claim. We emphasize that the main question here is whether, by doing bad arithmetic that yields only *periodic pseudo-orbits*, one is *guaranteed* by a relatively weak mathematical approximation theorem (the so-called shadowing lemma) to obtain the statistics of a particular class of *unstable nonperiodic orbits* at some coarsegrained level that is necessarily less than that set by the computer's precision. Also behind the above claim is the hidden assumption that, in spite of the fact that both simple and complex dynamical systems generate an infinity of *different* statistical distributions, only one particular statistical distribution is of interest in natural science, namely, the one that has been mislabeled 'the natural measure' (see Chapter 9).

One thing ought to be clear: it is dangerous to assume, without careful analysis, that the effect of roundoff/truncation on a deterministic machine, in some unexplained way, mimics randomness: one effect of the roundoff/truncation mechanism is to impose periodicity systematically upon computed orbits. Whether the resulting long-period pseudo-orbits generated by a particular computer can approximately exhibit some degree of *statistical independence* for a small number of coarsegrainings is a question that should be analyzed for a given map on a given machine: there is, as we shall see in section 11.2 below, no justification from mathematics for *assuming* it to be true. For cases where it turns out to be approximately true, the explanation cannot follow from continuum mathematics, but must be based upon an understanding of a particular computer's roundoff/truncation algorithm.

Before advancing to a discussion of the shadowing lemma, let us take a look at another tempting assumption that requires further analysis. Consider a dissipative system such as the Lorenz model, where the Jacobian is small, $J \ll 1$. In this case, volume elements in phase space contract rapidly. One might attempt to make the following argument: because of the rapid contraction of volume elements (meaning that the product $\delta x_0 \delta y_0 \delta z_0$ of uncertainties in the three initial conditions (x_0, y_0, z_0) of the Lorenz model contracts), computed orbits must lie near the Lorenz attractor and therefore give statistics like those on the attractor. In order to believe this argument we must assume that the long period pseudo-orbits obtained by floating-point computation provide reliable information about (*a*) 'the attractor' and (*b*) the statistics of 'generic' orbits on the attractor – that is, about the nonperiodic orbits on the attractor that are supposed to be 'typical' from the measure theoretic standpoint. Let us analyze a transparent model in order to show that the assumption is not on firm ground and that, without additional assumptions, it does not necessarily guarantee anything of the sort.

Consider, for example, the Bernoulli shift $f(x, D) = Dx \bmod 1$ with $D \geq 2$ an integer (for simplicity). In this case, the attractor is the entire unit interval. But, in contrast with the Lorenz model, where an initial condition not initially on the attractor (but lies in the basin of attraction) would in principle merely *approach* the attractor as $t \to \infty$, the Bernoulli shift, for *any* initial condition x_0 places the map precisely *on* 'the attractor' after exactly *one* iteration! Let us now see just how much mileage this gives us with respect to the stated aims (*a*) to mimic the statistics of nonperiodic orbits on the attractor, in particular, (*b*) to generate 'generic' invariant distributions from arbitrarily chosen initial conditions in computation. In particular, what does the shadowing lemma guarantee us under these circumstances?

Merely to start 'on the attractor' is of absolutely no help; 'the attractor' is made up of an uncountable infinity of repelling nonperiodic orbits and a countable infinity of repelling finite invariant sets (unstable periodic orbits). Therefore, what comes out of a computation depends upon (i) *where* you start on the attractor (this is completely clear for the Bernoulli shift, and there is no reason to expect the Lorenz model to be easier) and (ii) whether you calculate in such a way that your computed orbit is periodic or nonperiodic. Even if you compute in such a way that the calculated orbit is nonperiodic, there is no guarantee that your orbit (even as $n \to \infty$) yields the 'generic' invariant distribution of the map. That is because the distribution of iterates that you generate depends upon the choice of initial conditions. That is, it depends upon from *where* you start your iterations on 'the attractor'. The so-called shadowing lemma (Bowen–Anosov lemma) is often quoted as the reason why one need not take care to compute correctly when the dynamical system is chaotic. Let us now examine the shadowing lemma in order to find out whether this is true by asking for its computational content.

11.2 The shadowing lemma

Consider the exact orbit $y_0, y_1, y_2, \ldots, y_n$, where the iterates

$$y_n = f^{(n)}(y_0, D),\qquad(11.3)$$

follow from a precise initial condition y_0. By an α-pseudo-orbit we mean any numerical sequence $x_0, x_1, \ldots, x_n, \ldots$, where

$$|x_{n+1} - f(x_n, D)| < \alpha \qquad(11.4)$$

and the difference $x_{n+1} - f(x_n, D)$ is nonvanishing for at least one iteration of the map. The pseudo-orbit comes from making a bounded error per iteration of the map. That is what computers do. The pseudo-orbit approximates the true orbit if

$$|x_n - y_n| < \beta \qquad(11.5)$$

for some prescribed number of iterations. This approximation is called β-shadowing. The β-shadowing lemma says that, for hyperbolic systems, every α-pseudo-orbit β-shadows an exact orbit of the map. Never mind that, given α, there is no general prediction for β (or vice versa), and that most systems of interest in nature are not hyperbolic. Let us instead ask what is the implication for computation of β-shadowing.

We can discover a general property by looking at the simplest case: consider the Bernoulli shift $y \to Dy \bmod 1$ with $D \geq 2$ an integer and with a uniform error α per iteration. The pseudo-orbit therefore follows from

$$x_{n+1} = Dx_n + \alpha, \qquad \bmod 1, \tag{11.6}$$

which can be solved to yield

$$x_n + \alpha/(D-1) = D^n(x_0 + \alpha/(D-1)), \qquad \bmod 1, \tag{11.7}$$

Therefore the exact orbit obeys $y_n = x_n + \alpha/(D-1)$ and has the initial condition $y_0 = x_0 + \alpha/(D-1)$, so that $\beta = \alpha/(D-1)$. To mimic a digital computer, we must choose a rational initial condition x_0 and α must be rational. It follows that y_0 is rational, so the periodic pseudo-orbit shadows an exact periodic orbit of the Bernoulli shift. We therefore shadow an exact orbit of the Bernoulli shift, but it is *not* a nonperiodic orbit of 2^∞ points that is generated by the Bernoulli shift with an irrational initial condition. For the special case of the Bernoulli shift, one can prove, e.g., that the elements of long-period unstable orbits of the ternary shift map in binary arithmetic tile the unit interval and so are distributed more or less evenly. Consequently, pseudo-orbits from floating-point arithmetic in this case can reflect at least one coarsegrained property of one particular class of nonperiodic orbits correctly. Namely, the pseudo-orbits can mimic the uniform invariant density in this case. That this happens or does not happen is not at all a consequence of the shadowing lemma, however. Rather, it is due to the fact that the periodic points of the map have a certain *distribution* in the phase space. In general, for an arbitrary map, we do not know the relation of the distribution of points on unstable cycles to the infinity of different distributions that can be generated by the infinity of different nonperiodic orbits. The best that can be said in this direction is that methods have been developed to determine the unstable cycles and to arrange them hierarchically upon trees (see Cvitanović et al., 1988). It is interesting that, in this work, the authors try to discover the coarsegrained properties of *attractors* by studying *finite repelling* invariant sets. The study of the periodic orbits provides a universal approach to chaotic dynamics, because (as with the nonperiodic symbol sequences), the periodic symbol sequences are topological invariants. In contrast, metric properties of particular maps are not universal: they change from one map to another. Our main point is that the shadowing lemma alone, without further information about the symbolic dynamics, gives one no reason to expect that bad computer arithmetic provides a reliable way to extract correct coarsegrained statistics of nonperiodic orbits of a chaotic map. For which *class* of exact periodic orbits are the pseudo-orbit statistics supposed to be correct, and why, one must ask? In other words, the shadowing lemma without further assumptions has no implication at all for computation. While the shadowing lemma is not useful for understanding the results of computation, symbolic dynamics can be put to work to solve that task.

If one wants to analyze the statistics generated by a particular pseudo-orbit, one need only compute the natural partitioning of the map and construct the symbol sequence of the pseudo-orbit. The method of Chapter 9 can then be applied to discover the statistics that are generated by that pseudo-orbit. If the resulting statistics should turn out approximately to have the property of statistical independence, then we expect, without proof, that it is because of the distribution of points on periodic orbits of the map. Pseudo-orbit statistics for the same map can be expected to vary from one computer to another, because computers with different roundoff/truncation algorithms will produce entirely different pseudo-orbits. If a particular computer generates pseudo-orbits that more or less mimic the invariant density of a map, for some finite number of coarsegrainings of the map's invariant set, then that same computer's pseudo-orbits cannot mimic the multifractal statistical distributions that also can be generated by the same map. Let us now look at a few examples of statistics from pseudo-orbits.

Consider the logistic map $x_{n+1} = f(x_n) - 4x_n(1 - x_n)$, whose invariant density is $p(x) = (\pi(x(1 - x))^{1/2})^{-1}$. By computing the preimages of the maximum of f (the maximum is located at $y = \frac{1}{2}$), we can compute the natural intervals systematically for each generation (see section (9.1)). For the first generation, we find that $l_1 = l_2 = \frac{1}{2}$. For the second generation, we find that $l_1 = l_4 = 0.1464\cdots$ and $l_2 = l_3 = 0.3535\cdots$. For the third generation, the eight different intervals have (because of the map's symmetry) four different sizes, $l_1 = 0.0380\cdots, l_2 = 0.1083\cdots, l_3 = 0.1622\cdots$, and $l_4 = 0.1913\cdots$. Typical computer results of iterating the map in single-precision floating-point arithmetic yield visitation frequencies for the pseudo-orbits that more or less mimic the even distribution $P_i = 2^{-n}$ for $n = 1, 2,$ and 3, and the result seems easy to reproduce pretty much independently of the choice of initial conditions on the computer. This means that the invariant density is approximately simulated by pseudo-orbits for several generations of coarsegraining, but it also means that all of the multifractal distributions that the logistic map can generate are beyond the reach of pseudo-orbits. Whether, with entirely different roundoff/truncation algorithms, the forward iteration of a map in the computer's floating-point mode could simulate multifractal statistics, is an interesting unanswered question. Furthermore, the shadowing lemma, taken alone, cannot be used to explain that the map's invariant density should or will be simulated at all. For example, if we consider the numerically generated Lorenz map (see section 2.1), then we can also compute the natural coarsegraining for that map, but the resulting computer-imposed statistics for the first several generations of coarse-grainings do not show statistical independence in the form $P_i = \bar{P}_1^m \bar{P}_2^{n-m}$ for any values of $\bar{P}_1$ and $\bar{P}_2$. Finally, if one studies the asymmetric tent maps with $ab = a + b$ (we showed in Chapter 2 that the invariant density is uniform) then one again gets results (at least for the case where $a = 3$ and $b = \frac{3}{2}$) from floating-point arithmetic that are more or less akin to the invariant density-based theoretical result, $P_i = l_i$, for several generations of the natural coarsegraining $\{l_i\}$. Again, the multifractal distributions that can be produced by using algorithms that generate either initial conditions or symbol sequences are completely beyond reach. And, in every case, as the coarsegraining is refined, one soon reaches the point where the computer's pseudo-orbits no longer mimic the invariant density at all, which is the opposite of what is true when one generates statistics by using algorithms. We leave it for the reader to discover the point where the coarsegrained invariant density is no longer mimicked on his particular machine.

At first sight, one is tempted to be impressed by the fact that the use of floating-point arithmetic seems to reproduce the statistics corresponding to several different invariant densities, at least for the first few generations. We suggest that the result is not necessarily an accurate reflection of true chaotic behavior, for one very good reason: asymptotic distributions like invariant densities would be produced, with probability one, by certain infinite-length symbol sequences. In the limit of infinite symbol sequence length, where the limiting (nonfractal and multifractal) measures occur, the first n terms of the symbol sequence can be discarded, *because all that matters in generating the limiting invariant distribution is the infinite-length tail*. In computation, and in experiment, the infinite length tail is *totally inaccessible* and is therefore irrelevant, because our empirical statistics follow precisely from *the first n terms* of a symbol sequence. It is exactly for this reason that one should not expect to obtain statistics that accurately reflect limiting invariant distributions, *even if initial conditions were chosen carefully and computations were performed accurately*. In other words, when deterministic noise is absent, then the too-smooth results should be suspect (cf. Chapter 9 for a discussion of the effect of deterministic noise on statistics).

We end this section with the following observation: it *is* possible to use a pseudo-orbit of the logistic map $f(x) = 4x(1 - x)$ as a pseudorandom number generator, by using symbolic dynamics, and thereby to accomplish in a different way part of what the advocates of the shadowing lemma would like to accomplish. First, every symbol sequence formed by a pseudo-orbit over the natural partitioning is a correct symbol sequence of the fully chaotic logistic map. These particular symbol sequences are periodic, with long period, and we have observed that the symbols have approximately the even distribution $P_i = 2^{-n}$, at least for $n = 1, 2,$ and 3. Let $n_{max} = M$ be the last generation of coarsegraining, in any floating-point computation, where $P_i = 2^{-M}$ holds to within your desired accuracy. Next, slide an M-bit window along the symbol sequence and record the sequence of M-bit words. If you then replace each L by 0 and each R by 1, and place a binary point to the left of the first bit, you will have a sequence of numbers that satisfies the condition for an even distribution (they are distributed with uniform coarsegrained density in the unit interval). Hence, the numerical sequence satisfies the condition stated by Knuth for 'weak (effective) randomness' (Knuth, 1981). However, the condition for an even distribution, taken alone, is not adequate for producing a 'good' pseudorandom number generator. The reader should refer to Knuth for details. However, this means that a kind of shadowing works in this case in the restricted sense that we get a valid symbol sequence (to get an invalid one is impossible), even though the map is not uniformly hyperbolic. The reason why this happens has nothing at all to do with the shadowing lemma: *it is because the map's binary tree is complete*. Trees and symbolic dynamics can therefore be used to formulate a better version of the original shadowing idea that can prove to be computationally useful. The modified idea is as follows: consider the collection of symbol sequences of a given map. Then a pseudo-orbit is accepted as useful if it generates, over the map's natural partitioning, a correct symbol sequence, whether periodic or nonperiodic. The main point is that one gets correct statistics from such a pseudo-orbit, but not from pseudo-orbits that fail to satisfy this condition. Again, the main idea is that unstable periodic orbits of all periods are generated by the map $f(x) = 4x(1 - x)$, and while the details of the pseudo-orbit are wrong, it cannot help but generate a correct symbol sequence. Finally, the statistics that follow from the symbol sequence are determined by and are restricted to those that are compatible with the roundoff/truncation algorithm for the computer in question.

We turn now to the more fundamental question: how easy is it in a correctly performed computation to pick an initial condition that generates an orbit of a given map that reproduces that map's invariant density for arbitrarily fine but finite coarsegrainings? We know that different classes of initial conditions generate different invariant distributions. If we use a uniform measure on the unit interval, then almost all initial conditions yield a uniform invariant distribution for the binary Bernoulli shift and for the binary tent map. For the ternary tent map, which peaks above unity, and for a much larger class of symmetric maps that generate repelling fractals, almost all initial conditions on the fractal yield a uniform distribution of iterates (one can deduce this from the results of Chapter 9). These are examples of the orbits that we would call 'generic' for these maps *if* we could justify the assumption that it is possible to make a 'random draw' (random *choice*) of initial conditions from a set with μ^{∞} points (the continuum or a Cantor subset of the continuum). In such a case, we would, with probability (measure) one, draw an initial condition whereby the iteration of the map would yield a uniform distribution over its natural partitioning. Does this mean that, in *computation*, an *arbitrary* choice of initial condition is overwhelmingly likely to produce a computed orbit whose iterates are distributed uniformly on the natural partitioning of the map? We turn now to the analysis of this question. Is the interpretation of mathematical measure as a probability correct? Its correctness, or lack of same, depends upon (i) the availability of the continuum, and (ii) the possibility to make a *truly random choice*. We shall see that there is no basis for the justification of either assumption.

11.3 The problem with the continuum and 'measure-one'

In order to set the stage for the discussion, we review briefly the reasoning that leads to the conclusion that almost all irrational numbers, taken as initial data for a Bernoulli shift, yield a uniform coarsegrained invariant distribution $P_i = \mu^{-n}$ for every generation $l_n = \mu^{-n}$ of coarsegraining of the unit interval where $\mu \geq 2$ is an integer. The key is to use the normal numbers that were studied by Borel.

Choose any integer $\mu \geq 2$ as the base of arithmetic, and consider the Bernoulli shift $x \to \mu x \bmod 1$. All initial conditions that are irrational yield chaotic orbits because $\lambda = \ln \mu > 0$, and in base-μ, the map is just a base point shift on the nonperiodic string $x_0 = 0.\varepsilon_1\varepsilon_2\cdots\varepsilon_N\cdots$ where $\varepsilon_i = 0, 1, 2, \ldots$, or $\mu - 1$: $x_1 = 0.\varepsilon_2\varepsilon_3\cdots\varepsilon_N$, $x_2 = 0.\varepsilon_3\varepsilon_4\cdots\varepsilon_N\cdots$, and so on. Choose $l_i^{(1)} = \mu - 1$. If, in x_0, each digit $\varepsilon_i = 0, 1, 2, \ldots$, or $\mu - 1$ occurs as often as every other digit than $P_i = \mu^{-1}$. For the next stage of coarsegraining, $l_i^{(2)} = \mu^{-2}$, we are interested in the coarsegrained orbit given by $x_n = 0.\varepsilon_1(n)\varepsilon_2(n)\cdots$, which is just the two-digit description of the exact orbit. Here, we get an even probability distribution $P_i = \mu^{-2}$ for all the μ^2 intervals if, in x_0, the block 00 occurs as often as 01, which occurs as often as 10, ..., and which occurs as often as the block $(\mu - 1)(\mu - 1)$. In other words, the distribution of *pairs* of symbols in x_0 is even. If you carry out this demand for an even coarsegrained distribution for every generation, then as $n \to \infty$ your initial condition is an example of a normal number. Normal numbers are, by definition, irrational, and are discussed in Niven (1956) and Kac (1959). Borel (see Kac, 1959) discovered that the distribution of the digits of normal numbers is the first example of precisely defined objects in mathematics to which the laws of probability can be applied. Borel proved that almost all numbers that can be defined to exist in the continuum are normal. If we could draw at random from the continuum, then we would with certainty (measure one) get an initial condition whose iteration by *any* Bernoulli shift (with $\mu > 1$ an integer) would yield a uniform invariant density. That is because the base of arithmetic is irrelevant: 'almost all' numbers are normal to *every* integral base of arithmetic (Niven, 1956).

Because the digits $\varepsilon_i = 0, 1, 2, \ldots, \mu - 1$ are in one-to-one correspondence with a μ-sided die (the numbers in $[0, 1]$ in base-μ or the die-toss sequences can be used to label a complete tree of order μ), the fact that normal numbers occur with measure one in the continuum is analogous to the law of large numbers for the hypothetical repeated tosses of a fair, μ-sided die (gedanken experiment, as n goes to infinity). Here, we come close to the hypothetical question whether random orbits can exist for deterministic dynamical systems: if the digits of x_0 could be regarded as appearing randomly, then we would be led to believe that randomness and determinism could coexist. Here is the way it would have to work. Suppose that you have an automaton that gives you one digit at a time of x_0, so that every piece of information you get is an iteration of the lowest-order coarsegrained orbit $x_1 = 0.\varepsilon_1$, $x_2 = 0.\varepsilon_2, \ldots, x_n = 0.\varepsilon_n$, and so on. Suppose, now, that your automaton produces each digit ε_i by a rule that is equivalent to the toss of a fair, μ-sided die. There are two questions: how easy is it to produce a rule (algorithm) that generates a normal irrational number, and how closely does this rule mimic the apparent randomness of a die-toss? I say *apparent* randomness, because the toss of a die is in fact describable by the laws of mechanics.

We observe first that, although all possible die sequences are in one-to-one correspondence with all possible base-μ strings, the toss of a die can in no way be regarded as an *algorithm* for generating one of these strings. If you toss a die n times then you generate a string of length n, but you cannot toss a die infinitely many times: once you stop (and you have to) then μ^{∞} digits are left undetermined. That die-tossing may be used to *define* infinite-length strings is not the same as the *construction* (computation) of a string. To the extent that irrational numbers can be *computed*, digit by digit, they follow from *algorithms* that are purely deterministic: every digit is fixed in advance by the rule, beyond the reach of human invention and intervention (Turing, 1937). That excludes die-tossing. For example, there is no question but that the digits of $\sqrt{2}$, π, and e are fixed rigidly – we can compute the digits of each number, to within the limits set by computer time. There is no arbitrariness in a single digit of $\sqrt{2}$, π, e or any other irrational number that can be computed. Such strings are generated by arithmetic, pure determinism, and therefore are not random, but they can be 'effectively random' in the sense of good pseudorandom number generators.

We return to the question: how easy is it to produce an algorithm for a 'generic' member of the continuum? Although, with measure one, almost all numbers are normal, there is only one *known* algorithm that produces a normal number! The algorithm is: write down the natural numbers in any base-μ. In base-10, the number is

$$x = 0.123456789101112131415\cdots \qquad (11.8)$$

and in base-2 it is

$$x = 0.110111001011101111000\cdots, \qquad (11.8\text{b})$$

and we must admit that the step-by-step method by which this algorithm works is the absolute opposite of the toss of a fair die. The most orderly rule imaginable deterministically produces a statistical distribution of blocks of digits that is a model of total disorder. Randomness was not used and is not needed – we can generate *apparent* randomness deterministically.

It is unknown whether (11.8) is normal to base-2 or whether (11.8b) is normal to base-10. It is not known whether the digits of $\sqrt{2}$, e, or π in any base are normally distributed (see Waggoner (1985) for a numerical study of the statistical independence and frequency of occurrence of blocks of digits of π in base-10).

On the other hand, it is very easy to invent or to discover algorithms that generate the digit strings of nonnormal numbers. An obvious one is the binary string

$$x = 0.101001000100001\cdots. \tag{11.9}$$

Another is given by the Fibonacci sequence:

$$x = 0.0110100110010110\cdots, \tag{11.9b}$$

which, while nonperiodic, is quasi-periodic in the sense of a one-dimensional Penrose tiling (see Peyriere, 1986). Therefore, what is typical in the continuum can be notoriously hard to realize in computation. There is a reason for this: computable numbers form only a *countable* subset of the continuum. In computation, as in nature, there is no mathematical continuum: we are fundamentally restricted to a set of measure zero. Therefore, the finite approach to chaos based upon algorithms has the better possibility of providing a correct description of nature than does the analytic description of chaos that takes the continuum for granted.

11.4 Why measure-one behavior may be hard to find

Let us recall how the continuum was brought into existence historically in analysis. There were 'gaps' between the rationals: limits of infinite sequences of rationals are not, in general, themselves rational. For example, we can generate $\sqrt{2}$ as the limit of a convergent sequence of rationals by the method of continued fractions: writing $2 = (1 + x)^2$ yields $\sqrt{2} = 1 + x$, so $1 = x(2 + x)$ and

$$x = \frac{1}{2 + x}. \tag{11.10}$$

This can be iterated to yield

$$x = \cfrac{1}{2 + \cfrac{1}{2 + \cfrac{1}{2 + \cdots}}}, \tag{11.10b}$$

the continued fraction expansion for $\sqrt{2} - 1 = x$. If we truncate the expansion, we obtain the sequence of rational approximations $x_0 = \frac{1}{2}$,

$x_1 = \frac{2}{5}$, $x_2 = \frac{5}{12}$, ... As $n \to \infty$, $x_n = P_n/Q_n \to 0.416\cdots$, an infinite non-periodic expansion, although P_n and Q_n are themselves integers for finite n. The real number system becomes 'complete' when one includes, along with all possible rational numbers P/Q, all possible limits of sequences of the form $[P_n/Q_n]$. So, every irrational number can be *defined* to exist in mathematics as the limit of a convergent sequence of rational numbers. Given that 2^∞ irrationals can be 'defined to exist', how many can, in principle, be computed? The point is that definition, or 'existence', is a far cry from construction.

The last question was analyzed carefully by Turing, the father of modern computer science, who was inspired by Hilbert's famous problem: can every mathematical problem that is correctly posed be answered? A particular example is: can every number that can be defined to exist be computed by an algorithm? A general answer to Hilbert's question was given by Gödel: there are statements in mathematics (in logic) whose truth cannot be decided (incompleteness theorem). Turing's contribution was to show that there are numbers in mathematics that can be defined to exist but whose digit strings cannot be computed by any possible algorithm. By the same method that Cantor used to prove that the irrationals cannot be put into one-to-one correspondence with the integers (Cantor's diagonal argument – see Beckman (1980)) – that the rational gives birth to the irrational – Turing proved that the computable gives rise to the noncomputable. A computable number is one whose digits can be computed by a finite algorithm (infinitely many digits are determined by a finite-length 'computer program'). $\sqrt{2}$, e, and π are examples of computable numbers. Turing's result is shocking: *almost all* numbers that can be defined to exist cannot be computed by any possible algorithm: by the diagonal argument, the number of algorithms is countable. That means that computable numbers are countable, so that limits of sequences of computable numbers are generally noncomputable. So far as prediction by computation is concerned, there is no continuum. That is the core of the reason why it is dangerous to entertain the notion of a 'random draw' of initial conditions. Almost all numbers are normal, so that almost all normal numbers are therefore noncomputable. The noncomputable numbers are mathematical figments of the imagination: they cannot be 'used' to generate chaotic orbits. The extent to which normal numbers might be generated by discoverable algorithms is still unknown. That is why 'measure-one' arguments must not be relied upon as a guide for the results of computation. We can expect that the same is true in nature, that measure-one conditions do not in any way constrain the natural evolution of the world: there is absolutely no reason to *believe* that they do. That is why we have emphasized the discovery of coarsegrained phase spaces and coarsegrained orbits and have avoided the $n \to \infty$ ($l_n \to 0$) limit, except as an occasional convenience. In nature, the $n \to \infty$ ($l_n \to 0$) limit does not exist – it 'exists' only in the definitions of the pure mathematician. As an example, fractal scaling in nature, as Mandelbrot was so careful to emphasize in his book, always occurs between *finite* limits $l_{min} < l_n < l_{max}$, where $l_{max} <$ the size of the object and $l_{min} \gg$ an interatomic spacing.

Let us return to Turing's fascinating result and try to understand it a little better. In particular, we can think of the method of continued fractions as a finite algorithm (the computation of each rational approximate P_n/Q_n needs finitely many steps), and every number in the continuum can be written formally as a continued fraction expansion (see Niven, 1956). Why, then, cannot every irrational number be *computed* by the method of continued fractions. For example, the nonintegral part of every square root can be computed by continuous fractions.:

$$\sqrt{17} - 4 = \frac{(\sqrt{17} - 4)(\sqrt{17} + 4)}{\sqrt{17} + 4} = \frac{1}{\sqrt{17} + 4}, \qquad (11.11)$$

so that with $x = \sqrt{17} - 4$ we get

$$x = \frac{1}{8 + x},$$

which can be iterated to yield

$$x = \cfrac{1}{8 + \cfrac{1}{8 + \cdots}}. \qquad (11.11c)$$

Clearly, the nonperiodic infinite string for x can be *computed* by higher- and higher-order truncations of (11.11c) to as many digits of $\sqrt{17} - 4$ as we want, within the limits set by computer time (that, in the end, is limited by the speed of light).

Why, then, can we not generate the digits of every irrational number by continued fraction approximants? Just as the existence of a formal one-to-one correspondence between numbers written as binary strings and the tosses of a fair coin does not mean that even one single irrational number can be *computed* by coin tossing (they are merely in one-to-one correspondence with coin tossing), the formal one-to-one correspondence between numbers in the continuum and continued fraction expansions does not guarantee that all irrational numbers can be *computed* by the method of continued fractions. The reason for this is that, along with the rule, the continued fraction rule, you have to have a 'seed', or initial condition. The 'seed' that allows us to compute $\sqrt{2} - 1 = x$ is the digit 2 in

$$x = \frac{1}{2 + x}. \tag{11.10}$$

The meaning of Turing's proof is that, for almost all numbers that can be defined to exist, no seed can be found, even in principle. The corresponding noncomputable numbers therefore have nothing to do with deterministic chaos or any other branch of physics.

We are reminded of Max Born's advice to get rid of the idea of the continuum in physics. Turing's result gives weight to that advice, and Borel expressed a similar opinion from the viewpoint of applied probability theory. The beauty of the method of multifractals for finite partitions of phase space is that it frees us from the expectation that we should look for evidence of 'generic' chaotic orbits in nature and in computation, and to encourages us search for the other statistical distributions that may be more useful for the description of nature anyway.

There is another interesting point. Turing's result is an example of Gödel's incompleteness theorem. An inference from Gödel's theorem is that there is nothing that you can do or express with functions that cannot be encoded as arithmetic. The encoding of the statistics of infinitely many functional compositions of a chaotic map, for infinitely many different maps in the class, onto a single nonperiodic binary symbol sequence (i.e., onto a single string of digits for a single irrational number), seems to come close to that idea: there is no chaos in infinitely many functional compositions that is not already contained in the algorithm for the corresponding irrational number. Deterministic chaos reduces to the laws of arithmetic via symbolic dynamics.

Finally, indeterminism *can* be found in nature: radioactive decay times are a good example. In quantum mechanics, the deterministic time evolution rule governs probability amplitudes rather than particle trajectories. One can write down a probability amplitude ψ for each possible classical particle trajectory, and the amplitude for a particle to go from one point in space–time to another is the sum over all possible amplitudes for mutually exclusive alternatives (see Feynman and Hibbs, 1965). Because probability is defined by $|\psi|^2$ in a quantum mechanics, interference phenomena occur. In other words, the particle trajectories are not statistically independent but are what Feynman called interfering alternatives. There is no need to rely upon (as Bohr and others did) the idea of a classical wave: it is possible to understand quantum particle behavior entirely from the rule that adds probability amplitudes rather than probabilities, for mutually exclusive alternatives (see Feynman and Hibbs, 1965).

There are attempts to study 'quantum chaos' by the WKB approximation to nonintegrable Hamiltonian systems, but it is too early to try to discuss the results of those efforts. In fact, it is not yet clear what is meant by the term quantum chaos, as nonintegrable quantum systems and their classical limits have yet to be explored adequately (for recent results, see Srivastava and Müller, 1990).

11.5 The idea of the universal computer

The confusion that is generated by the question of whether random (therefore undetermined) orbits of a deterministic chaotic map can occur is avoided when we realize that we are restricted to computable maps of computable numbers. The hypothetical noncomputable numbers are truly random in the sense that there is no algorithm from which their digit strings can be constructed from the laws of arithmetic. Bernoulli shifts, tent maps, symmetric and asymmetric logistic maps are all examples of computable maps, so long as the corresponding control parameters are computable numbers. If we use as control parameters and initial conditions algorithms for computable numbers, then we can compute coarse-grained orbits with controlled accuracy. It seems sensible to base deterministic chaos (and the rest of physics) not upon the continuum but upon the part of the continuum that is accessible in computation, because what is noncomputable in mathematics cannot be used for the prediction of nature.

If, instead, we should base our considerations upon the so-called algorithmically complex numbers, we would be led in a different direction. An algorithmically complex number is one that requires an infinite-length computer program for its construction. Almost all algorithmically complex numbers are defined by decimal expansions that satisfy the requirements for randomness (see Martin-Löf, 1966). The 'use' of these numbers in deterministic-chaotic maps, such as the Bernoulli shift (see Ford, 1988) or tent map, would lead to the conclusion that random orbits are possible. The catch is that algorithmically complex numbers are, by their very definition, not computable. The intellectual gyrations that are avoided by the elimination of noncomputable numbers in deterministic chaos are analogous to those that are avoided in quantum mechanics when one follows the Dirac–Feynman interpretation rather than the Copenhagen interpretation. But in our case, something more than interpretation is at stake, namely, the question how correctly to formulate deterministic chaos: in quantum theory, Born and Heisenberg argued strongly that physics should not be based upon nonobservable concepts – because of this, Max Born argued for the elimination of the continuum concept from physics. By restricting to computable numbers in classical dynamics, we take a small step in that direction. It means that formal Hilbert space theory cannot be the final foundation for quantum mechanics, because Hilbert space is built upon the generalization to function spaces of the idea of the continuum, the completeness of the real number system (a space is complete when all the limits of all convergent sequences in the space also belong to the space). But this introduces noncomputability into the foundations of quantum mechanics, because almost all functions that can be defined to exist are noncomputable (see Turing, 1937).

The restriction to computable numbers and functions leads to a rigid determinism in classical mechanics, and that is as it should be: indeterminism in a *closed* dynamical system can, in the end, arise only at the level of quantum mechanics, owing to the Heisenberg uncertainty principle. Particle motions in quantum mechanics are not deterministic: one can at best write down a probability amplitude that a particle follows a given path from x_0 to x_n, and the total probability amplitude is then the sum of all amplitudes for different classical paths connecting the two points.

Classical dynamics, in spite of the possibility of chaos, is rigidly deterministic: true randomness cannot come out of arithmetic, but in nature 'least-significant digits' can be jiggled in an undetermined way only at atomic and smaller-length scales. However, chaos is pseudo-random in a very interesting way that nonchaotic deterministic dynamics is not: it seems to provide us with a *deterministic* realization of an old idea of Fechner (see the article by Heidelberger in Kruger *et al.*, 1987), whereby the past provides no basis for the prediction of the future owing to the continual creation of 'new initial conditions' that have never been experienced before (each present state of a chaotic dynamical system is a nontrivial initial condition for all future states).

We have seen that, for chaotic systems, coarsegrained probabilities (empirical probabilities) can be computed from a symbol sequence, and so statistical properties can be predicted. But is it true that the statistical properties of simple dynamical systems ('simple machines') can *always* be predicted? The answer seems to be no, because there is machine behavior that is more complicated than that of a merely chaotic one: that of the so-called 'universal computer'. In other words, there is a higher level of complexity of purely deterministic machines that is possible without the inclusion of either quantum mechanics or classical ideas of randomness.

The theme of complex behavior from simple dynamical systems continues: just as deterministic chaos for the simplest iterative maps occur, universal computational capability is in principle possible for simple (although not the simplest) automata. The 'Turing machine' is an example of a universal computer: the Turing machine is an abstract machine that can compute any computable number or computable function. No real computer operating in a fixed-precision mode is a universal computer (infinite storage capacity is necessary in the definition of a universal computer), but real digital computers can be made to simulate universal computers by programming methods and increased storage capacity. The interesting thing for us is that very simple mathematical models of machines, cellular automata, can also have universal computational capability.

It is easy to see that infinite storage capacity is required of a universal computer. Consider, for example, the computation of the digits of $\sqrt{2}$ by the school-child algorithm:

$$
\begin{array}{r}
1. \quad 4\cdots \\
\hline
\sqrt{2.\ 00\ 00\cdots} \\
\hline
1 \\
28)1\ 00 \\
\vdots
\end{array}
$$

Only finitely many repetitions of a simple rule are required to know $\sqrt{2}$ to any finite number of decimal places, but a strip of paper of infinite length (and an infinite-time computation) would be required of the machine that carries out the gedanken computation to infinitely many digits. Turing machines are based upon the question of what is computable by *algorithms* without regard to the amount of time required for the computation. These machines set a limit on real computations that is analogous to the limit set upon real machines by the second law of thermodynamics, where 'reversible engines' have the highest possible efficiency for doing work. In reality, one must consider relaxation times for irreversible processes in nature, and the analog of this in our context is the amount of time needed for a particular computation. This leads to the classification of problems according to NP-completeness, which is a more useful one than merely the question of whether the problem is

Turing-computable (see Wilf, 1988). What follows next is a brief discussion of very *simple* mathematical models of machines that are equivalent to Turing machines: machines that can, given unlimited computation time and storage space, compute any number or function that is computable. Although these machines can compute all dynamical systems (including chaotic ones) that are computable, the machines are themselves dynamical systems. The trick in using them is to be able to program (encode and decode) them, and that is not necessarily so simple.

An automaton or a model of an automaton can be thought of as a processor. Sequential processors are in practice complicated machines that can be programmed relatively simply. We can think of the program as the initial condition, and the processor as a dynamical system whose output corresponds to the final state or to the time evolution of the system. The idea can be stood on its head: we can instead construct (in reality or as models) highly parallel computers, made up of a large number of simple processors coupled together. In this case, the step-by-step dynamics can be simple, but writing down meaningful 'initial conditions' (programs) is not simple. Here is the idea in its simplest form (see Fig. 11.2): consider a large number of identical processors coupled as nearest neighbors in a chain. Each processor is set in an initial internal state – one of a finite number of discrete states – and this corresponds to the choice of program. The processors are turned on simultaneously ($n = 0$). At the next discrete time step ($n = 1$), each processor receives discrete information from each of its two nearest neighbors and 'computes', that is to say, goes through a finite number of steps and ends in a definite final state that is determined (i) by its initial state and (ii) the information that it got from its two nearest neighbors. It is easy to make a simple-minded mathematical model of this. Let $\varepsilon_j(n)$ be a discrete variable, $\varepsilon_j(n) = 0$ or 1, e.g., and let

$$\varepsilon_j(n + 1) = \varepsilon_{j+1}(n) + \varepsilon_{j-1}(n) \bmod 2. \tag{11.12}$$

Fig. 11.2   The simplest idea of a parallel-processing system: many simple processors (the black boxes) are coupled together to form a nearest-neighbor network. Each separate processor is set in one of its possible initial states and then all processors compute simultaneously. Information is then exchanged with nearest neighbors, and each processor uses this new information to go mechanically into a new state. The new state plays the role of the initial state for the next computation whenever the whole process is repeated.

Here n is the discrete time variable, j is the discrete space variable that corresponds to the location of a 'processor', and the processor can be found in one of two states: $\varepsilon_j(n) = 0$ or 1. In the nearest-neighbor rule (11.12), the 'program' corresponds to specifying the initial two-sided string $\cdots\varepsilon_{-N}(0)\cdots\varepsilon_{-2}(0)\varepsilon_{-1}(0)\varepsilon_0(0)\varepsilon_1(0)\varepsilon_2(0)\cdots\varepsilon_N(0)\cdots$ (one can consider an infinite cellular space, or else boundary conditions can be imposed). Computation of this program then corresponds to the sequence of strings $\cdots\varepsilon_{-1}(1)\varepsilon_0(1)\varepsilon_1(1)\ldots,\ldots,\cdots\varepsilon_{-1}(n)\varepsilon_0(n)\varepsilon_1(n)\cdots$. The main idea of parallel processing is that the independent processors are coupled (here, in nearest-neighbor style), and that each processor receives information and computes simultaneously with all others. This cellular automaton is not capable of universal computation, but it can generate a fractal pattern: from the initial condition $\varepsilon_i(0) = \delta_{i,0}$ one gets the Sierpinski gasket (see Fig. 11.3a, which is to be compared with the usual geometrical

Fig. 11.3 Sierpinski gasket as generated by (a) programming the cellular automaton (11.12) in the initial state $\varepsilon_i(0) = \delta_{i,0}$, and (b) by the geometrical process of removing successive open middle-thirds from the triangles.

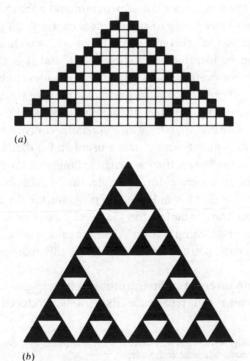

(a)

(b)

construction that is obtained by removing 'middle thirds' and is shown as Fig. 11.3b). In this dynamical system, as with the logistic map at FDC, information is created with each iteration of the map. In general, the patterns generated by cellular automata can be stable- or unstable-periodic, depending upon the particular cellular automaton and upon the initial conditions. Also, they can fall into the category of yielding undecidable results, which is a level deeper than chaos in complexity.

Another example of a cellular automaton is given by the 'Game of Life' (see Berlekamp, Conway, and Guy, 1982; Gardner, 1983). Just as the logistic map can be interpreted as a population growth model on the unit interval, the game of life is a birth–competition–death model on an infinite cellular space in the plane. Assign to each cell a binary variable $\varepsilon_{ij} = 0$ (dead) or 1 (alive). Each cell has eight nearest neighbors. A dead cell at time n is alive at time $n + 1$ if (at time n) exactly three of its nearest neighbors are alive (birth). A cell that is alive at time n is dead at time $n + 1$ if four or more of its nearest neighbors were alive (local over-population). A cell that is alive at time n but has no more than one live nearest neighbor is dead at time $n + 1$. However, a live cell at time n survives at time $n + 1$ if it has exactly 2 or 3 live neighbors at time n (Fig. 11.4). Fixed points, limit cycles and many sorts of surprises are possible in this game. For an *arbitrary* initial program and an unbounded cellular space, there is no known way to say in advance what will be the outcome of the computation carried out by this model machine. Special configurations can be identified as logical gates, and it is argued that the Game of Life is an example of a universal computer, although it is not known how to program it to do a particular computation.[2]

Automata (including cellular automata) are usually proven to be 'universal' by showing that they are *formally* equivalent to a Turing machine, but this is merely an *existence* proof and is therefore dangerous: unless one can actually *construct* a machine language for the game of life that permits one to *compute*, for example, the logistic map (the logistic map plus initial condition would be the program for the machine), then it is of no use to know whether the Game of Life is or is not abstractly capable of 'universal computation'. Here, again, we face the question of existence versus construction, which in this context means real computation.

Von Neumann invented cellular automata in response to the question whether a machine can reproduce itself, and produced just such an

---

[2] Berlekamp, Conway, and Guy (1982) discuss what is required in order to turn the Game of Life into a formally universal computer.

example. Examples of diffusion and also phase transitions have been discovered in *simple* cellular automata (see Grassberger, 1988 and Kinsel, 1985), and phenomena so seemingly diverse as sandpiles and earthquakes recently have been modeled by cellula automata (see Bak, Tang, and Wiesenfeld, 1988).

It is tempting to try to imagine the brain as a highly parallel electrochemical computer, and cellular automata have been used to make corresponding models. The first such was the McCullough–Pitts model (see Minsky, 1967) which led, by the inclusion of ideas of temperature and noise, to the Hopfield model (see McCauley in Haken, 1987 or Pynn and Skjeltorp, 1985). There is much literature in this field – the Hopfield model is essentially a pattern-recognition system that can be generalized to perform computations of a sort. Here, there is as yet no overlap with

Fig. 11.4 Evolution of an initial pattern in a two-dimensional cellular space according to the rules of 'Life'. The arrows show the directions of (discrete) time, and the result of the chosen initial condition (program!) is a 'space–time 4-cycle', because the same pattern is repeated after every four iterations but is shifted downward along the diagonal.

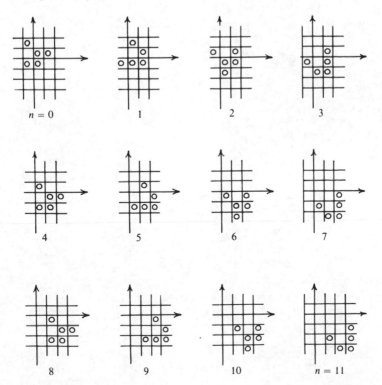

chaotic dynamics: because of the relation of the Hopfield model to the spin-glass problem, and its use as a pattern-recognition machine, the concentration so far has been upon finding fixed-point solutions in a many-degree-of-freedom system with many competing low-energy states. It would be interesting to continue our discussion in this direction, but that is beyond the scope of the present book.

Bibliography

CHAPTER 1

ARNOL'D, V. I., *Ordinary Differential Equations*, MIT Press, Cambridge, Mass. (1973).

ARNOL'D, V. I., *Mathematical Methods of Classical Mechanics*, Springer, Berlin (1978).

ARNOL'D, V. I., *Geometric Theory of Ordinary Differential Equations*, Ch. 1, Springer, Berlin (1983).

BENDER, C. M. and ORSZAG, S. A., *Advanced Mathematical Methods for Scientists and Engineers*, McGraw-Hill, New York (1978).

BIRKHOFF, G. D., *Hydrodynamics*, Dover, New York (1950).

CAMPBELL, J. E., *Continuous Groups*, Chelsea, New York (1966).

CARTAN, E., *Sur la Structure des groupes de transformations finis et continus*, Thèse, Paris (1894; 2nd edn 1933).

DAVIS, H. T., *Nonlinear Integral and Differential Equations*, Dover, New York (1960).

GOLDSTEIN, H., *Classical Mechanics*, Addison-Wesley, Reading, Mass. (1980).

HAKEN, H., *Advanced Synergetics*, Springer, Berlin (1983).

INCE, E. L., *Ordinary Differential Equations*, Dover, New York (1986).

JORDAN, D. W. and SMITH, P., *Introduction to Nonlinear Ordinary Differential Equations*, Clarendon, Oxford (1977).

MARSDEN, J. E. and McCRACKEN, M., *The Hopf Bifurcation and Its Applications*, Springer, Berlin (1976).

STRUBLE, R. A., *Nonlinear Differential Equations*, McGraw-Hill, New York (1962).

WYBOURNE, B. G., *Classical Groups for Physicists*, Wiley, New York (1974).

CHAPTER 2

FORD, J., in *Fundamental Problems in Statistical Mechanics III*, ed. E. G. D. Cohen, North-Holland/American Elsevier, New York (1975).

GHIL, S. M. and CHILDRESS, S., *Topics in Geophysical Fluid Dynamics: Atmospheric Dynamics, Dynamo Theory and Climate Dynamics*, Springer, Berlin (1987).

GLEICK, J., *Chaos*, Viking, New York (1987).

GROSSMAN and THOMAE, S., *Z. Naturforsch.* **32a**, 1351 (1977).

GUCKENHEIMER, J. and HOLMES, P., *Nonlinear Oscillations. Dynamical Systems and Bifurcations of Vector Fields*, Springer, Berlin (1983).

KNUTH, D. E., *Seminumerical Algorithms*, Vol. II in *The Art of Computer Programming*, Addison-Wesley, Reading, Mass. (1981).

LICHTENBERG, A. J. and LIEBERMAN, M. A., *Regular and Stochastic Motion*, Springer, Berlin (1983).

LORENZ, E. N., Deterministic nonperiodic flow, in *Universality in Chaos*, ed. P. Cvitanović, Hilger, Bristol (1984).

McCAULEY, J. L., *Z. Naturforsch.* **42a**, 547 (1987).

McCAULEY, J. L., *Physica Scripta* **T20** (1988).

NIVEN, I., *Irrational Numbers*, Carus Math. Monogr. No. 11, Mathematical Association of America (1956).

RICHTMYER, R. D., *Principles of Advanced Mathematical Physics*, Vol. 2, Springer, Berlin (1981).

RÖSSLER, O., *Phys. Lett.* **57A**, 397 (1976).

SCHUSTER, H. G., *Deterministic Chaos*, Chs. 1, 2, Physik-Verlag, Weinheim (1984).

SPARROW, C., *The Lorenz Equations*, Springer (1982).

CHAPTER 3

ARNOL'D, V. I., *Mathematical Methods of Classical Mechanics*, Springer (1978).

ARNOL'D, V. I., *Geometric Theory of Ordinary Differential Equations*, Ch. 3, Sect. 13, Springer, Berlin (1983).

ARNOL'D, V. I. and AVEZ, A., *Ergodic Problems in Classical Mechanics*, Chs. 1 and 2 (1986).

BALESCU, R., *Statistical Mechanics of Equilibrium and Non-Equilibrium* (cf. appendix on ergodic theory), Wiley-Interscience, New York (1975).

BERRY, M. V. and FORD, J., in *Topics in Nonlinear Dynamics*, ed. S. Jorna, AIP, New York (1975).

BRUSH, S. G., *Kinetic Theory*, Vol. II, Pergamon, Oxford (1966).

DUFF, G. F. D., *Partial Differential Equations*, University of Toronto Press, Toronto (1962).

EISENHART, L. P., *Continuous Groups of Transformations*, Dover, New York (1961).

FORD, J., in *Fundamental Problems in Statistical Mechanics III*, ed. E. G. D. Cohen, North-Holland/American Elsevier, New York (1975).

GIBBS, J. W., *Collected Works*, Vol. II, Ch. XII, Longmans Green, London (1928).

GOLDSTEIN, H., *Classical Mechanics*, 2nd edn, Addison-Wesley, Reading, Mass. (1980).

HÉNON, M., in *Chaotic Behavior of Deterministic Systems*, ed. G. Loos, R. H. Helleman and R. Stora, North-Holland, Amsterdam (1983).

KADANOFF, L. P., *Phys. Rev. Lett.* **47**, 1641 (1981).

KLEIN, F., *On Riemann's Theory of Algebraic Functions and Their Integrals*, transl. by Frances Hardcastle, Dover, New York (1963).

LEBOWITZ, J. L. and PENROSE, O., Modern ergodic theory, *Physics Today* **26** (1973).

LICHTENBERG, A. J. and LIEBERMAN, M. A., *Regular and Stochastic Motion*, Ch. 5, Springer, Berlin (1983).

LIE, S. and ENGEL, F., *Theorie der Transformationsgruppen*, Vols 1 and 2, Teubner, Leipzig (reprinted 1930).

MacKAY, R. S., in *Universality in Chaos*, ed. P. Cvitanović, Hilger, Bristol (1984).

MONIN, F. S. and YAGLOM, A. M., *Statistical Fluid Mechanics*, MIT Press, Cambridge, Mass. (1973).

NIVEN, I., *Irrational Numbers*, Carus Math. Monogr. No. 11, Mathematical Association of America (1956).

SALETAN, E. J. and CROMER, H. A., *Theoretical Mechanics*, Wiley, New York (1974).

SUDARSHAN, E. C. G., *Classical Dynamics: A Modern Perspective*, Wiley, New York (1974).

TABOR, M., *Chaos and Integrability in Nonlinear Dynamics*, Wiley, New York (1989).

VON NEUMANN, J., *Mathematical Foundations of Quantum Mechanics*, Princeton University Press (1955).

WEIGERT, S., Dissertation, University of Basel (1991).

WHITTAKER, E. T., *Analytical Dynamics*, Cambridge University Press, Cambridge (1965).

ZASLAVSKII, G. M. and CHIRIKOV, B. V., *Sov. Phys. Usp.* **14**, 549 (1972).

ZASLAVSKY, G. M., SAGDEEV, R. Z., USIKOV, D. A., and CHERNIKOV, A. A., *Weak Chaos and Quasi-regular Patterns*, Cambridge Nonlinear Science Series 1, Cambridge University Press, Cambridge (1991).

CHAPTER 4

BARNSLEY, M., *Fractals Everywhere*, Academic Press, London (1988).

GELBAUM, B. R. and OLMSTED, J. H., *Counterexamples in Analysis*, Holden-Day, San Francisco, London, Amsterdam (1964).

GRASSBERGER, P. and PROCACCIA, I., *Physica* **13D**, 34 (1984).

HALSEY, T. C., JENSEN, M. H., KADANOFF, L. P., PROCACCIA, I., and SHRAIMAN, B. I., *Phys. Rev.* **A33**, 1141 (1986).

McCAULEY, J. L., *Physica Scripta* **T20** (1988).

MANDELBROT, B. V., *The Fractal Geometry of Nature*, Freeman, San Francisco (1982).

NIVEN, I., *Irrational Numbers*, Carus Math. Monogr. No. 11, Mathematics Association of America (1956).

NORTHRUP, E. P., *Riddles in Mathematics*, The English University Press, London (1945).

PALMORE, J. and McCAULEY, J. L., *Int. J. Mod. Phys.* **B3**, 1447 (1989).

PEITGEN, H. O. and RICHTER, P. H., *The Beauty of Fractals*, Springer, Berlin (1986).

SCHUSTER, H. G., *Deterministic Chaos*, Physik-Verlag, Weinheim (1984).

VISCEK, T., *Fractal Growth Phenomena*, World Scientific, Singapore (1989).

CHAPTER 5

ALSTRØM, P., Dissertation, Universitetet av København (1987).

ARNOL'D, V. I., *Trans. Am. Math. Soc.* **46**, 213 (1965).

ARNOL'D, V. I., *Geometric Methods in the Theory of Ordinary Differential Equations*, Springer, New York (1983).

BOHR, T., BAK, P., and JENSEN, M. G., *Physica Scripta* **T9**, 50 (1985); *Phys. Rev.* **A30**, 1960 (1984); *Phys. Rev.* **A30**, 1970 (1984).

CVITANOVIĆ, P., JENSEN, M. H., KADANOFF, L. P., and PROCACCIA, I., *Phys. Rev. Lett.* **55**, 343 (1985).

CVITANOVIĆ, P. and SØDERBERG, B., *Physica Scripta* **32**, 263 (1985).

FEIGENBAUM, M., KADANOFF, L. P., and SHENKER, S., *Physica* **5D**, 370 (1982).

GUCKENHEIMER, J. and HOLMES, P., *Nonlinear Oscillations, Dynamical Systems, and Bifurcations of Vector Fields*, Springer, Berlin (1983).

KADANOFF, L. P., *Phys. Rev. Lett.* **47**, 1641 (1981).

OSTLUND, S., RAND, D., SETHNA, J., and SIGGIA, E., *Physica* **8D**, 303 (1983).

SCHUSTER, H. G., *Deterministic Chaos*, Physik-Verlag, Weinheim (1984).

SHENKER, S., *Physica* **5D**, 405 (1982).

CHAPTER 6

COLLET, P. and ECKMANN, J. P., *Iterated Maps of the Interval as Dynamical Systems*, Birkhäuser, Basel, Boston, Stuttgart (1980).

CVITANOVIĆ, P. (ed.), *Universality in Chaos*, Hilger, Bristol (1984).

DERRIDA, B., in *Bifurcation Theory in Mathematical Physics and Related Topics*, pp. 137–54, ed. C. Bardos and D. Bessis, Reidel, Dordrecht (1980).

FEIGENBAUM, M. J., *J. Stat. Phys.* **19**, 25 (1978).

FEIGENBAUM, M. J., *Los Alamos Science* **1**, 4 (1980).

GLEICK, J., *Chaos*, Viking, New York (1987). Translations into German and Norwegian are also available.

GROSSMANN, S. and THOMAE, S., *Z. Naturforsch.* **32A**, 1353 (1977).

HU, B., *Phys. Rep.* **31**, 233 (1982).

HU, B., in *Proc. of the 1986 Summer School on Statistical Mechanics*, ed. C. K. Hu, Academia Sinica, World Scientific, Singapore (1987).

KADANOFF, L. P., *Physica* **2**, 263 (1966).

KADANOFF, L. P. et al., *Rev. Mod. Phys.* **39**, 395 (1967).

LIBCHABER, A. et al., in *Universality in Chaos*, ed. P. Cvitanović, Hilger, Bristol (1984).

MAY, R. M., *Nature* **262**, 459 (1976).

METROPOLIS, M. L., STEIN, M. L., and STEIN, P. R., *J. Combinatorial Theory* (A) **15**, 25 (1973).

SCHUSTER, H. G., *Deterministic Chaos*, VCH Verlagsgesellschaft, Weinheim (1988).

VON NEUMANN, J., *J. Res. Nat. Bur. Std. Appl. Math. Ser.* **3**, 36 (1951).

WILSON, K. G., *Phys. Rev.* **B4**, 3174 (1971).

WILSON, K. G., *Physica* **73**, 119 (1974).

WILSON, K. G. and KOGUT, J., *Phys. Rep.* **C12**, 75 (1974).

CHAPTER 7

ARTUSO, R., AURELL, E., and CVITANOVIĆ, P., *Nonlinearity* 3, 325, 361 (1990).

ARTUSO, R., CVITANOVIĆ, P., and KENNY, B. G., *Phys. Rev.* **A39**, 268 (1989).

CVITANOVIĆ, P., GUNARATNE, G. H., and PROCACCIA, I., *Phys. Rev.* **A38**, 1503 (1988).

FEIGENBAUM, M. J., *Los Alamos Science* **1**, 4 (1980).

HALSEY, T. C., JENSEN, M. H., KADANOFF, L. P., PROCACCIA, I., and SHRAIMAN, B. I., *Phys. Rev.* **A33**, 1141 (1986).

JENSEN, M. H., KADANOFF, L. P., and PROCACCIA, I., *Phys. Rev.* **A36**, 1409 (1987).

McCAULEY, J. L., *Int. J. Mod. Phys.* **B3**, 821 (1989).

McCAULEY, J. L., *Phys. Rep.* **189**, no. 5 (1990).

PYNN, R. and RISTE, T. (eds), *Time-dependent Effects in Disordered Materials*, Plenum, New York (1987).

TEL, T., *Z. Naturforsch.* **43A**, 1154 (1988).

VISCEK, T., *Fractal Growth Phenomena*, World Scientific, Singapore (1989).

CHAPTER 8

BOHR, T. and RAND, D., *Physics* **25D**, 387 (1987).

CVITANOVIĆ, P., in *Nonlinear Evolution and Chaotic Phenomena*, ed. G. Gallowotti and P. Zweifel, Plenum, New York (1988).

FEIGENBAUM, M. J., *J. Stat. Phys.* **46**, 919 (1987).

FEIGENBAUM, M. J., in *Nonlinear Evolution and Chaotic Phenomena*, eds G. Gallowotti and P. Zweifel, Plenum, New York (1988).

FEIGENBAUM, M. J., *Nonlinearity* **1**, 577 (1988).

FEIGENBAUM, M. J., *J. Stat. Phys.* **52**, 527 (1988).

FEIGENBAUM, M. J., JENSEN, M. H., and PROCACCIA, I., *Phys. Rev. Lett.* **57**, 1503 (1986).

HALSEY, T. C., JENSEN, M. H., KADANOFF, L. P., PROCACCIA, I., and SHRAIMAN, B. I., *Phys. Rev.* **A33**, 114 (1986).

JENSEN, M. H. and BOHR, T., *Phys. Rev.* **A36**, 4904 (1987).

JENSEN, M. H., KADANOFF, L. P., LIBCHABER, A., PROCACCIA, I., and STAVANS, J., *Phys. Rev. Lett.* **59**, 2798 (1985).

McCAULEY, J. L., *Int. J. Mod. Phys.* **B3**, 821 (1989).

McCAULEY, J. L., *Phys. Rep.* **189**, No. 5 (1990).

OTT, E., WITHERS, W. D., and YORKE, J. A., *J. Stat. Phys.* **36**, 687 (1984).

PALMORE, J. and McCAULEY, J. L., *Int. J. Mod. Phys.* **B3**, 1447 (1989).

RUELLE, D., *Statistical Mechanics, Thermodynamic Formalism*, Addison-Wesley, Reading, Mass. (1978).

TAKAHASHI, Y. and OONO, Y., *Prog. Theor. Phys.* **71**, 85 (1984).

VUL, E. B., SINAI, Y. G., and KHANIN, K. M., *Usp. Mat. Nauk.* **39**, 3 (1984).

CHAPTER 9

BOREL, E., *Rend. Circ. Mat. Palermo* **27**, 247 (1909).

BOREL, E., *Les Nombres Inaccessibles*, Gauthier-Villars, Paris (1952).

BOREL, E., *Elements of The Theory of Probability*, Engl. trans. by J. E. Freund, Prentice-Hall, Englewood Cliffs, NJ (1965).

CHAMPERNOWNE, D. G., *J. Lond. Math. Soc.* **8**, 254 (1933).

CHHABRA, A., JENSEN, R. V., and SREENIVASAN, K. R., *Phys. Rev.* **A40**, 4593 (1989).

CVITANOVIĆ, P., GUNARATNE, G., and PROCACCIA, I., *Phys. Rev.* **A38**, 1503 (1988).

FARMER, J. D., in *Evolution of Ordered and Chaotic Patterns in Systems Treated by the Natural Sciences and Mathematics*, ed. H. Haken, Springer, Berlin (1982).

FARMER, J. D., OTT, E., and YORKE, J. A., *Physica* **7D**, 153 (1983).

FEIGENBAUM, M. J., *J. Stat. Phys.* **21**, 699 (1979).

FEIGENBAUM, M. J., *Los Alamos Science* **1**, 4 (1980).

FEIGENBAUM, M. J., *J. Stat. Phys.* **52**, 527 (1988).

FEIGENBAUM, M. J., JENSEN, M. H., and PROCACCIA, I., *Phys. Rev. Lett.* **57**, 1503 (1986).

FEIGENBAUM, M. J., KADANOFF, L. P., and SHENKER, S., *Physics* **5D**, 370 (1982).

GROSSMANN, S., in *Multicritical Phenomena*, eds R. Pynn and A. T. Skjeltorp, Plenum, New York (1984).

GROSSMANN, S., in *Nonequilibrium Cooperative Phenomena in Physics and Related Fields*, ed. M. G. Verlande, Plenum, New York (1984).

GROSSMANN, S. and THOMAE, S., *Z. Naturforsch.* **32a**, 1353 (1977).

GYÖRGI, G. and SZEPFALUSY, P., *Z. Phys.* **B55**, 179 (1984).

HALSEY, T. H., JENSEN, M. H., KADANOFF, L. P., PROCACCIA, I., and SHRAIMAN, B. I., *Phys. Rev.* **A33**, 114 (1986).

JENSEN, M. H., KADANOFF, L. P., and PROCACCIA, I., *Phys. Rev.* **A36**, 1409 (1987).

KAC, M., *Statistical Independence in Probability, Analysis, and Number Theory*, Carus Math. Monogr. No. 12, Mathematical Association of America (1959).

KNUTH, D. E., *The Art of Computer Programming II: Semi-Numerical Algorithms*, Addison-Wesley, Reading, Mass. (1981).

LICHTENBERG, A. J. and LIEBERMAN, M. A., *Regular and Stochastic Motion*, Springer, New York (1983).

McCAULEY, J. L., *Z. Naturforsch.* **42a**, 547 (1986).

McCAULEY, J. L., in *Computational Systems – Natural and Artificial*, ed. H. Haken, Springer, Berlin (1987).

McCAULEY, J. L., *Int. J. Mod. Phys.* **B3**, 821 (1989).

McCAULEY, J. L., *Z. Phys.* **81**, 215 (1990).

NIVEN, I., *Irrational Numbers*, Carus Math. Monogr. No. 11, Mathematical Association of America (1956).

PALMORE, J. and McCAULEY, J. L., *Int. J. Mod. Phys.* **B3**, 1447 (1989).

RAND, D., *The Singularity Spectrum for Hyperbolic Cantor Sets and Attractors*, unpublished.

SCHUSTER, H. G., *Deterministic Chaos*, VCH-Verlagsgesellschaft, Weinheim (1988).

TURING, A., *Proc. Lond. Math. Soc.* (2) **42**, 230 (1937).

CHAPTER 10

ANSELMET, F., COGNE, Y., HOPFINGER, E. J., and ANTONIO, R. A., *J. Fluid Mech.* **140**, 63 (1984).

BENZI, R., PALADIN, G., PARISI, G., and VULPIANI, A., *J. Phys.* **A17**, 3521 (1984).

BOHR, T., in *Spontaneous Formation of Space–Time Structures and Criticality*, ed. D. Sherrington and T. Riste, Kluwer, Dordrecht (1991).

CHATE, H. and MANNEVILLE, P., *Physica* **D32**, 409 (1988).

CHATE, H. and MANNEVILLE, P., *Physica* **D37**, 33 (1989).

CHATE, H. and MANNEVILLE, P., *J. Stat. Phys.* **56**, 357 (1989).

EFFINGER, H. and GROSSMAN, S., *Z. Phys.* **B66**, 289 (1987).

FRISCH, U. and ORSZAG, S. A., *Physics Today*, April, p. 24 (1990).

FRISCH U. and PARISI, G., *Turbulence and Predictability of Geophysical Flows and Climate Dynamics*, Varenna Summer School LXXXVIII (1983).

FRISCH, U., SULEM, P., and NELKIN, M., *J. Fluid Mech.* **87**, 719 (1987).

GOLDSTEIN, S., *Modern Developments in Fluid Dynamics*, Dover, New York (1965).

HEISENBERG, W., *Z. Phys.* **124**, 628 (1948).

JENSEN, M. H., in *Spontaneous Formation of Space–Time Structures and Criticality*, eds D. Sherrington and T. Riste, Kluwer, Dordrecht (1991).

KOLMOGOROV, A. N., *C. R. Acad. Sci. USSR* **30**, 301 (1941).

McCAULEY, J. L., *Int. J. Mod. Phys.* **B 4**, 1517 (1990).

McCAULEY, J. L., *Z. Phys.* **81**, 215 (1990).

McCAULEY, J. L., *Introduction to Multifractals in Dynamical Systems Theory and Fully Developed Fluid Turbulence*, *Phys. Rep.* **189**, No. 5 (1990).

McCAULEY, J. L., in *Spontaneous Formation of Space–Time Structures and Criticality*, eds D. Sherrington and T. Riste, Kluwever, Dordrecht (1991).

MANDELBROT, B., *The Fractal Geometry of Nature*, Freeman, San Francisco (1983).

MENEVEAU, C. and SREENIVASAN, K. R., *Phys. Rev. Lett.* **59**, 1424 (1987).

NELKIN, M., *J. Stat. Phys.* **54**, 1 (1989).

ONSAGER, L., *Nuovo Cim.* Suppl. VI, Nos 2, 3 (1949).

PALADIN, G. and VULPIANI, A., in *Fractals in Physics*, eds L. Pietronero and E. Tosati, pp. 447–51, North-Holland, Amsterdam (1986).

PRANDTL, L., *Essentials of Fluids Dynamics*, Eng. trans. of *Führer durch die Strömungslehre*, Hafner, New York (1952).

SAUCIER, A., Dissertation, University of Toronto (1991).

SCHLICHTING, H., *Boundary Layer Theory*, McGraw-Hill, New York (1979).

TENEKES, H., *Physics Today* **57**, No. 1 (1974).

TENEKES, H. and LUMLEY, J. L., *A First Course in Turbulence*, MIT Press, Cambridge, Mass. (1972).

CHAPTER 11

AXEL, F., ALLOUCHE, J. P., KLEMAN, M., MENDES-FRANCE, M., and PEYRIERE, J., *J. Physique colloque* **C3**, suppl. 7, **47**, C3–181 (1986).

HAO BAI-LIN, *Elementary Symbolic Dynamics*, World Scientific, Singapore (1989).

BAK, P., TANG, C., and WIESENFELD, K., *Phys. Rev.* **A38**, 304 (1988).

BECKMAN, F. S., *Mathematical Foundations of Programming*, Addison-Wesley, Reading, Mass. (1980).

BENELTIN, G., CASARTELLI, G. M., GALGANI, L., GIORGILLI, A., and STRELCYN, J. M., *Nuovo Cim.* **B44**, 183 (1978).

BERLEKAMP, E. R., CONWAY, J. H., and GUY, R. K., *Winning Ways for Your Mathematical Plays*, Academic Press, London (1982).

BOREL, E., *Les Nombres Inaccessibles*, Gauthier-Villars, Paris (1952).

CRUTCHFIELD, J. P. and PACKARD, N., *Int. J. Theor. Phys.* **21**, 433 (1982).

CVITANOVIĆ, P., GUNARATNI, G., and PROCACCIA, I., *Phys. Rev.* **A38**, 1503 (1988).

DEKKING, M., MENDES-FRANCE, M., and VAN DER POORTEN, A., *Math. Intell.* **4**, 130 (1982).

FEYNMAN, R. P., LEIGHTON, R., and SANDS, M. *The Feynman Lectures on Physics: Quantum Mechanics* (Vol. III), Addison-Wesley, Reading, Ma. (1965).

FEYNMAN, R. P. and HIBBS, A. R., *Quantum Mechanics and Path Integrals*, McGraw-Hill, New York (1965).

FORD, J., *Physics Today* **36**, 40 (1988).

GARDNER, M., *Wheels of Life and Other Amusements*, Freeman, New York (1983).

GRASSBERGER, P., *Z. Naturforsch.* **43a**, 671 (1988).

HODGES, A., *Alan Turing: The Enigma*, pp. 91–110, Simon & Schuster, New York (1980).

HOPKIN, D. and MOSS, B., *Automata*, Elsevier/North-Holland, New York (1976).

KAC, M., *Statistical Independence in Probability, Analysis and Number Theory*, Carus Math. Monogr. No. 12, Mathematical Association of America (1959).

KINSEL, W., *Z. Phys.* **B58**, 229 (1985).

KNUTH, D. E., *The Art of Computer Programming II: Semi-Numerical Algorithms*, Addison-Wesley, Reading, Mass. (1981).

KRUGER, L., DASTON, J., and HEIDELBERGER, M., (eds), *The Probabilistic Revolution*, Vol. I, MIT Press, Cambridge, Mass. (1987). In this collection of essays, one can also find a brief description of Borel's argument against the use of the continuum in applications of probability theory.

LANFORD III, D. E., in *Chaotic Behavior in Deterministic Systems*, eds Loos, Helleman, and Stora, North-Holland, Amsterdam (1983).

McCAULEY, J. L. and PALMORE, J. I., *Phys. Lett.* **A115**, 433 (1986).

McCAULEY, J. L., *Zeitschr. für Naturforsch.* **42a**, 547 (1987).

McCAULEY, J. L., in *Computation Systems, Natural and Artificial*, ed. H. Haken, Springer, Berlin (1987).

MARTIN-LÖF, P., *Inf. Control.* **9**, 602 (1966).

MINSKY, M. L., *Computation, Finite and Infinite Machines*, Prentice-Hall, London (1967).

NIVEN, I., *Irrational Numbers*, Carus Math. Monogr. No. 11, Mathematical Association of America (1956).

OZORIO DE ALMEIDA, A. M., *Hamiltonian Systems, Chaos and Quantization*, Cambridge University Press, Cambridge (1988).

PEYRIERE, J., *J. de Physique* **C3** (Suppl 7) 47, 41 (1986).

PYNN, R., and SKJELTORP, A. J., *Scaling Phenomena in Disordered Systems*, Plenum, New York (1985).

SRIVASTAVA, N. and MÜLLER, G., *Phys. Lett.* **147A**, 282 (1990).

SRIVASTAVA, N. and MÜLLER, G., *Z. Phys.* **B8**, 137 (1990).

TURING, A., *Proc. Lond. Math. Soc.* (2) **42**, 230 (1937).

VON NEUMANN, J., *Theory of Self-Reproducing Automata*, University of Illinois Press, Urbana (1966).

WAGGONER, S., *Math. Intell.* **7**, 65 (1985).

WILF, H. S., *Algorithms and Complexity*, Prentice-Hall, Englewood Cliffs, NJ (1988).

WOLFRAM, S., *Los Alamos Science* **9**, 2 (1983).

WOLFRAM, S., *Physica* **10D**, 1 (1984).

Index